P9-DMY-543

DISCARDED
FROM
UNH LIBRARY

Ion solvation

Ion solvation

Yizhak Marcus

Department of Inorganic and Analytical Chemistry
The Hebrew University of Jerusalem
Israel

A Wiley-Interscience Publication

JOHN WILEY & SONS LIMITED
Chichester · New York · Brisbane · Toronto · Singapore

Library of Congress Cataloging in Publication Data:

Marcus, Y.
Ion solvation.
'A Wiley–Interscience publication.'
Includes indexes.
1. Solvation. 2. Ionic solutions. I. Title.
QD543.M383 1985 541.3'72 85-6436
ISBN 0 471 90756 1

British Library Cataloguing in Publication Data:

Marcus, Yizhak
Ion solvation.
1. Ions 2. Solvation
I. Title
541.3'72 QD561

ISBN 0 471 90756 1

Printed and bound in Great Britain

Contents

Preface

During my work of research on chemical interacting systems, I seem to succumb every eight years or so to the urge to sit down and write a book. The first effort, *Ion Exchange and Solvent Extraction of Metal Complexes* (with Aviezer Steven Kertes) was published in 1969; the second one, *Introduction to Liquid State Chemistry*, was published in 1977, and here is the third one. As the title gets shorter and the volume slimmer, my main purpose does not change. On the one hand it is to look back on what has been accomplished in a given field, organize it in an orderly fashion, and present it in a comprehensible and unified manner. On the other hand, I use this opportunity to locate gaps in our knowledge, and either fill these with new research while the book is being written, or to do this to the best of my ability in the course of time. Many new research ideas have thus come to me from my attempts to present current factual knowledge and theoretical interpretation in an organized form. However, this mode of operation may have slowed down the progress of the writing, and may have caused chapters written early in the course of writing to be somewhat less up to date than those written at the later stages.

The scope of university research nowadays, I regret to record, is not conductive to the carrying out of extensive and systematic sets of measurements of high accuracy on the properties of systems. Both from the standpoint of student interest and from that of the necessary financial support, work on systems that are not of immediate practical importance, but which could become so in time, is not encouraged. Still, as pointed out in the final chapter of this book, the applications of ion solvation in many fields of chemistry and other disciplines depends on the availability of reliable data on such systems. Furthermore, theories, models, and interpretations require adequate data to operate on, in order to be tested and to provide the insight on the interactions and processes that is being sought. I have therefore endeavoured in this book to present as many reliable data as seem to be relevant, without trying to be exhaustive, and to provide these with appropriate annotations. This monograph can therefore serve as a source-book within the scope of its subject. (This corrects an oversight in my previous book, where the sources of the data included have not been mentioned, so that for some of them that were needed for the present work, I had to do the literature search over again.) I hope that the long lists of references and the extensive tables do not detract too much from the readability of the book. I preferred to have the tables right at the place where the data are discussed or where they can be employed by the reader as an illustration to the points discussed, rather than have them relegated to appendixes.

The significance of the subject of ion solvation and the organization of the present book dealing with it are described in detail in the introductory chapter and need not to be repeated here. It is sufficient to say that the central role that water plays in solution chemistry is recognized by the allocation of an entire chapter to ion hydration. Also in other chapters, in many cases, separate tables are devoted to the presentation of numerical data concerning hydration. However, the mistake of disregarding solvation in and by other solvents was avoided, and a major fraction of the book is devoted to solvation in nonaqueous and mixed solvents. Since, however, the essential concepts have already been discussed with regard to hydration, only the changes occurring on *transfer* from water to the nonaqueous or mixed solvent need to be discussed at this stage. Some more fundamental reasons for doing so are provided in the text.

It is a pleasant duty to acknowledge the support obtained during some of the period when the book was being written from the Gmelin Institute of the Max Planck Society in Frankfurt, Germany. Professor E. Fluck and Dr W. Lippert at the head of the Institute and my other colleagues there provided the atmosphere and the facilities that helped me to do the library research and some of the writing, while being engaged also on work for the Institute. Being in Frankfurt meant also absence from my activities and family in Jerusalem. Leave of absence on a sabbatical from the Hebrew University permitted this stay in Frankfurt on the one hand, the patience and forbearance of my wife and daughters—as was the case during the writing of my previous book—permitted it on the other. This is hereby acknowledged and deeply appreciated.

Yizhak Marcus
Jerusalem
May 1985

Chapter 1

Introduction

1.1 The significance and phenomenology of ion solvation

Much of chemistry pertains to and is conducted in liquid solutions and involves ionic species. Examples from nature are the oceans, these vast bodies of multicomponent aqueous salt solutions, and physiological fluids, which may reflect the distant marine origin of life on earth, and contain ionized or ionizable solutes together with nonelectrolytes in an essentially aqueous medium. Mixed aqueous–organic or totally organic reaction media have been utilized by organic chemists to synthesize many valuable drugs and other compounds by reaction paths that involve ionic intermediates or products. Analytical chemists have since long ago employed organic solvents to precipitate salts from aqueous solutions in which they are too soluble for a gravimetric determination. More recently, mixed aqueous–organic solvents and also nonaqueous solvents have been used for the study of electrochemical processes and for the carrying out of electroanalytical determinations. The technology of liquid–liquid distribution, i.e., solvent extraction, has been employed for hydrometallurgical processing on an industrial scale, metal ions being removed from an aqueous solution into a second, essentially immiscible, liquid phase.

In all these and many other systems the ions interact with one another and with nonelectrolytes, when present, and usually these interactions are of prime interest to the chemist. However, the extent of these interactions and the rate of any reactions in which the ions take a part depend strongly on the nature of the solvent present, i.e., on the ion—solvent interactions. These, in turn, are generally sufficiently strong so that the properties relevant for any further interactions or reactions become those of the solvated ion, rather than those of the bare, i.e., unsolvated ion. In fact, 'bare' ions are encountered only in the gas phase but never in condensed media. It is a question of semantics, of course, whether the ions in a molten salt are considered to be bare. They are always surrounded by other ions, preponderantly of the opposite charge, with which they interact strongly. In the normal situation, which is the subject of this book, where a large excess of a nonionic solvent is present, each ion is surrounded by the solvent molecules, with which it interacts, by which it is solvated. Otherwise, some or all of the solvent may be removed from the near environment of an ion, but it must then be exchanged for some other solvent molecules or some other ligands.

The role of the solvent, affecting the interactions and reactions that ions undergo in solution, has often been ignored in the past. Only in recent years has it been given due attention, as the publication of a number of monographs dedicated partly or entirely to this subject attests.[1-4] For the present purposes 'ion solvation' will be the term that describes all ion–solvent interactions. A detailed knowledge of it is a prerequisite for intelligent and skilful exploitation of the benefits that the free choice of the solvent as a reaction medium can provide. Conversely, if the solvent is prescribed, an understanding of ion solvation can help the chemist to optimize the other variables in his reaction system to effect his purpose.

The mere fact that a solution exists, that contains an electrolyte in ionized, or at least partly ionized, form, indicates that ion solvation has taken place. The pure electrolyte may be a solid ionic salt, such as NaCl, $(C_4H_9)_4NClO_4$, or $C_{12}H_{25}NH_3{}^+C_7H_{15}CO_2{}^-$. It may, for instance, also be a gaseous ionophore, such as HCl or NH_3, that does not contain ions. In the case of the solid salt, the electrostatic lattice energy that holds the ions together (with superimposed other interactions) must be overcome in order for the salt to dissolve and form a dilute solution where the ions are far from each other. In the case of the gaseous solutes, covalent bonds have to be broken (in the case of NH_3 dissolving in water, it is a covalent bond of H_2O that is mainly relevant) in order for ions to be formed, and a loss of translational entropy must be compensated for in order for the gas to dissolve. In addition, the mutual interactions of the solvent molecules that make it a coherent condensed phase at room temperature must be worked against, in order to accomodate the ions in the solution. All these equivalents of energy that have to be invested for an ionic solution in the solvent to be formed have to be recovered from the direct interactions of the finally resulting ions with the molecules of the solvent. Thus, ion solvation becomes a prerequisite for the existence of electrolyte solutions. Of course, nothing is said at this stage about the nature of these interactions, that are designated by the general term 'ion solvation'.

If one wishes to form an idea of ion solvation unencumbered by other interactions, one must envisage a single ion in an infinite amount of solvent. Generally, however, it is necessary to have ions of opposite signs, constituting a neutral component, dissolved in the solvent at some high dilution, in order to study ion solvation experimentally in the laboratory. When such measurements are extrapolated to infinite dilution one arrives at the state to which standard thermodynamic quantities refer. Because of the absence of ion–ion interactions in this state, these standard thermodynamic quantities must be additive in the contributions of the individual kinds of ions. Theoretical examination of the interactions of the ions with the solvent that constitute 'ion solvation' are based, indeed, on models involving one given ion and its surroundings. The model employed may be electrostatic and relate to ion–dipole, ion–quadrupole, ion–induced dipole, etc., interactions. On the other hand, the model may be quantum-mechanical and relate to the overlap of orbitals of the ion with those of certain atoms in the solvent molecules. The model may, finally, deal with certain empirical properties of the ions and the solvent molecules, but in all cases it is one ion that is considered, interacting with the surrounding solvent. It is, therefore, profitable to discuss individual ion solvation, provided that the resulting quantities are added in the proportions appropriate for the neutral component before comparison is made with the experimentally obtained data of thermodynamic nature (solubilities, heats of solution, densities, etc.).

Certain non-thermodynamic properties of individual kinds of ions can, however, be measured in the laboratory and related directly or indirectly to their solvation. These include transport properties, such as mobilities under electrical field gradients (equivalent conductivities) or concentration gradients (diffusivities). They include also some spectroscopic properties, such as the ultraviolet–visible spectra of transition metal ions or nuclear magnetic resonance spectra involving nuclei of certain metal ions (e.g. ^{7}Li, ^{27}Al, or ^{205}Tl). These properties may also include certain effects of ions on molecular spectra, such as proton magnetic resonance or infrared spectra, of the solvent. It is not straightforward to relate the one type of information to the other, but consistent descriptions of the ion solvation and predictions of the behaviour of the electrolyte solution must be the outcome of valid applications of the theoretical interpretations in all these areas.

In most elementary texts hydration is the only form of ion solvation encountered. Since water is nearly unique among solvents by being highly structured, special effects are encountered in hydration that are in general absent when ions are solvated by other solvents. An unbalanced view of solvation may be the result of this exclusive consideration of solvation by water. On the other hand, hydration is of great practical as well as theoretical interest, and the special phenomena observed for hydration should be studied carefully. For instance, 'negative hydration', a term much used by Russian chemists, refers to the water-structure-breaking properties of certain kinds of ions, and is irrelevant to other, unstructured, solvents. 'Hydrophobic interactions' is another concept that is practically unique to aqueous solutions, and applies to such ions as $(C_4H_9)_4N^+$. The enhanced intermolecular structure observable in the water molecules surrounding such an ion in the solution strongly affects the properties of the solution. Probably of greater significance than such a highly symmetrical ion, but more difficult to study, hence much less investigated and understood, are the ionic side chains of proteins. These carry carboxylate groups, $-CO_2^-$, primary ammonium groups, $-NH_3^+$, or protonated nitrogen atoms that are members of heterocyclic rings, bonded to aliphatic or aromatic moieties that undergo hydrophobic interactions in an aqueous environment. A consideration of these is an area of biophysics that is outside the scope of the present book (see Refs. 5, 6 for their discussion).

The study of ion solvation in nonaqueous and mixed aqueous–organic solvents has proliferated in recent years, first because techniques for studying it have become available, and secondly because of foreseeable applications. The techniques that have been evolved are mainly electrochemical and spectroscopic; applications are, for instance, in the field of high-energy-density batteries and hydrometallurgy. An important issue in this respect is the comparison of different solvents with one another (and with a reference solvent, such as water). The efficacy of solvents for the solvation of ions may be beneficial, but it may also be detrimental, as it decreases the reactivity of the ions required for some chemical reaction. Organic chemists have, for instance, established the great advantage of using aprotic solvents for reactions involving anions, since these are not appreciably solvated by such solvents. In another context it is important to refer electrochemical redox potentials and pH values to a unique scale and reference point, rather than have a multitude of unrelated scales. Analytical chemists and electrochemists have endeavoured for many years to achieve such a unification of these scales, and a solution to the problem seems to be now at hand.

Solvents can be characterized in many ways. For the present purpose the gross division into nonpolar and polar solvents is useful. The former category plays little role in ion solvents, because of the very small solubility of most electrolytes in solvents belonging to it. Polar solvents are often subdivided into protic and aprotic ones, both subcategories being significant for the solvation of ions. The commonly quoted measures of solvent polarity, the dipole moment and the relative permittivity (dielectric constant) are not sufficient for the characterization of the properties relevant for ion solvation. The electron pair acceptance and donation abilities of the solvents and the hydrogen bond donation and acceptance abilities are of great importance in this respect. The goal of a useful theory of ion solvation should be the prediction of ion solvation behaviour for any ion–solvent combination, given small sets of independently measurable properties of the ions and the solvents. This has now been attempted, with moderate success so far.

Solvent mixtures are encountered in 'real life' situations more frequently than very pure single solvents. In order to master such situations it is necessary first of all to study the properties of solvent mixtures, not least those of mixtures rich in a relatively not very reactive solvent that contain as a minor constituent a highly reactive solvent, such as water. It is not to be expected that the solvation properties of mixed solvents should be additive. Solvent molecules would compete for inclusion in the near envionment of an ion, its first solvation shell. This situation should first be described in quantitative terms (selective solvation), then explained in terms of the ion and solvents involved, and lastly be predictable from the properties of these. Some little progress has been made along these lines, and it is worthwhile to consider what is now known.

Ion solvation proper is most definite at infinite dilution, where only ion–solvent interactions are superimposed on the perennial solvent–solvent interactions that occur in the pure solvent (or mixture of solvents). However, 'real life' situations involve finite concentrations of the electrolyte solute in the solution. They may even involve quite high concentrations, so that the ratio of the number of solvent molecules present per formula unit of the electrolyte is relatively small (<10). As the concentration of the electrolyte is permitted to increase gradually, the ion–solvent interactions must adjust themselves to the presence of increasingly important ion–ion interactions. These changes occur perceptibly only at moderate concentrations, unless ion pairing occurs. In the latter case the solvent and an ion of the opposite charge compete for sites in the solvation shell of a given ion. When sufficiently high concentrations are reached, methods for the investigation of structure in liquids, such as neutron diffraction, can be employed. These have been applied almost exclusively to aqueous solutions, for understandable reasons, but they do yield information concerning the relative importance of hydration and ion–ion interactions, which is, however, also obtainable from spectroscopic investigations.

Ion solvation has been surveyed above from several aspects, and it is the purpose of this book to delve deeper into the facts that are already known about these aspects, and the interpretations that have been given to these facts. It is hoped that the insight gained will lead readers to useful applications in their own fields of investigation.

1.2 List of symbols and abbreviations

Roman type characters are used for chemical species and units of physical quantities, whereas Greek characters and *italic* characters are used for physical quantities expressible in numerical terms, except for universal physical constants, that are in **bold type**. Partial molar quantities generally have a bar ¯ over the symbol, but extensive and molar quantities are not generally distinguished. Mathematical symbols have their usual meanings, but specifically, $\langle\ \rangle$ around a symbol designates its average value for all the configurations of a system. The following list includes the main symbols employed, their SI units, and, where required, the equation where a symbol is defined or first used.

List of principal symbols, Latin characters

A	coefficient in the Debye–Hückel expression for the activity coefficient
AN	acceptor number (Gutman)
a	distance of closest approach of ions (m)
a_{x}	thermodynamic activity of x
(aq)	state of infinitely dilute aqueous solution
B	coefficient in the Debye–Hückel expression for the activity coefficient
B	B-coefficient of the Jones–Dole viscosity equation (5.29) ($dm^3\ mol^{-1}$)
B_{x}	binding energy of x to the rest of the system
b	coefficient of expression for ε(E) (equation (3.35)) ($m^2\ V^{-2}$)
b	parameter in Bjerrum's ion pairing theory (equation (8.4))
C_{p}	molar heat capacity at constant pressure ($J\ K^{-1}\ mol^{-1}$)

c_x	molar concentration of x (moles of solute per unit volume, n_x/V, $\rho_x/\mathbf{N}$, mol dm^{-3})
D	Debye unit of dipole strength ($3.33564 \cdot 10^{-30}$ C·m)
D	diffusion coefficient (m^2 s^{-1} mol^{-1})
D_x, D(x, y/z)	distribution ratio of x (between liquids y and z)
DN	donor number (Gutmann)
d	density (kg m^{-3})
d, d_{xy}	interatomic distance (between x and y) (m)
E	energy, molar energy (J mol^{-1})
E	electrical field strength (V m^{-1})
E	molar expansibility (K^{-1} mol^{-1})
E	electromotive force of electrochemical cell (V)
E^0	standard electrode potential (V)
E_j	liquid junction potential (V)
$E_{1/2}$	polarographic half-wave potential (V)
E_T	electron pair acceptance polarity index (kcal mol^{-1})
$\mathbf{e}$	elementary charge ($1.6021 \cdot 10^{-19}$ C)
$\mathbf{e}_0$	rest energy of electron ($8.184 \cdot 10^{-14}$ J)
e_{xy}	interaction energy of neighbouring x and y (J)
$\mathbf{F}$	Faraday's constant ($9.6487 \cdot 10^4$ C mol^{-1})
F	number of solvent hydrogen bonds broken by anion (equation (4.17))
f	ultrasonic frequency (s^{-1})
f_r	relaxation frequency ($1/2\pi$ times the relaxation time of the system, s^{-1})
f_x	coherent scattering amplitude by x
G	Gibbs free energy, molar Gibbs free energy (J mol^{-1})
g	geometrical factor for hydrogen bonding (equation (3.40))
g	dipole orientation parameter (Kirkwood) (equation (6.1))
g	preferential solvation parameter (equation (7.27))
(g)	gas phase
g_{xy}	pair correlation function (probability of finding y, given x)
H	enthalpy, molar enthalpy (J mol^{-1})
H_0	acidity function (Hammett)
h	solvation or hydration number
$I^{\pm}$, I^z	generalized ion
I	ionic strength ($\frac{1}{2}\Sigma c_i z_i^2$, mol dm^{-3})
$I(\theta)$	intensity of radiation scattered through angle θ
i	generalized ion
K	equilibrium constant
K	molar compressibility (Pa^{-1} mol^{-1})
K_a, K_b	acid, base dissociation constant in water (dm^3 mol^{-1}, mol dm^{-3}, respectively)
K_{ai}	auto-ionization ion product (mol^2 dm^{-6})
K_{ass}	association constant of (contact) ion pair (dm^3 mol^{-1})

K_i	stepwise replacement, association constant (dimensionless, $dm^3 mol^{-1}$)
K_{ip}	ion pair association constant from Bjerrum's theory ($dm^3 mol^{-1}$)
K_W	K_{ai} of water ($mol^2 dm^{-6}$)
$\bar{K}$	mean equilibrium constant for a series of steps
$\mathbf{k}$	Boltzmann's constant ($1.3805 \cdot 10^{-23}$ J K^{-1})
k	rate constant (s^{-1} for unimolecular reaction, $dm^3 mol^{-1} s^{-1}$ for bimolecular reaction)
k	wave-number of radiation used for diffraction (m^{-1})
k	packing factor (equation (4.26))
k_r	rate constant for solvent release from solvation shell (s^{-1})
k_x	rate constant in solvent x (see k for units)
L	length of cubic box used in Monte Carlo computer simulation calculations (m)
(l)	liquid phase
M^{z+}	metal ion of charge z^+
M_x	molar mass of x (kg mol^{-1})
m_x	molal concentration (moles of x per unit mass of solvent, mol (kg solvent)$^{-1}$)
$\mathbf{N}$	Avogadro's number ($6.0223 \cdot 10^{23} mol^{-1}$)
N_x	number of x particles in system
N_{xy}	number of nearest x–y neighbouring pairs
n	number of solvent molecules or other ligands bound to an ion
n_C	number of carbon atoms in alkyl chain
n_D	refractive index for yellow sodium D line
n_x	amount of the substance x (mol)
$\bar{n}_x$	number of moles of x imbibed by unit mass of dry ion exchange resin (mol kg^{-1})
P	pressure (Pa)
p	vapour pressure (Pa)
$P(x)$	generalized numerical property of x (equation (6.27))
Q	classical canonical partition function
Q	equilibrium quotient (for subscripts see K)
$Q(b)$	integral in Bjerrum's theory of ion pairing (equation (8.5))
Q_y^x	ion exchange selectivity quotient for x over y
q_x	internal partition function of an x particle
$\mathbf{R}$	molar gas constant (8.3143 J $K^{-1} mol^{-1}$)
R_0	fixed coordinates and orientations at a fixed point of a system
R^N	one configuration of coordinates and orientations of N particles
R_W	ratio of moles of water per mole of salt
R_x	molar refractivity of x ($m^3 mole^{-1}$)
r	distance between centres of particles
r	radius of sphere representing an ion (see subscripts for descriptors)

r	ratio of lattice parameters in two solvents (equation (7.43))
S	generalized solvent
S	entropy, molar entropy (J K^{-1} mol^{-1})
$S(k)$	structure factor in k-space
S_κ,S_V	slopes of partial molar compressibility, volume expressions (coefficients of terms in $c^{1/2}$)
s_x	solubility of x (molar, mol dm^{-3})
(s)	solid phase
(ss)	standard state
T	(absolute) temperature (K)
T_1, T_2	longitudinal, transverse relaxation times of NMR signal (s)
t	(centigrade) temperature (°C)
t_b	normal boiling temperature (at 0.101325 MPa, °C)
t_m	freezing, melting temperature (°C)
U	potential energy of interactions in a system (J)
U_{hs}	pair potential for hard sphere particles (J)
u	speed of ultrasound (m s^{-1})
u_{hb}	molar hydrogen bond energy (J mol^{-1})
u_x	minimum value of pair potential (depth of potential well) for interactions between x particles (J)
V	volume, molar volume (m^3 mol^{-1})
W	a reference solvent, water
W	work (per molecule) (J)
$W(x \mid y)$	coupling work of an x particle to the system y (J)
w	weighting factor (equation (7.43))
w_x	mass fraction of x
X	generalized solute
x_x	mole fraction of x
Y	generalized molar thermodynamic function (G, H, S, V, C_p . . .)
y	packing fraction of solvent ($\pi\sigma_S^3N/6V_S$)
y_x	molar activity coefficient of x
Z	atomic number
Z	configurational partition function
Z	coordination number, number of nearest neighbours
Z	lattice parameter
z_i, z	charge number of ion (taken alegebraically)

List of principal symbols, Greek characters

α	hydrogen bond donation ability (Kamlet and Taft)
α	absorption coefficient for ultrasonic energy
α	fraction of electrolyte dissociated into ions
α_P, α_{PS}	isobaric thermal expansibility (of solvent S) ((∂ ln $V/\partial T)_P$, K^{-1})
α_x	polarizability of x (m^3)
α_x	'real' chemical potential of x (J mol^{-1})

β	propensity to accept a hydrogen bond (Kamlet and Taft)
$\gamma_{\pm}$	mean ionic molal activity coefficient
γ	surface tension of solvent S ($N\ m^{-1}$)
Δ	partition function at constant pressure
δ	chemical shift of NMR signal (ppm, against an external reference)
δ	solubility parameter (Hildebrand, $J^{1/2}\ cm^{-1/2}\ mol^{-1/2}$)
ε_0	absolute permittivity of free space ($8.8542 \cdot 10^{-12}\ C^2\ J^{-2}\ m^{-1}$)
ε	relative permittivity
η	coefficient of viscocity ($Pa \cdot s$)
θ	scattering angle (between incident and scattered radiation)
θ_x	quadrupole moment of molecules of x ($C \cdot m^2$)
κ	specific conductance ($S\ m^{-1}$)
κ_S	adiabatic compressibility (Pa^{-1})
κ_T, κ_{TS}	isothermal compressibility (of solvent S) ($(\partial \ln V/\partial P)_T$, Pa^{-1})
Λ	equivalent conductivity of an electrolyte ($S\ m^{-1}\ dm^3\ mol^{-1}$)
Λ_x	momentum partition function of an x particle
λ_i	equivalent conductivity of ion i ($S\ m^{-1}\ dm^3\ mol^{-1}$)
μ_x	dipole moment of x (D)
μ_x	chemical potential of x ($J\ mol^{-1}$)
ν	wave-number (cm^{-1})
ν	stoichiometric coefficient (number of ions per formula of ionizing electrolyte)
π^*	polarity index (Abboud, Kamlet and Taft)
σ_x	number density of x (N_x/V, m^{-3})
σ	molecular collision diameter (m)
σ	chemical shift of NMR signal (ppm, against an internal reference)
σ_x	diameter of a sphere representing particle x
σ_M, σ_X	softness parameter (Marcus) of cation, anion
τ	mean residence time (s)
ϕ_x	volume fraction of x
χ	surface potential against a gas phase (V)
χ_x	molecular (diamagnetic) susceptibility of x

List of principal subscripts and superscripts

ad	pertaining to the process of adsorption
cav	pertaining to the process of cavity formation
conv	conventional (thermodynamic quantity)
cov	covalent (radius)
dip	contribution from dipole interactions
disp	contribution from dispersion (London) forces
E	excess thermodynamic quantity of mixture
el	contribution from electrostatic interactions
els	pertaining to electrostriction
f	pertaining to the process of formation

F	of freezing, fusion
hydr	pertaining to the process of hydration
i	of an ion i
i c	crystal (radius) of ion
ih	hydrated (radius) of ion
ind	contribution from induction
intr	intrinsic value of solute
i S	Stokes (radius) of ion
N	normalized
neut	pertaining to a neutral species
S	of the solvent S
s	pertaining to a standard buffer solution
soln	pertaining to the process of solution
solv	pertaining to the process of solvation
solv(x, y)	pertaining to the solvation of the solute x by the solvent y
str	structural contribution
t	of transfer
$t(x, y \rightarrow z)$	pertaining to the transfer of solute x from solvent y to solvent z
V	of vaporization
vdW	van der Waals (radius or volume)
*	standard state of pure substance, liquid
$\circ$	standard thermodynamic function (standard state understood)
∞	standard state of infinite dilution
$\cdot$	pertaining to a single particle
$\neq$	of activation
ϕ	(presuperscript) apparent molar

Chemical substances are generally referred to in the text by their names, occasionally by their formulas. In Tables and as subscripts, however, abbreviations are often used. In addition to the standard abbreviations, such as Me = methyl, Et = ethyl, Pr = propyl, Bu = butyl, Pn = pentyl, Hx = hexyl for aliphatic straight-chain radicals, and Ph = phenyl, a few others are used (e.g., pic = picrate anion), except for the following list of abbreviations and mnemotechnics employed for solvents.

List of the principal abbreviations for the names of solvents

AcOH	acetic acid (CH_3COOH)
Ac_2O	acetic anhydride ($CH_3C(O)OC(O)CH_3$)
BuOAc	n-butyl acetate ($C_4H_9OC(O)CH_3$)
BuOH	1-butanol (C_4H_9OH)
2-BuOH	2-butanol ($C_2H_5CH(OH)CH_3$)
i-BuOH	2-methyl-1-propanol ($(CH_3)_2CHCH_2OH$)
t-BuOH	2-methyl-2-propanol ($(CH_3)_3COH$)
BzCN	phenylacetonitrile, benzyl cyanide ($C_6H_5CH_2CN$)
BzOH	banzyl alcohol ($C_6H_5CH_2OH$)

1,1DClE	1,1-dichloroethane (CH_3CHCl_2)
1,2DClE	1,2-dichloroethane (CH_2ClCH_2Cl)
DEA	N,N-diethylacetamide ($CH_3C(O)N(C_2H_5)_2$)
DEF	N,N-diethylformamide ($HC(O)N(C_2H_5)_2$)
DMA	N,N-dimethylacetamide ($CH_3C(O)N(CH_3)_2$)
DME	1,2-dimethoxyethane ($CH_3OC_2H_4OCH_3$)
DMF	N,N-dimethylformamide ($HC(O)N(CH_3)_2$)
DMSO	dimethylsulphoxide ($CH_3S(O)CH_3$)
DMThF	N,N-dimethylthioformamide ($HC(S)N(CH_3)_2$)
EC	1,3-dioxolan-2-one, ethylene carbonate ($\overline{\llcorner OC_2H_4OC(O) \lrcorner}$)
en	1,2-diaminoethane, ethylene diamine ($H_2NC_2H_4NH_2$)
$En(OH)_2$	2-hydroxyethanol, ethylene glycol (HOC_2H_4OH)
EtOH	ethanol (C_2H_5OH)
EtOAc	ethyl acetate ($C_2H_5OC(O)CH_3$)
Et_2O	diethyl ether ($C_2H_5OC_2H_5$)
FA	formamide ($HC(O)NH_2$)
HMPT	hexamethyl phosphoric triamide ($OP(N(CH_3)_2)_3$)
HxOH	1-hexanol ($C_6H_{13}OH$)
MeCN	acetonitrile, methyl cyanide (CH_3CN)
Me_2CO	acetone ($CH_3C(O)CH_3$)
$MeNO_2$	nitromethane (CH_3NO_2)
MeOAc	methyl acetate ($CH_3OC(O)CH_3$)
MeOH	methanol (CH_3OH)
NMA	N-methylacetamide ($CH_3C(O)NHCH_3$)
NMF	N-methylformamide ($HC(O)NHCH_3$)
NMP	N-methylpropanamide ($C_2H_5C(O)NHCH_3$)
NMPy	N-methylpyrrolidinone ($\overline{\llcorner C_3H_6N(CH_3)C(O) \lrcorner}$)
OcOH	1-octanol ($C_8H_{17}OH$)
PC	4-methyl-1,3-dioxolan-2-one, propylene carbonate ($\overline{\llcorner OCH(CH_3)CH_2OC(O) \lrcorner}$)
PhCl	chlorobenzene (C_6H_5Cl)
PhCN	benzonitrile (C_6H_5CN)
$PhNH_2$	aniline ($C_6H_5NH_2$)
$PhNO_2$	nitrobenzene ($C_6H_5NO_2$)
PhOH	phenol (C_6H_5OH)
PnOH	1-pentanol ($C_5H_{11}OH$)
Pr_2O	di-n-propyl ether ($C_3H_7OC_3H_7$)
PrOH	1-propanol (C_3H_7OH)
iPrOH	2-propanol ($(CH_3)_2CHOH$)
py	pyridine (C_5H_5N)
TBP	tri-n-butyl phosphate ($(C_4H_9O)_3PO$)
TEA	triethylamine ($(C_2H_5)_3N$)
TEP	triethyl phosphate ($(C_2H_5O)_3PO$)
TFE	2,2,2-trifluoroethanol (CF_3CH_2OH)
THF	tetrahydrofuran ($\overline{\llcorner C_4H_8O \lrcorner}$)

TMP	trimethyl phosphate ($(CH_3O)_3PO$)
TMS	S,S-tetrahydrothiophenedioxide, tetramethylene sulphone, sulpholane ($\llcorner C_4H_8S(O)_2 \lrcorner$)
TMU	N,N,N′,N′-tetramethylurea ($(CH_3)_2NC(O)N(CH_3)_2$)
W	water (H_2O)

References

1. J. F. Coetzee and C. D. Ritchie, eds., *Solute–Solvent Interactions*, Dekker, New York (1969).
2. S. Petrucci, ed., *Ionic Interactions*, Academic Press, New York, Vol. I and in particular Vol. II (1971).
3. J. J. Lagowski, ed., *The Chemistry of Non-Aqueous Solvents*, Academic Press, New York, Vol. I (1966), Vol. II (1967), Vol. III (1970), Vol. IV (1976), Vols. Va and Vb (1978).
4. J. Burgess, *Metal Ions in Solution*, Ellis Horwood, Chichester (1978).
5. B. E. Conway, *Ionic Hydration in Chemistry and Biophysics*, Elsevier, Amsterdam (1981).
6. A. Ben-Naim, *Hydrophobic Interactions*, Plenum Press, New York (1980).

Chapter 2

Ion solvation in the gas phase

The most direct information concerning the nature and strength of the interaction of a given ion with a given solvent is obtained when the interaction takes place in a dilute gaseous phase. The interaction of the ion with one or more solvent molecules to form so called solvation clusters can be studied under such conditions by appropriate experimental methods. It can also be established by calculation, in favourable cases *ab initio*, in others at least semi-empirically. The information obtained from the former kind of studies is the bonding Gibbs free energy or its enthalpy (and entropy) for increasing solvation numbers. That obtained from the second kind of studies is the geometry and bond distances that minimize the energy of the system for a given solvation number, and the resulting bonding energy. This should be in accord with the measured bonding enthalpy for the solvation number considered.

The question of the relevance of this information to the study of the solvation of ions in dilute liquid solutions has been raised. Neither measurements nor calculations can be made beyond solvation numbers of about 8. The Gibbs free energy or enthalpy for the gas phase solvation may be plotted against the solvation number, and the curve extrapolated to such large solvation numbers, where the whole of the effect has been taken into account. Since it is found that the dependence on the solvation number is not necessarily monotonic, the extrapolation is not reliable. In certain models for ion solvation, however, that consist of multi-layer solvation shells, the direct interaction that takes place in the innermost shell should be that obtained from gas-phase studies. The models then take care of further interactions in various ways, as specified in Section 3.6.

2.1 Experimental data

A great deal of the experimental determinations of the solvation Gibbs free energies of ions in the gas phase is due to Kebarle and his co-workers, who have also reviewed this work and that of others.[1] The method employed involves the mass-spectrometric measurement of the equilibrium concentrations of ionic solvates, present at low concentrations in a large excess of gaseous solvent molecules at a pressure relatively high for mass-spectrometric conditions (500 to

1500 Pa). The solvent molecules are in thermal equilibrium with the surroundings of the ion chamber, and sufficient time is permitted also for the ions and solvates to attain equilibrium, without losing too many of them by reactions with the walls of the vessel or with ions of opposite sign. The latter conditions are ensured by the short free path lengths of the ions diffusing through the relatively abundant solvent molecules, and by the low abundance of the ions. Various arrangements of the mass spectrometer have been used to achieve this. A pulsed beam electron source combined with a high pressure ion source is very effective. The positive ions must be obtained by thermionic emission rather than by electron beam irradiation. The ions and solvated ions are bled from the high pressure region through an orifice, into an evacuated part of the apparatus for mass-spectrometric determination. Some problems, such as the break-up of ion-solvent clusters, may be traced back to this source. The pulsed electron beam, trapped ion cell ion cyclotron resonance (ICR) method obviates this particular source of error. The pressures utilized in this method are of the order of 10^{-4} Pa.

The direct results of the measurements are equilibrium constants and their temperature coefficients. These lead to values of $\Delta G^0_{n-1,n}$ and of $\Delta H^0_{n-1,n}$ for the reactions

$$I^{\pm}S_{n-1}(g) + S(g) \rightleftharpoons I^{\pm}S_n(g) \tag{2.1}$$

where $I^{\pm}$ is a cation or an anion and S a solvent molecule.

Extensive information for a variety of solvents is available for the solvation of an ion with just one solvent molecule (i.e., $n = 1$). Representative data are shown in Table 2.1. Information on solvation with further solvent molecules (i.e., $n \geqslant 2$) is available mainly for the solvent $S = H_2O$, and is shown in Table 2.2. The standard states in all cases are ideal gases at 1 atm pressure (1 atm = 101 325 Pa), and the temperature at which the data are valid is 298 K. The changes in entropy for the individual steps[2] do not depend much on n, and for the solvent water they contribute $-T\Delta S^0_{n-1,n}$ of 25 to 30 kJ mol^{-1}, making $\Delta G^0_{n-1,n}$ less negative by that amount relative to the values of $\Delta H^0_{n-1,n}$ tabulated in Table 2.2.

The trend in the results in Table 2.2 is that there is only a moderate decline in the enthalpies of the individual steps (except, in some cases, from the first to the second), until a fairly large cluster is built up, when a more abrupt drop is noted in some cases (e.g., NH_4^+–NH_3, F^-–CH_3CN, Cl^-–CH_3CN).

Excluded from the ions considered in Tables 2.1 and 2.2 is the bare proton. So called 'proton affinities' have been determined for many bases in the gas phase. A representative selection of values is shown in Table 2.3. The data are normalized relative to the value -846.4 kJ mol^{-1} for ΔH^0 of the reaction[3]

$$H^+(g) + NH_3(g) \rightleftharpoons NH_4^+(g)$$

The proton affinities, i.e. the enthalpies for the solvation of a bare proton with a base (i.e. a solvent molecule), are seen to be much larger than the corresponding

Table 2.1 Thermodynamic quantities for the reaction $I^{\pm}(g) + S(g) \rightleftharpoons I^{\pm}S(g)$

$I^{\pm}$	S	$-\Delta G^0$* (k J mol^{-1})	$-\Delta H^0$ (k J mol^{-1})	$-\Delta S^0$ (J K^{-1} mol^{-1})	Ref.
Li^+	H_2O	(113)	142	96	a
	CH_2Cl_2		121		b
	c-C_6H_{12}		100		b
	C_6H_6		155		b
	CH_3OH		159		b
	CH_3NO_2		163		b
	CH_3CHO		172		b
	$HCOOCH_3$		172		b
	CH_3COOCH_3		184		b
	CH_3COCH_3		188		b
	$HCON(CH_3)_2$		209		b
	NH_3		159		b
	CH_3CN		180		b
	C_5H_5N		184		b
Na^+	H_2O	(73)	100	92	a
K^+	H_2O	48	75	90	s
		(46)	71	83	a, c
	NH_3	(49)	74	84	c
	CH_3OCH_3	(56)	87	104	c
	$C_2H_5OC_2H_5$	(62)	93	103	c
	CH_3NH_2	(53)	80	90	c
	$(CH_3)_2NH$	(55)	82	90	c
	$(CH_3)_3N$	(55)	84	98	c
	$C_6H_5NH_2$	(65)	95	99	c
	C_5H_5N	(64)	87	78	c
	CH_3CN	75	102	90	c
	$H_2NC_2H_4NH_2$	(80)	108	93	d
	$CH_3OC_2H_4OCH_3$	(96)	129	112	d
Rb^+	H_2O	(41)	67	88	a
	CH_3CN	64	87	76	c
Cs^+	H_2O	(35)	59	81	a
	CH_3CN	57	80	78	c
H_3O^+	H_2O	(96)	138	141	g
		(104)	134	102	e, f
NH_4^+	H_2O	(47)	72	84	h
	NH_3	65	90	84	p
		(73)	105	109	h
F^-	H_2O	(75)	97	73	i
	CH_3CN		67		j
Cl^-	H_2O	34	55	69	i
	CH_3OH	41	59	62	j, k
	n-C_4H_9OH	46			l
	t-C_4H_9OH	46	59	43	k
		(46)	80	113	m
	C_6H_5OH	62, 60	81	65	k, l
	FC_6H_4OH	64			l
	ClC_6H_4OH	72			l
	NCC_6H_4OH	96			l

(*continued*)

Table 2.1—*continued*

$I^{\pm}$	S	$-\Delta G^0$* (k J mol^{-1})	$-\Delta H^0$ (k J mol^{-1})	$-\Delta S^0$ (J K^{-1} mol^{-1})	Ref.
	HCOOH	106	156	166	k
		78			l
		(86)	117	105	m
	CH_3COOH	66	90	81	k
		62			l
	CH_3COCH_3	29	57		l
	$CHCl_3$	45	64	62	k
	CH_3CN	40	56	51	j, l
Br^-	H_2O	(29)	53	79	i
I^-	H_2O	24			t
		(23)	43	68	i
OH^-	H_2O	(79)	105	87	n
			~150		q
		71	94	80	r
CN^-	H_2O	(38)	58	66	n
NO_2^-	H_2O	34	60	88	n, o
NO_3^-	H_2O	28	52	80	n, o

* Values in ( ) calculated from $\Delta H^0 - 298.15\Delta S^0$, the others are given in the papers cited. [a] I. Džidič and P. Kebarle, *J. Phys. Chem.*, **74**, 1466 (1970); [b] R. H. Staley and J. L. Beauchamp, *J. Am. Chem. Soc.*, **97**, 5920 (1975); [c] W. R. Davidson and P. Kebarle, *J. Am. Chem. Soc.*, **98**, 6125 (1976); [d] W. R. Davidson and P. Kebarle, *Canad. J. Chem.*, **54**, 2594 (1976); [e] P. Kebarle, S. K. Searls, A. Zolla, J. Scarborough and M. Arshadi, *J. Am. Chem. Soc.*, **89**, 6393 (1967); [f] J. J. Solomon and F. H. Field, *J. Am. Chem. Soc.*, **97**, 2625 (1975); [g] M. Meot-Ner and F. H. Field, *J. Am. Chem. Soc.*, **99**, 998 (1977); [h] J. D. Payzant, A. J. Cunningham and P. Kebarle, *Canad. J. Chem.*, **51**, 3242 (1973); [i] M. Arshadi, R. Yamdagni and P. Kebarle, *J. Phys. Chem.*, **74**, 1475 (1970); [j] R. Yamdagni and P. Kebarle, *J. Am. Chem. Soc.*, **94**, 2940 (1972), and R. Yamdagni, J. D. Payzant and P. Kebarle, *Canad. J. Chem.*, **51**, 2507 (1973); [k] R. Yamdagni and P. Kebarle, *J. Am. Chem. Soc.*, **93**, 7139 (1971); [l] P. Kebarle, W. R. Davidson, M. French, J. B. Cumming and T. B. McMahon, *Disc. Faraday Soc.*, **64**, 220 (1977, publ. 1978); [m] M. A. French and P. Kebarle, unpublished, 1977, quoted in: P. Kebarle, *Ann. Rev. Phys. Chem.*, **28**, 445 (1977); [n] J. D. Payzant, R. Yamdagni and P. Kebarle, *Canad. J. Chem.*, **49**, 3309 (1971); [o] F. C. Fehsenfeld and E. F. Ferguson, *J. Chem. Phys.*, **61**, 3181 (1974); [p] M. R. Arshadi and J. H. Futrell, *J. Phys. Chem.*, **78**, 1482 (1974); [q] M. DePaz, A. G. Giardini and L. Friedman, *J. Chem. Phys.*, **52**, 687 (1970); [r] M. Arshadi and P. Kebarle, *J. Phys. Chem.*, **74**, 1483 (1970); [s] S. K. Searls and P. Kebarle, *Canad. J. Chem.*, **47**, 2619 (1969); [t] P. Kebarle, M. Arshadi and J. Scarborough, *J. Chem. Phys.*, **47** 817 (1968).

solvation enthalpies of other ions with the same solvents. The degree of covalency achieved in the bonding of the proton is evidently much larger than for the other ions, for which an electrostatic model proves to be adequate.

2.2 Theoretical calculations

Consider a single (monoatomic) ion and a single dipolar solvent molecule approaching it from an infinitely large distance. The potential energy of the system will decrease, once the particles come into the range of the interparticle (ion–molecule) forces. When the ion and the dipolar solvent molecule get too

Table 2.2 The standard enthalpy changes for consecutive solvation steps in the gas phase $I^{\pm}S_{n-1}(g) + S(g) \rightleftharpoons I^{\pm}S_n(g)$, ΔH_n^0/k J mol^{-1}

$I^{\pm}$	S	$n = 1$	2	3	4	5	6	Ref.
Li^+	H_2O	142	108	87	69	58	51	a
	NH_3	162	139	88	69	46		k
Na^+	H_2O	100	83	66	58(57)	51(49)	45	a(b)
	NH_3	122	95	72	62	45		k
K^+	H_2O	75	67	55	49	45	42	a
	NH_3	84	68	57	49			m
	CH_3CN	102	86	76	57	48		j
Rb^+	H_2O	67	57	51	47	44		a
	NH_3	78	64	55	48	43		m
Cs^+	H_2O	57	52	47	44			a
H_3O^+	H_2O	132	82	73	64	54	49	c, d, p
NH_4^+	H_2O	72	62	56	51	41		e
	NH_3	105	73	58	52	31		e
F^-	H_2O	97	69	57	56	55		f
	CH_3CN	67	54	49	44	22		g
Cl^-	H_2O	55	53	49	46			f
	CH_3OH	59	54	51	47	44		h
	CH_3CN	56	51	44	26			g
Br^-	H_2O	53	51	48	46			f
	CH_3CN	54	49	42	23			g
I^-	H_2O	43	41	39	38			f
	CH_3CN	50	44	39				g
OH^-	H_2O	105	69	63	59	59		i
NO_2^-	H_2O	64(60	57(54)	49(54)	49			l(i)
NO_3^-	H_2O	61	60	58				l
HCO_3^-	H_2O	66	62	57	56			n

[a] I. Džidič and P. Kebarle, *J. Phys. Chem.*, **74**, 1466 (1970); [b] I. N. Tang and A. W. Castleman, Jr., *J. Chem. Phys.*, **57**, 3638 (1972); [c] P. Kebarle, S. K. Searles, A. Zolla, J. Scarborough, and M. Arshadi, *J. Am. Chem. Soc.*, **89**, 6393 (1967); [d] A. J. Cunningham, J. D. Payzant, and P. Kebarle, *J. Am. Chem. Soc.*, **94**, 7627 (1972); [e] J. D. Payzant, A. J. Cunningham, and P. Kebarle, *Canad. J. Chem.*, **51**, 3242 (1973); [f] M. Arshadi, R. Yamdagni, and P. Kebarle, *J. Phys. Chem.*, **74**, 1475 (1970); J. R. Yamdagni and P. Kebarle, *J. Am. Chem. Soc.*, **94**, 2940 (1972); [h] R. Yamdagni, J. D. Payzant, and P. Kebarle, *Canad. J. Chem.*, **51**, 2507 (1973); J. D. Payzant, R. Yamdagni, and P. Kebarle, *Canad. J. Chem.*, **49**, 3309 (1971); [i] W. R. Davidson and P. Kebarle, *J. Am. Chem. Soc.*, **98**, 6125 (1976); [k] A. W. Castleman, Jr., P. M. Holland, D. M. Lindsay, and K. I. Peterson, *J. Am. Chem. Soc.*, **100**, 6039 (1978); [l] N. Lee, R. G. Keese and A. W. Castleman, Jr., *J. Chem. Phys.*, **72**, 1089 (1980); [m] A. W. Castleman, Jr., *Chem. Phys. Lett.*, **53**, 560 (1978); [n] R. G. Keese, N. Lee, and A. W. Castleman, Jr., *J. Am. Chem. Soc.*, **101**, 2599 (1979); [p] Y. Lau, S. Ikuta, and P. Kebarle, *J. Am. Chem. Soc.*, **104**, 1462 (1982) obtained lower values starting with $n = 4$.

close together, or when the orientation of the dipole to the ionic charge causes electrostatic repulsion, the potential energy will increase again. Obviously there exists at least one distance and mutual orieintation where the potential energy is minimal. There may exist a few equivalent minima, depending on the symmetry of the solvent molecule, and possibly an absolute minimum in addition to several local ones. Forces other than the electrostatic ion–dipole ones also play their role.

Table 2.3 The proton affinities of bases, i.e. ΔH^0 for the reaction $H^+(g) + B(g) \rightleftharpoons HB^+(g)$

B	$-\Delta H^0/$ kJ mol^{-1}	Ref.	B	$-\Delta H^0/$ kJ mol^{-1}	Ref.	B	$-\Delta H^0/$ kJ mol^{-1}	Ref.
H_2O	713	a	$HCOOC_2H_5$	800	a, b	CH_3SCH_3	827	a
	708	b	$HCOOC_3H_7$	803	a, b	$C_2H_5SC_2H_5$	848	a
CF_3CH_2OH	720	a	CH_3COOCH_3	818	a, b	CH_3CN	782	a
CHF_2CH_2OH	744	a	CF_3COOCH_3	754	a	C_2H_5CN	796	a
CCl_3CH_2OH	749	a	$CH_3COOC_2H_5$	828	a, b	(NH_3	846	e)*
CH_3OH	762	a, b	$CF_3COOC_2H_5$	767	a	CH_3NH_2	884	e
C_2H_5OH	781	a, b	$CH_3COOC_3H_7$	831	b	$C_2H_5NH_2$	895	e
CH_3OCH_3	794	a, b	C_6H_6	754	c	$C_3H_7NH_2$	902	e
$CH_3OC_2H_5$	812	a	$C_6H_5CH_3$	789	c	$C_4H_9NH_2$	904	e
Tetrahydrofuran	822	a	$C_6H_5C_2H_5$	784	d	$(CH_3)_2NH$	912	e
Tetrahydropyran	825	a		793	c	$(CH_3)_3N$	929	e
$C_2H_5OC_2H_5$	826	a, b	C_6H_5F	748	c	$(C_2H_5)_2NH$	932	e
Dioxane	805	a	C_6H_5Cl	749	c	$(C_2H_5)_3N$	958	e
HCHO	731	a	$C_6H_5NO_2$	800	c	$(C_3H_7)_3N$	966	e
CH_3CHO	775	a, b	C_6H_5CN	810	c	C_5H_5N	913	e
C_2H_5CHO	786	a, b	C_6H_5OH	810	c	4-F-C_5H_4N	896	e
C_3H_7CHO	794	a	C_6H_5CHO	827	c	4-Cl-C_5H_4N	900	e
CH_3COCH_3	811	a, b	$C_6H_5OCH_3$	828	c	4-CN-C_5H_4N	869	e
HCOOH	745	b	$C_6H_5NH_2$	869	c	4-NO_2-C_5H_4N	863	e
CH_3COOH	785	b		873	e	$C_6H_5N(CH_3)_2$	921	e
$CH_2ClCOOH$	763	b	H_2S	728	a	$(CH_3)_2SO$	871	e
CF_3COOH	713	b	CH_3SH	776	a	$HCON(CH_3)_2$	874	e
$HCOOCH_3$	786	a, b	C_2H_5SH	792	a	$CH_3CON(CH_3)_2$	895	e

[a] J. W. Wolf, R. H. Staley, I. Koppel, M. Taagepera, R. T. McIver, J. L. Beauchamp, and R. W. Taft, *J. Am. Chem. Soc.*, **99**, 5417 (1977); [b] R. Yamdagni and P. Kebarle, *J. Am. Chem. Soc.*, **98**, 1320 (1976); [c] Y. K. Lau and P. Kebarle, *J. Am. Chem. Soc.*, **98**, 7452 (1976); [d] W. J. Hehre, R. T. McIver, J. A. Pople, and P. V. R. Schleyer, *J. Am. Chem. Soc.*, **96**, 7162 (1974); [e] R. W. Taft, in *Proton Transfer Reactions*, E. F. Caldin, V. Gold, eds., Chapman and Hall, London (1975), p. 31. * Reference value.

A spherically symmetrical solvent molecule, having a centrally located point dipole, and that is solvating a cation, is shown in Figure 2.1a. Free rotation around the axis through the centres of the ion and the solvent molecule is expected. If the charge of the ion is 'flipped' from positive to negative, the solvent molecule will 'flip' around to present the ion with the positive end of its dipole, but the potential energy of the system will remain the same (provided the solvent does not have an electrical quadrupole moment). Such an idealized system does not exist in practice, but might be approached (with respect to rotation around the ion–solvent axis) with solvents such as acetonitrile, pyridine, or triethylamine, where the nitrogen atom has a single lone pair of electrons.

When the solvent has an appreciable electrical quadrupole moment, or when its donor atom is oxygen with two lone pairs of electrons, a configuration like that in Figure 2.1b may have the lowest potential energy. Rotation around the ion–solvent molecule axis is then strongly hindered. Also, the potential energy depends on the sign of the charge on the ion. This is also the case when the

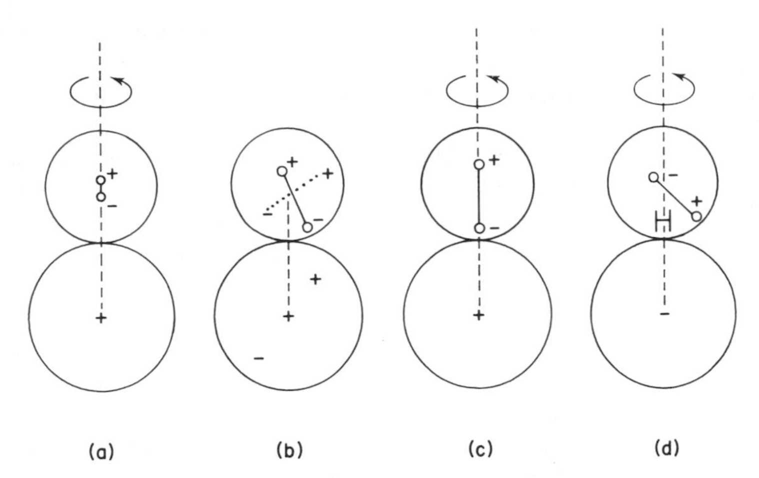

Figure 2.1 Some possible mutual configurations of ions and solvent molecules. (a) A point dipole in the solvent oriented towards a cation: rotation of the solvent around the dipole axis is free. (b) The solvent has a large quadrupole moment, so that the induced charges fix the orientation at an angle to the dipole axis. (c) A cation solvated by an aprotic dipolar solvent. (d) Hydrogen bonding, as it affects the solvation of an angion by a relatively weakly polar protic solvent. (From Y. Marcus, *Introduction to Liquid State Chemistry*, Wiley, Chichester, 1977, by permission.)

dipole of the solvent molecule is not centrally located: aprotic solvents tend to have the negative end of their dipole (i.e., the partial excess negative charge) nearer the surface of the solvent molecule than is the positive end. Such solvents therefore solvate a cation more readily than an anion of equal size and charge; see Figure 2.1c. Dimethylsulphoxide is an example of such a solvent.

Protic solvents are capable of the formation of a hydrogen bond between the solvent molecule and an anion. They then seek a configuration where the hydrogen atom is on the straight line connecting the electronegative atom to which it is bonded with the anion, as shown in Figure 2.1d, in balance with the configuration where the dipole points directly towards the anion. It depends on the relative strength of the ion–dipole interaction on the one hand and the strength of the hydrogen bond on the other, where this balancing point will be. Water and alcohols orient themselves in this fashion. Rotation around the anion–solvent axis is not hindered very strongly in this case.

The potential energy of each configuration can be calculated by the methods of quantum chemistry, and thus the minimal potential energy and the favoured geometry can be found by a program that systematically considers all the possible configurations. When more than one solvent molecule approaches the ion, the mutual interactions of the solvent molecules are, of course, also taken into account. These include both attractive interactions and repulsive ones, due to crowding around the ion.

Schuster *et al.*[4] reviewed the methods for the calculation of the solvation of small ions and the results obtained thereby. There are essentially two methods: the *ab initio* calculation and the semi-empirical one (using mainly a CNDO-type approach).

In the *ab initio* calculation, the difference in energy ΔE between a solvate $I^{\pm}S_n$ and the separate constituents is determined as a function of the geometry (bond angles and lengths d). The (absolute value of the) solvation energy is then maximized with respect to these parameters. The solvent molecules may be regarded as rigid, having the same geometry as in the free state, or their geometry may also be adjusted. The latter procedure, it was found, had little effect for small solvent molecules like H_2O. The calculation generally employs the *s*elf-*c*onsistent-*f*ield (SCF) version of *m*olecular-*o*rbital (MO) theory to solve the Schrödinger equation for the solvate and for the constituents separately, neglecting electron correlation. Charge transfer between the ion and the solvent is found to be minimal. The main problem seems to be the proper choice of the basis set of orbitals, whether *G*aussian *t*ype (GTO) or *S*later *t*ype (STO) *o*rbitals, each of the latter expanded into a number of GTO's, and how extensive this basis set is. After all, ΔE is obtained as a small difference between very large numbers and the errors must be minimized. With a given basis set, however, the full molecular Hamiltonian is used in the calculation.

The semi-empirical calculations generally employ an approach due to Pople *et al.*,[5] which involves *c*omplete *n*eglect of *d*ifferential *o*verlap (CNDO) of the basis set orbitals. With a given basis set of orbitals this decreases drastically the number of integrals that have to be calculated. The version usually applied is called CNDO/2, in which empirical parameters (based on ionization potentials and electron affinities) are used for atomic matrix elements, and certain terms (leading to excessive bonding between nonbonded atoms) are omitted from the Hamiltonian. The energy of the system can then be expressed as a sum of one-atom and two-atom terms. In essence, the success of the method depends on a cancellation of errors, which is not adequate when the geometries are chosen far from the correct (equilibrium) one. Thus a local minimum of the total energy may suggest a wrong geometry, and much care must be used when the method is applied not to fall into this kind of trap.

Rrepresentative results obtained with these two calculation methods for the monosolvates $I^{\pm}S$ ($n = 1$) are shown in Table 2.4. The $-\Delta E$ values may be compared with the corresponding experimental $-\Delta H^0$ data in Table 2.1. The range of values shown in Table 2.4 applies to the results of different authors, different modifications of the general methods, different basis sets of orbitals, and mainly different geometries. The maximal (in the absolute sense) value of the solvation energy does not correspond necessarily to proper minimization of the energy of the system, since it may correspond to unrealistic bond distances.

Solvate clusters, $I^{\pm}S_n$ with $n > 1$, have also been subjected to calculations. Certain highly symmetrical geometries (e.g. linear for $n = 2$, planar triangular for $n = 3$, tetrahedral for $n = 4$, trigonal bipyramidal for $n = 5$, octahedral for $n = 6$ and archimedean antiprismatic for $n = 8$) have generally been prescribed

Table 2.4 Calculated bond energies and distances for monosolvates of ions, according to various *ab initio*[4] and semi-empirical CNDO methods[6]

		Range of *ab initio* results		Range of CNDO results	
Ion	Solvent	$-\Delta E$/ kJ mol^{-1}	d/nm	$-\Delta E$/ kJ mol^{-1}	d/nm
Li^+	H_2O	130–198	0.181–0.230	188–216	0.235–0.242
	CH_3OH			185	0.180
	C_2H_5OH			226	0.175
	$(CH_3)_2O$			215	0.175
	H_2CO	180–188	0.177–0.184	159	0.185
	$(CH_3)_2CO$			211	0.180
	$HCOOH$			431	
	CH_3COOCH_3			326	0.210
	NH_3	168–207	0.190	154	0.185
	HCN	155–164	0.194–0.196	177	0.191
	CH_3CN				
	$HCONH_2$	234	0.171		
	$CH_3CON(CH_3)_2$			347	0.210
	HF	113–128	0.172		
Na^+	H_2O	100–170	0.199–0.225	86–146	0.292–0.311
	CH_3OH			127	0.212
	C_2H_5OH			156	0.212
	$(CH_3)_2O$	163		143	0.212
	H_2CO	124	0.229	110	0.217
	$(CH_3)_2CO$			147	0.217
	$HCOOH$			295	0.305
	CH_3COOCH_3	162	0.199	251	0.300
	NH_3	143–157	0.228	102	0.217
	CH_3CN			120	0.228
	$CH_3CONHCH_3$	208	0.195		
	$CH_3CON(CH_3)_2$			264	0.300
	HF	85–95	0.206		
K^+	H_2O	70–117	0.240–0.265		
	CH_3OH			96	0.254
	C_2H_5OH			120	0.249
	$(CH_3)_2O$	108		110	0.249
	H_2CO	85	0.282	85–215	0.232–0.259
	$(CH_3)_2CO$			115	0.254
	CH_3COOCH_3	107	0.240		
	NH_3			71	0.265
	CH_3CN			89	0.270
	$CH_3CONHCH_3$	147	0.235		
Rb^+	H_2O	69[a]		67	0.281
	CH_3OH			85	0.275
	C_2H_5OH			107	0.275
	$(CH_3)_2O$			97	0.270
	CH_2O			74	0.286
	$(CH_3)_2CO$			100	0.275
	NH_3			60	0.291
	CH_3CN			89	0.296
Cs^+	H_2O	61[a]			
NH_4^+	H_2O	67–156	0.240–0.265	108	
	NH_3	67–177			
$CH_3NH_3^+$	H_2O	143	0.240	90	
$(CH_3)_2NH_2^+$	H_2O			79	
$(CH_3)_4N^+$	H_2O	23–43			

(continued)

Table 2.4—*continued*

Ion	Solvent	Range of *ab initio* results		Range of CNDO results	
		$-\Delta E$/ k J mol^{-1}	d/nm	$-\Delta E$/ k J mol^{-1}	d/nm
$(C_2H_5)N(CH_3)_3^+$	H_2O	14–38			
$(C_2H_5)_2N(CH_3)_2^+$	H_2O	36	0.36		
$(C_2H_5)_3N(CH_3)^+$	H_2O	34			
Be^{2+}	H_2O	586–592	0.150–0.156	910	0.170
	NH_3	615–637	0.169		
	HF	361–382	0.149		
Mg^{2+}	H_2O	335	0.195		
	NH_3	402–420	0.200		
	HF	242–258	0.180		
Ca^{2+}	H_2O	222	0.240		
Al^{3+}	H_2O	753	0.175		
F^-	H_2O	72–127	0.230–0.264	95–339	0.222–0.238
	CH_3OH			32–82	0.243–0.370
	C_2H_5OH			34–81	0.238–0.365
	$(CH_3)_2O$			26–35	0.302–0.434
	CH_2O			57–58	0.286–0.291
	$(CH_3)_2CO$			34–63	0.286–0.365
	NH_3			31–64	0.254–0.286
	CH_3CN			57–71	0.281–0.296
Cl^-	H_2O	47–79	0.304–0.330	52–99	0.274–0.307
	CH_3OH			34–54	0.307–0.434
	C_2H_5OH			35–55	0.302–0.429
	$(CH_3)_2O$			31–46	0.370–0.450
	CH_2O			38–51	0.365–0.376
	$(CH_3)_2CO$			39–71	0.365–0.423
	HCOOH			160	
	NH_3			35–38	0.339–0.349
	CH_3CN			51–56	0.360–0.365
ClO_4^-	HCOOH			68	
Br^-	H_2O	53[b]	0.304[b]	47–120[c]	0.268–0.318
	CH_3OH			33–52	0.318–0.455
	C_2H_5OH			34–54	0.312–0.450
	$(CH_3)_2O$			31–46	0.392–0.471
	CH_2O			36–48	0.386–0.397
	$(CH_3)_2CO$			38–70	0.386–0.445
	NH_3			32–38	0.360–0.370
	CH_3CN			50–54	0.381
I^-	H_2O	44[b]	0.334[b]	32–82[c]	0.299[c]–0.392
	CH_3OH			32–41	0.381–0.503
	C_2H_5OH			32–44	0.365–0.492
	$(CH_3)_2O$			32–47	0.434–0.519
	CH_2O			33–45	0.429–0.450
	$(CH_3)_2CO$			38–69	0.434–0.492
	NH_3			26–36	0.418–0.429
	CH_3CN			46–52	0.429–0.434

[a] K. G. Spears and S. H. Kim, *J. Phys. Chem.*, **80**, 673 (1976); [b] K. G. Spears, *J. Phys. Chem.*, **81**, 186 (1977); [c] V. B. Volkov and D. A. Zhogolev, *Chem. Phys. Lett.*, **49**, 591 (1977).

for the semi-empirical CNDO calculations, in order to stay around realistic values of bond distances. The results for the solvent S = water for a representative series of ions are shown in Table 2.5. The table also contains results from recent calculations made by means of an empirical electrostatic method, calibrated with the experimental data available. For the anions the geometry chosen was that with a linear hydrogen bond, as in $F^- \ldots H{-}O$. For the larger anions, Br^- and I^-, an ion–dipole structure may be preferable. Calculations have also been made for so called chain hydrates, i.e. where a second water molecule is attached to the first one through hydrogen bonding, rather than directly to the ion, and for clusters where the first coordination shell is some symmetrical arrangement (e.g. tetrahedral or octahedral), and further water molecules are attached in a second coordination sphere, Such structures may have more favourable energies than those where all the water molecules are attached directly to the ion.

Table 2.5 Calculated hydration energies for multihydrated ions.[k]

		Ab initio		CNDO/2		Electrostatic	
Ion	n	$-\Delta E$/k J mol^{-1}	$d_{I\text{-}O}$/nm	$-\Delta E$/k J mol^{-1}	$d_{I\text{-}O}$/nm	$-\Delta E$/k J mol^{-1}	$d_{I\text{-}O}$/nm
Li^+	1	155[a]	0.185	188[c]	0.238	154[f]	
	2	271[b]	0.187	364[c]	0.239	279[f]	
	3	371[b]	0.190	531[d]	0.239	364[f]	
	4	443[b]	0.194	681[c]	0.243	425[f]	
	5	472[b]	0.202			442[f]	
	6	521[b]	0.208	906[c]	0.249	467[f]	
	8			1075[e]			
Na^+	1	113[a]	0.220	136[c]	0.294	101[f]	
	2	200[b]	0.223	267[c]	0.295	191[f]	
	3	279[b]	0.225	249[d]	0.313	263[f]	
	4	343[b]	0.228	519[c]	0.295	320[f]	
	6			749[c]	0.298	390[f]	
	8			961[e]	0.298		
K^+	1	75[a]	0.265			75[f]	
	2	140[b]	0.269			144[f]	
	3	149[b]	0.269			203[b]	
	4					253[b]	
NH_4^+	1	156[g]	0.240	108[d]			
	2	238[g]	0.245	197[d]			
	3	277[g]	0.255	272[d]			
	4	295[g]	0.260	339[d]			
Be^{2+}	1	586[a]	0.150[h]	910[e]	0.170		
	2	1225[a]	0.150[h]	1723[e]	0.172		
	3	1272[a]	0.150[h]				
	4			2857[e]	0.180		
	6			3473[e]	0.189		
	8			3729[e]	0.199		
F^-	1	95[b]	0.256	90[c]	0.214	101[i]	0.214
	2	182[b]	0.255	165[c]	0.216	187[i]	0.222
	3	256[b]	0.258			252[i]	0.232
	4	307[b]	0.267	291[c]	0.224	304[i]	0.240
	6			415[c]	0.268		

(continued)

Table 2.5—*continued*

Ion	n	*Ab initio*		CNDO/2		Electrostatic	
		$-\Delta E$/k J mol^{-1}	d_{I-O}/nm	$-\Delta E$/k J mol^{-1}	d_{I-O}/nm	$-\Delta E$/k J mol^{-1}	d_{I-O}/nm
Cl^-	1	50[b]	0.330	79[c]	0.238	60[i]	0.286
	2	96[b]	0.331	145[c]	0.241	115[i]	0.288
	3	137[b]	0.334			168[i]	0.292
	4			294[c]	0.237	205[i]	0.296
	5			418[c]	0.242	239[i]	0.302
	6					267[i]	0.308
Br^-	1			120[j]	0.273	53[i]	0.304
	2			219[j]	0.277	102[i]	0.308
	3			305[j]	0.278	146[i]	0.312
	4			378[j]	0.281	184[i]	0.316
	5					215[i]	0.320
	6			500[j]	0.285	243[i]	0.324
	8			589[j]	0.290		
I^-	1			82[j]	0.299	44[i]	0.394
	2			154[j]	0.301	86[i]	0.336
	3			218[j]	0.303	123[i]	0.338
	4			276[j]	0.306	156[i]	0.342
	5					185[i]	0.346
	6			374[j]	0.308	210[i]	0.350
	8			455[j]	0.311		

[a] P. A. Kollman and I. D. Kuntz, *J. Am. Chem. Soc.*, **94**, 9236 (1972); [b] H. Kistenmacher, H. Popkie and E. Clementi, *J. Chem. Phys.*, **58**, 5627 (1973); **61**, 799 (1974); [c] K. Bange and Å. Støgård, *Acta Chem. Scand.*, **27**, 2683 (1973); [d] R. A. Burton and J. Daly, *Trans. Faraday Soc.*, **67**, 1219 (1971); [e] P. Russeger, H. Lischka, and P. Schuster, *Theor. Chim. Acta*, **24**, 191 (1972); [f] K. G. Spears and S. H. Kim, *J. Phys. Chem.*, **80**, 673 (1976); [g] A. Pullman and A. M. Armbruster, *Intern. J. Quant. Chem. Symp.*, **8**, 169 (1974); [h] not varied; [i] K. G. Spears, *J. Phys. Chem.*, **81**, 186 (1977); [j] V. B. Volkov and D. A. Zhogolev, *Chem. Phys. Lett.*, **49**, 591 (1977); [k] F. F. Abraham, M. R. Mruzik, and G. M. Pound, *Faraday Disc. Chem. Soc.*, **61**, 34 (1976) presented results of Monte Carlo calculations of stepwise hydration energies of Li^+, Na^+, K^+, F^-, and Cl^- up to $n = 6$.

An examination of the results presented in Tables 2.4 and 2.5 shows that agreement between different methods, or within one method when different data bases are used, is not particularly good. A comparison of the calculated $-\Delta E$ with the experimental $-\sum_1^n \Delta H_n^0$ from Table 2.2 shows only fair agreement with the *ab initio* results and rather good agreement with the CNDO/2 results.

The discrepancies for the latter may be due to the appreciable charge transfer (0.2 to 0.4 units) from the solvent to the ion, which the method predicts. This seems, however, to be unrealistically high: no more than 0.05 charge units are transferred,[4,6] except for Be^{2+}, where the value may be as high as 0.33, indicating some covalency.[4] (The electrostatic calculations shown in Table 2.5 were calibrated by means of the experimental data, hence agreement is expected). The conclusion must be, that so far the calculation methods have not progressed sufficiently to be completely reliable.

Certain trends shown by the calculated results should be noted, though, which are in agreement with the experimental data when available. The increments in energy as solvent molecules accumulate in the solvate cluster decrease with increasing n, and should reach the limit of the condensation energy of a solvent molecule to the bulk solvent when n tends to infinity. The distance between the ion and the donor atom of the solvent molecule increases gradually as n increases. For the monohydrated cations the hydrate has C_{2v} symmetry, all the atoms lying in a plane. Ion–dipole interactions and closed shell repulsion dominate the energy, hence the charge and size effects noted in Table 2.4. The solvent practically does not change its geometry between its free state and its state in the solvate.

2.3 Comparison with bulk solvation

The quantities derived for the interaction of an ion with a few solvent molecules may be related to the corresponding quantities for the transfer of the ion from the gas phase to an infinite amount of liquid solvent, its solvation. Consider the two processes of adding solvent molecules one by one in the gas phase to an ion on the one hand, and to a solvent molecule on the other. The enthalpy changes for the two processes would consist of additive terms $\Delta H^0_{n-1,n}(I^{\pm})$ and $\Delta H^0_{n-1,n}(S)$, respectively. As the number n of solvent molecules increases, the clusters increase in size. The effect of the central particle, whether

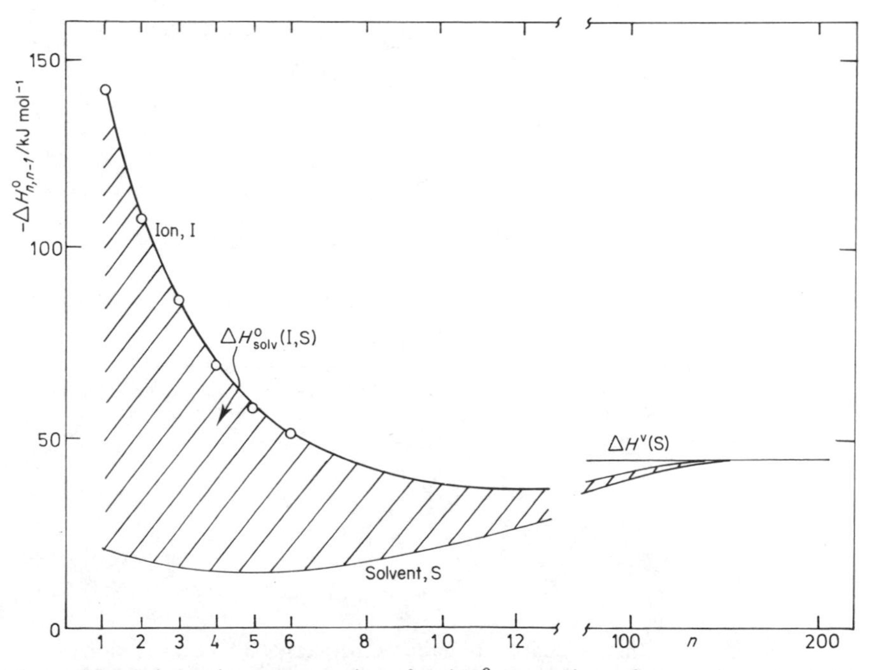

Figure 2.2 Schematic representation of $-\Delta H^0_{n-1,n}$ against n for gas phase clustering around an ion (circles: experimental points for Li^+ with H_2O) and around a solvent molecule. The horizontal asymptote is the enthalpy of evaporation of the solvent (the relative value for H_2O shown). The area between the curves represents the enthalpy of solvation of the ion

ion or solvent molecule, on the energetics of the addition of another solvent molecule diminishes. At the limit of $n \to \infty$ the difference vanishes:

$$\lim_{n\to\infty} [\Delta H^0_{n-1,n}(\mathrm{I}^{\pm}) - \Delta H^0_{n-1,n}(\mathrm{S})] = 0 \tag{2.2}$$

The same energy is, thus, involved in condensing a solvent molecule on to an infinity dilute solution of an ion as on to the bulk solvent. A schematic presentation of the enthalpy of condensing solvent molecules, as a function of the number $n - 1$ of solvent molecules already around an ion or a solvent molecule, is shown in Figure 2.2. The curves tend asymptotically towards $-\Delta H^{\mathrm{V}}(\mathrm{S})$, the bulk enthalpy of vaporization of the solvent. If n is taken as a continuous variable, then the area between the curves is the solvation enthalpy of the ion. Actually it is the sum:

$$\Delta H^0_{\mathrm{solv}}(\mathrm{I}^{\pm}, \mathrm{S}) = \sum_{n=1}^{\infty} [\Delta H^0_{n-1,n}(\mathrm{I}^{\pm}) - \Delta H^0_{n-1,n}(\mathrm{S})] \tag{2.3}$$

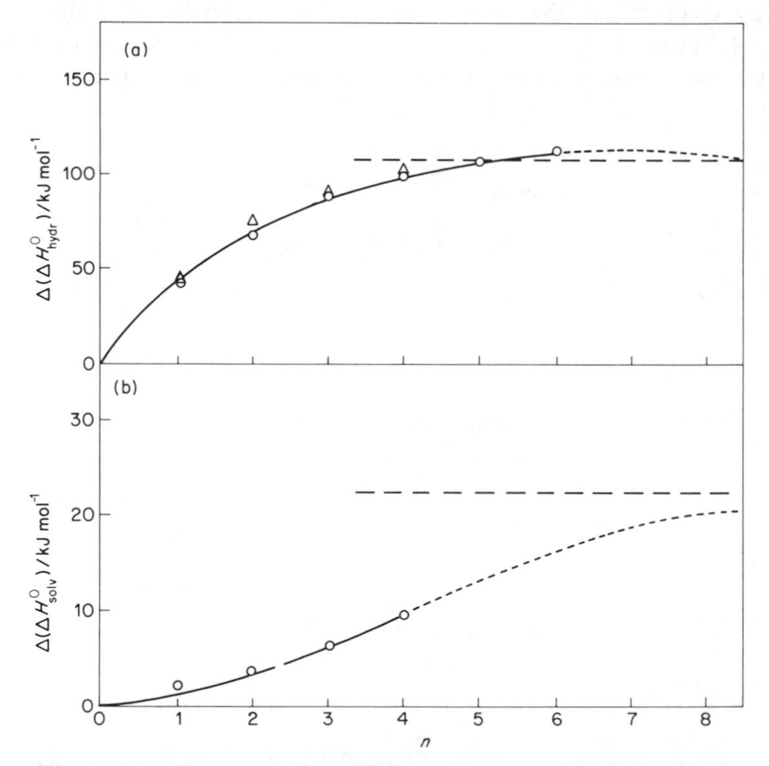

Figure 2.3 The cumulative differences between: (a) the hydration enthalpies of Li^+ and Na^+, (b) the solvation enthalpies for CH_3CN with Cl^- and Br^-, in the gas phase, compared with the difference between the enthalpies of hydration and solvation in bulk water and CH_3CN, respectively (dashed lines). Experimental values (circles) from Table 2.2, calculated values (triangles) from Table 2.5

that gives the enthalpy of solvation, and corresponding sums give its Gibbs free energy and entropy.

It may be expected that a more rapid convergence will be obtained if the different energetics of the clusters around two ions, rather than around an ion and a solvent molecule, are considred. This will also avoid the uncertainties in the solvation energetics of single ions, since the difference between those of two ions of the same charge can be obtained unequivocally from the measured values for electrolytes. Such differences are shown for two cases in Figure 2.3, and it is seen that indeed

$$\lim_{n\to\infty} [\Delta H^0_{n-1,n}(I_1) - \Delta H^0_{n-1,n}(I_2)] = \Delta H^0_{\mathrm{solv}}(I_1, S) - \Delta H^0_{\mathrm{solv}}(I_2, S) \quad (4)$$

Since experimental or calculated data for $\Delta H^0_{n-1,n}$ are not available for $n > 8$, a rigorous test of equation 4 is impossible. It is seen, however, how far gas phase data, in particular those obtained from calculations, can be used to predict solvation energetics for ions in bulk solvents.

References

1. P. Kebarle, W. R. Davidson, M. French, J. B. Cumming, and T. B. McMahon, *Disc. Faraday Soc.*, **64**, 220–9 (1977); P. Kebarle, *Ann. Rev. Phys. Chem.*, **28**, 445–76 (1977).
2. Y. Marcus and A. Loewenschuss, *Ann. Reports C* (Royal Soc. Chem.), **1984** (publ. 1985), Ch. 6.
3. R. W. Taft, in *Proton Transfer Reactions*, E. F. Caldin and V. Gold, eds., Chapman and Hall, London (1975), p. 31.
4. P. Schuster, W. Jakubetz, and W. Marius, *Topics Curr. Chem.*, **60**, 1–108 (1975).
5. J. A. Pople, D. P. Santry, and G. A. Segal, *J. Chem. Phys.*, **43**, S 129 (1965); J. A. Pople and G. A. Segal, *ibid.*, **43**, S 136 (1965); **44**, 3289 (1966); J. A. Pople and D. L. Beveridge, *Approximate Molecular Orbital Theory*, McGraw-Hill, New York (1970).
6. A. V. Bandura and N. P. Novoselov, *Teor. Eksp. Khim.*, **14**, 174–82 (1978); *Theor. Exp. Chem.*, **14**, 135–41 (1978).

Chapter 3

Interaction models for ion solvation

The solvation of individual ions in the gas phase can be studied experimentally and evaluated theoretically, as has been discussed in Chapter 2. The solvation of single ions in bulk solvents can be studied experimentally from several aspects by certain structural, spectrochemical and chemical kinetic methods (see Chapters 4 and 5). Thermodynamics, however, can be applied only to chemical components, so that thermodynamic data cannot be obtained directly for individual ions. This has already been noted in the comparison of the gas phase solvation with bulk solvation in solvents in Section 2.3. Extra-thermodynamic assumptions have been applied to thermodynamic solvation data for components, in order to divide them into the contributions from individual cations and anions. These are dealt with in Chapters 5 and 6. At present, the theoretical approach to this problem is based mainly on the consideration of suitable models of ion solvation. Increasing use of computer simulation techniques is forseen. These have so far provided some information, but not very conclusive results.

The ion solvation theories apply statistical mechanical concepts and methods to models of the interaction of an ion with the surrounding solvent, including any changes in the mutual interactions of the solvent molecules, that are induced by the presence of the ion. The sum of the contributions from all these interactions to the thermodynamic functions is then compared with the experimental data. A certain danger exists, that not all the relevant interactions have been taken into account, or that conversely, contributions from certain interactions have been counted more than once in the summation, though from different aspects.

Since the calculated results for individual ions can be combined in such a way as to give values for chemical components, a direct comparison with valid thermodynamic data can be made. The model calculations must withstand this test not only for the Gibbs free energy, which is relatively insensitive, but also for derived functions, such as the entropy, the enthalpy, the volume change, etc., of solvation. The present chapter examines the application of models for obtaining these thermodynamic functions for ion solvation.

3.1 Statistical thermodynamic considerations

Statistical thermodynamics relates the properties of individual particles (molecules, atoms, ions, etc.) to the properties of bulk phases. The methods can be applied to mixtures and solutions, and of concern here is a solution of ions (species X) in a solvent consisting of molecules (species S). Nothing is said at this stage about the specific interactions that occur in this case, and attention is focused on the general relationships arising from this situation, following the development by Ben-Naim.[1]

For a fluid mixture of N_S particles of species S and N_X particles of species X, the classical canonical partition function is

$$Q(T, V, N_S, N_X) = q_S^{N_S} q_X^{N_X} (N_S! N_X! \Lambda_S^{3N_S} \Lambda_X^{3N_X})^{-1} Z(T, V, N_S, N_X) \quad (3.1)$$

Here q is the internal partition function of a single particle, Λ is the momentum partition function of a single particle, integrated over all its possible momenta, and Z is the configurational partition function for specified temperature, volume and compsition. This is the integral over all the possible configurations of the $N = N_S + N_X$ particles, of the total potential energy of interaction among them, $U(R^N)$ (where (R^N) is a symbol for one configuration of the coordinates and orientations R of the N particles) divided by $\mathbf{k}T$.

Practical studies are conducted at constant pressure, rather than at a specified volume, hence the isothermal–isobaric partition function is required:

$$\Delta(T, P, N_S, N_X) = C \int_0^{\infty} Q(T, V, N_S, N_X) \exp(-PV/\mathbf{k}T)\, dV \quad (3.2)$$

Here C, which has the dimension of a reciprocal volume, is a normalization constant, the value of which need not be of concern, since differences in $\ln \Delta$ are always employed, so that it cancels out. The Gibbs free energy of the system is given by

$$G(T, P, N_S, N_X) = -\mathbf{k}T \ln \Delta(T, P, N_S, N_X) \quad (3.3)$$

The chemical potential of the solute is given by

$$\mu_X^{\cdot} = (\partial G/\partial N_X)_{T,P,N_S} = G(T, P, N_S, N_X + 1) - G(T, P, N_S, N_X) \quad (3.4)$$

where $\mu^{\cdot}$ designates the chemical potential per particle, in distinction from the chemical potential per mole, $\mu = \mathbf{N} \cdot \mu^{\cdot}$ generally used in thermodynamics. The second equality in equation 3.4 reflects the extensive character of the Gibbs free energy, and is valid at the thermodynamic limit of very large N_S. The chemical potential $\mu_X^{\cdot}$ is thus the change in the Gibbs free energy of the system, when a single particle of the solute X is added to a large number of particles of S plus any number of particles X.

It is convenient for the purpose of the evaluation of $\mu_X^{\cdot}$ in terms of equations 3.1, 3.2, and 3.3 to consider a slightly modified process, namely that of the addition of the particle of X to a fixed point R_0 (fixed coordinates and orientation) in the system. This process is identified by the descriptor (fp) in the following. Since all the points in an isotropic and homogeneous liquid are equivalent, the quantity $\mu^{\cdot}_{X}(\mathrm{fp})$ is not a function of the location of and exact orientation at the point R_0. If the functions $G(N_S, N_X + 1)$, $G(N_S, N_X + 1, R_0)$, and $G(N_S, N_X)$ are written out in terms of equations 3.1 to 3.3 (constant T and P are implicit in the equations, and are not specified for the sake of a simple notation), the following differences can be seen. First, the factor $(N_X + 1)!$ appearing in the first case is replaced by $N_X!$ in the other two, since in the first case there are $N_X + 1$ indistinguishable particles of X, in the second case only N_X, since the $(N_X + 1)$th particle is distinguishable by its fixed location at R_0. Secondly, the factor $q_X^{N_X+1}\Lambda_X^{-3(N_X+1)}$ appearing in the first case is replaced by $q_X^{N_X}\Lambda_X^{-3N_X}$ in the other two, since, again, in the second case the added particle is devoid of orientational and translational degrees of freedom and does not contribute to the momentum of the system. Thirdly, the integration in the first case is over the configurations of $N_S + N_X + 1$ particles, whereas in the other two it is over the configurations of only $N_S + N_X$ particles, since the localization of the added one in the second case makes it necessary to integrate over the configurations of only the remaining $N_S + N_X$ ones. This yields an extra factor of V in the first case, absent in the other two. Finally, the quantity $B_X(R^N, R_0)$ is defined as

$$B_X(R^N, R_0) = U(R^{N_S+N_X+1}, R_0) - U(R^{N_S+N_X}) \tag{3.5}$$

and is the interaction energy of the particle of the solute X located at R_0 with the rest of the system at some given configuration.

The difference between the G functions in equation 3.4 can now be evaluated by a standard procedure, yielding

$$\mu^{\cdot}_{X}(\mathrm{fp}) = -\mathbf{k}T \ln \langle \exp(-B_X(R^N, R_0)/\mathbf{k}T)\rangle_0 \tag{3.6}$$

$$\mu^{\cdot}_{X} = \mathbf{k}\mathrm{T} \ln (\rho_X\Lambda_X^3 q_X^{-1}) - \mathbf{k}T \ln \langle \exp(-B_X(R^N, R_0)/\mathbf{k}T)\rangle_0 \tag{3.7}$$

In equations 3.6 and 3.7, $\langle\ \rangle_0$ represents an average over all the configurations of the particles in the system, except for the one at the fixed point R_0. In equation 3.7 $\rho_X = N_X/V$ is the number density of the particles of the solute. The origin of the first term in $\mu^{\cdot}_{X}$, that in $N_X\Lambda_X^3/Vq_X$, has been indicated in the discussion above of the G functions for the three cases. A combination of equations 3.6 and 3.7 gives

$$\mu^{\cdot}_{X} = \mu^{\cdot}_{X}(\mathrm{fp}) + \mathbf{k}T \ln (\rho_X\Lambda_X^3/q_X) \tag{3.8}$$

which can be interpreted as follows.

When a particle of the solute is added to the solution, the process can be envisaged as taking place in two steps. First the particle is added to a fixed point R_0 and allowed to intereact with its surroundings. A clear relationship is established at this stage with considerations of the pair distribution function and pair potential energies, since the point R_0 can be selected as the origin of the coordinates. It is assumed that in this process the internal partition function q_X remains unchanged in spite of the interactions and may be factored out. In the second step the particle is liberated from its fixed location and orientation. It becomes then indistinguishable from the other N_X solute particles (introducing the factor N_X), the whole volume becomes accessible to it (introducing the factor V^{-1}), and it has its full translational and orientational degrees of freedom (introducing the factor $\Lambda_X^3 q_X^{-1}$). The entire process thus consists of a 'coupling' step and a 'liberation' step.

In view of the interpretation given here to $\mu_X^{\cdot}(\text{fp})$, it is convenient to use the following notation[1a]

$$\mu_X^{\cdot}(\text{fp}) = W(\text{X} \mid (\text{S} + \text{X}), x_X) \tag{3.9}$$

where the right-hand side is read as the coupling work of a particle of solute X to the rest of the system, consisting of species S and X at the composition specified by the mole fraction

$$x_X = N_X/(N_S + N_X)$$

The coupling work depends, thus, on the composition, since B_X depends on it.

The discussion up to this point has been quite general, and has not specified any particular concentration range.[1b] For the purpose of the evaluation of the thermodynamic functions of ion solvation, it is necessary to consider extremely dilute solutions. Equations 3.8 and 3.9 can then be combined to give

$$\begin{aligned}\mu_X^{\cdot}(N_S \gg N_X) &= [W(\text{X} \mid \text{S}) + \mathbf{k}T \ln (\Lambda_X^3 q_X^{-1})] + \mathbf{k}T \ln \rho_X \\ &= \mu_X^{\cdot\infty} + \mathbf{k}T \ln \rho_X\end{aligned} \tag{3.10}$$

The particles of the solute are surrounded practically by particles of the solvent only in these dilute solutions; hence the coupling work is written as $W(\text{X} \mid \text{S})$. The intensive quantities in the square brakets in equation 3.10 are collected together as $\mu_X^{\cdot\infty}$ which depends on the natures of species S and X, but not on the concentration of the solute X, provided it is sufficiently low for X to be surrounded by S particles only. It is now appropriate to change into ordinary thermodynamic notation, by multiplying $\mu_X^{\cdot}$ with $\mathbf{N}$ to give

$$\begin{aligned}\mu_X(\text{at low } c_X) &= [\mathbf{N}\mu_X^{\cdot\infty} - \mathbf{R}T \ln \mathbf{N}] + \mathbf{R}T \ln c_X \\ &= \mu_{X(c)}^{\infty} + \mathbf{R}T \ln c_X\end{aligned} \tag{3.11}$$

where $\mu_{X(c)}^{\infty}$ is the standard chemical potential on the molar scale and $c_X = \rho_X/\mathbf{N}$ is the molar concentration. A comparison of equation 3.11 with 3.10 shows that the standard chemical potential is in fact the limit

$$\mu_{X(c)}^{\infty} = \lim_{c_X \to 0} (\mu_X - \mathbf{R}T \ln c_X) \tag{3.12}$$

in the same way as $W(\text{X} \mid \text{S})$ is the limit of $W(\text{X} \mid (\text{S} + \text{X}))$ as $N_X/N_S \to 0$. Otherwise, an activity coefficient term

$$\mathbf{R}T \ln y_X = W(\text{X} \mid (\text{S} + \text{X})) - W(\text{X} \mid \text{S})$$

is added. It clearly depends on the concentration through the probability of finding particles of X in the immediate surroundings of a given X particle, rather than only S ones, as at infinite dilution.

The purpose of this exercise, which has led to the known and seemingly trivial equations 3.11 and 3.12, is on the one hand to relate the standard chemical potential $\mu^{\infty}_{X(c)}$ to the coupling work $W(X \mid S)$ through

$$\mu^{\infty}_{X(c)} = \mathbf{N}W(X \mid S) - \mathbf{R}T \ln (\mathbf{N}\Lambda_X^{-3} q_X) \tag{3.13}$$

and on the other hand to show that the statistical thermodynamic derivation, with its inherent interpretation in terms of the coupling work and the liberation work, equation 3.8, is straightforward and simple when the number density or the molar concentration scale is used. It is instructive to consider the form that the equations take when other common concentration scales are employed.

The mole fraction scale is defined by

$$x_X = N_X/(N_S + N_X) \approx \rho_X/\rho_S^* \tag{3.14}$$

where the second, approximate, equality is obtained by dividing the numerator and the denominator by V and neglecting in very dilute solutions ρ_X besides ρ_S, which becomes then ρ_S^*, the number density of the pure solvent S. Substitution of equation 3.14 in equation 3.10 yields

$$\mu^{\cdot}_X(\text{low } x_X) = [\mu_X^{\cdot\infty} + \mathbf{k}T \ln \rho_S^*] + \mathbf{k}T \ln x_X \tag{3.15}$$

The extra term it contains in the square brackets makes equation 3.15 more complicated than equation 3.10, in particular if the temperature and pressure derivatives are taken in order to obtain the partial entropy and volume of the solute, since ρ_S^* is both temperature and pressure dependent. Thus, contrary to the common belief, that the use of the mole fraction scale provides relief from the difficulties of temperature and pressure dependencies of the molar concentration scales, it addes to them.

The same is true of the molal scale, defined as

$$m_X = N_X/M_S N_S = \rho_X/M_S\rho_S^* \tag{3.16}$$

where M_S is the molar mass of the solvent S (in kg). Substitution of equation 3.16 in equation 3.10 yields

$$\mu^{\cdot}_X(\text{low } m_X) = [\mu_X^{\cdot\infty} + \mathbf{k}T \ln (M_S\rho_S^*)] + \mathbf{k}T \ln m_X \tag{3.17}$$

which is, again, the more complicated equation and the more sensitive one for the reasons stated above. Multiplication by $\mathbf{N}$ leads from equations 3.15 and 3.17 to the familiar forms

$$\begin{aligned} \mu_X(\text{at low } x_X \text{ or } m_X) &= \mu^{\infty}_{X(x)} + \mathbf{R}T \ln x_X \\ &= \mu^{\infty}_{X(m)} + \mathbf{R}T \ln m_X \end{aligned} \tag{3.18}$$

which look simple, but include in the standard chemical potentials on the mole fraction and molality scales in addition to the quantities of equation 3.13 also the term $\mathbf{R}T\ln\rho_S^*$ or $\mathbf{R}T\ln M_S\rho_S^*$, respectively. Hence the conclusion of Ben-Naim,[1a] that statistical mechanics arguments favour the use of the molar concentration scale, whereas no choice can be made on purely thermodynamic grounds, since equations 3.11, 3.12, and 3.18 all look alike.

No restriction has been made up to this point as to the nature of the solute species X. From the statistical thermodynamic aspect it may as well be an individual ionic species. This is notwithstanding the fact that the addition of an individual ionic species to a solvent cannot be carried out experimentally. In any case, the chemical potentials themselves cannot be measured, even for neutral solutes, only differences in chemical potentials being measurable.

The difference between the standard chemical potentials of an ionic species in the two states: the ideal gas state and the infinitely dilute solution in the solvent, is the quantity of interest in the present context. It represents the standard molar Gibbs free energy of solvation of the ion:

$$\Delta G^0_{\text{solv}}(\text{ion X in solvent S}) = \mu^{\infty}_{X(c)}(\text{X in S}) - \mu^{\infty}_{X(c)}(\text{X in gas}) \qquad (3.19)$$

The first term on the right-hand side of equation 3.19 is given by equation 3.13 and the second term by

$$\mu^{\infty}_{X(c)}(\text{X in gas}) = \mathbf{R}T\ln(\Lambda_X^3 q_X^{-1}\mathbf{N}^{-1})$$

so that the simple relationship results:

$$\Delta G^0_{\text{solv}}(\text{X in S}) = \mathbf{N}W(\text{X}\mid\text{S}) \qquad (3.20)$$

The standard molar Gibbs free energy of solvation of an ion is thus the coupling work of this ion at a fixed point to the surrounding solvent, per mole of ions.[1b] The other thermodynamic functions of solvation are the corresponding derivatives:

$$\Delta S^0_{\text{solv}} = -\mathbf{N}(\partial W(\text{X}\mid\text{S})/\partial T)_P$$

$$\Delta H^0_{\text{solv}} = \mathbf{N}[(W(\text{X}\mid\text{S}) - T(\partial W(\text{X}\mid\text{S})/\partial T)_P)]$$

$$\Delta V^0_{\text{solv}} = \mathbf{N}(\partial W(\text{X}\mid\text{S})/\partial P)_T), \quad \text{etc.}$$

3.2 Cavity formation

In order to examine the coupling work of an ion with the surrounding solvent more closely, consider first a system comprising a sphere of macroscopic size (diameter σ_X) in the solvent and the process of its removal. This may be envisaged to proceed in two steps. In the first all the interactions between the sphere and the solvent are 'switched off', the sphere becoming a 'hard sphere' (hs). In the second step the hard sphere is removed from the solvent, which is permitted to collapse into the cavity that has been left, and assume its normal properties also for this element of volume. In the reverse process, that of the introduction of the sphere into the solvent, the first step would then be the creation of the appropriate cavity. This involves the doing of work against the surface tension of the solvent, γ_S:

$$W_{cav} = \pi\sigma_X^2\gamma_S \quad \text{(for large } \sigma_X) \tag{3.21}$$

This expression has found use in the description of the solubility of inert gases in molten salts, but it is not expected to yield realistic values of W_{cav}, if σ_X is permitted to decrease and assume microscopic dimensions.

If a cavity is to be formed in a solvent for the introduction of an ion, the ion is first conceptually transformed into a 'hard particle', devoid of interactions. It is then replaced by a 'hard sphere' of corresponding volume, with an assigned diameter σ_X. This step prevents any difficulties which may arise from an irregular shape an ion may have. The work that has to be expended to create this cavity is then calculated. Simple-minded considerations lead to the expectation that this work will be larger than the pressure–volume work done against the atmospheric pressure, $P(\pi/6)\sigma_X^3$, which is independent of the solvent altogether, but smaller than the work required for the removal of a solvent molecule (assuming it to have the same size as the ion) from the liquid, its heat of vaporization per molecule minus $\mathbf{k}T$.

A successful approach to this problem that has gained wide acceptance is that of Pierotti,[2] based on the scaled particle theory of Reiss *et al.*[3] The cavity is considered as a region in space, from which the centres of the solvent molecules, regarded as spheres of diameter σ_S, are excluded, Figure 3.1. The radius of this cavity is $\frac{1}{2}(\sigma_S + \sigma_X)$. Two parameters determine the amount of work sought. One is the ratio of the diameters of the solute and solvent particles, both considered as hard spheres: σ_X/σ_S. The other is the packing fraction of the solvent or its reduced number density: $y = (\pi/6)\sigma_S^3\mathbf{N}/V_S$, where V_S is the molar volume of the solvent, an experimental quantity. The Gibbs free energy for the formation of the cavity, $\Delta G_{cav} = \mathbf{N}W_{cav}$ is given by[2]

$$\begin{aligned}\Delta G_{cav} = \mathbf{R}T\{&-\ln(1-y) + 3y(1-y)^{-1}(\sigma_X/\sigma_S)\\ &+ [3y(1-y)^{-1} + (9/2)y^2(1-y)^{-2}](\sigma_X/\sigma_S)^2\}\\ &+ PV_S y(\sigma_X/\sigma_S)^3\end{aligned} \tag{3.22}$$

The last term is the pressure–volume term, and the first results from the expansion of the expression from the scaled particle theory in a power series,

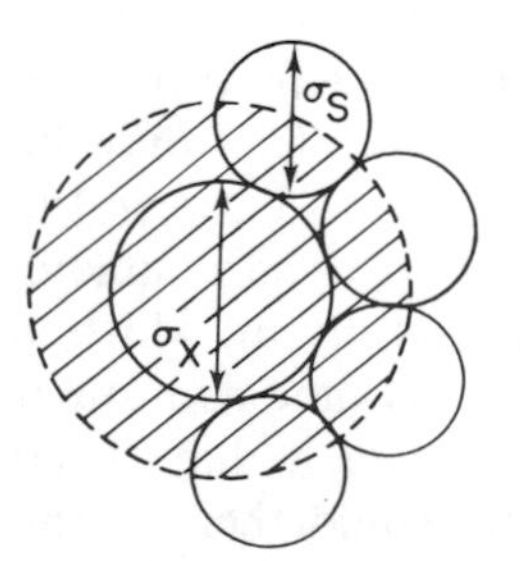

Figure 3.1 A two-dimensional representation of the cavity induced in the solvent S by the solute X. The centres of the molecules of the solvent are excluded from the shaded area, the radius of which is one half of the sum of the diameters σ_X of the solute and σ_S of the solvent molecules

truncated after the term in the second power of the radius of the cavity, which in a sense corresponds to equation 3.21. The corresponding enthalpy change is[2]

$$\Delta H_{\text{cav}} = y\alpha_{\text{pS}}\mathbf{R}T^2(1-y)^{-3}\{(1-y)^2 + 3(1-y)(\sigma_X/\sigma_S) + 3(1+2y)(\sigma_X/\sigma_S)^2\} + PV_S y(\sigma_X/\sigma_S)^3 \qquad (3.23)$$

where α_{pS} is the thermal expansibility of the solvent at constant pressure. The corresponding entropy change is obtained from the difference $\Delta S_{\text{cav}} = (\Delta H_{\text{cav}} - \Delta G_{\text{cav}})/T$. Other derivative thermodynamic functions are obtained by further derivation, e.g., of ΔG_{cav} with respect to the pressure to obtain ΔV_{cav}, etc.

The applicability of equations 3.22 and 3.23 depends on the possibility to specify values for σ_X of the solute and of σ_S of the solvent, in addition to the necessary availability of V_S and α_{pS} data for the solvent. The value of σ_X of a monoatomic ion or for such ions as have approximately spherical shape is generally set equal to twice the crystal radius. For non-ionic solutes or non-spherical ions estimates based on the molecular dimensions, the bond lengths and the geometry are used.

The quantity σ_S is more difficult to estimate, since solvent molecules are generally non-spherical, even not approximately. A procedure based on the solubility of the noble gases has proved helpful for this estimation.[2] If the diameters σ_X of the noble gases are plotted against their polarizabilities α_X and the curve is extrapolated to $\alpha_X = 0$, a 'hard sphere noble gas' of diameter $\sigma_X^0 = 0.255$ nm is thereby defined. If the solubilities of the noble gases (see Table 3.1), expressed as Henry's law constants, are plotted against their diameters σ_X and extrapolated to σ_X^0, their 'hard sphere' solubility is obtained, for the particular solvent employed. The reciprocal of this Henry's law constant is the mole fraction $x_X^0 = (V_S/\mathbf{R}T)\exp(-\Delta G_{\text{cav}}/\mathbf{R}T)$, giving an 'experimental' estimate of the hard sphere Gibbs free energy of cavity formation for this solvent and $\sigma_X = \sigma_X^0$. From this and equation 3.22, σ_S of this solvent can be computed numerically.

A simpler method employs a relationship suggested by Reiss *et al.*[3], involving experimental data of the solvent in question only, and being applicable to non-hydrogen-bonding solvents. The expression is

$$\Delta H_S^V = \mathbf{R}T + \alpha_{\text{pS}}\mathbf{R}T^2[(1+2y)^2/(1-y)^3] \qquad (3.24)$$

where H_S^V is the heat of evaporation of the solvent. It can be shown[4] that by rewriting equation 3.24 and making a power series approximation, the packing fraction can be obtained from

$$Y = [(H_S^V/\mathbf{R}T) - 1]/\alpha_{\text{pS}}T$$

$$y = -0.0469 + 0.4418(\log Y) - 0.0520(\log Y)^2 \qquad (3.25)$$

and the assigned diameter of the solvent molecule from

$$(\sigma_S/\text{nm}) = 0.14692 y^{1/3}(V_S/\text{cm}^3\,\text{mol}^{-1})^{1/3} \qquad (3.26)$$

A still different manner to estimate y (and σ_S from equation 3.26) employs isothermal compressibility data, κ_T, according to Mayer[5]:

$$A = (\mathbf{R}T\kappa_{TS}/V_S)^{1/2}$$

$$y = (A + 1) - [A(A + 3)]^{1/2} \tag{3.27}$$

Equations 3.25 and 3.27 imply a temperature dependence of the packing fraction y of the solvent, hence of its σ_S, that should be taken into account. The ΔH_S^V, V_S, α_{pS}, and κ_{TS} data for about 110 solvents are presented, as far as they are known, for the fixed temperature of 298.15 K in Table 6.1. Values of σ_S for these solvents are listed in Table 6.4. A comparison of alternative values of σ_S for some 50 solvents has led Kim[6] to the correlation

$$(\sigma_S/\text{nm}) = (0.1363 \pm 0.0019)(V_S/\text{cm}^3\,\text{mol}^{-1})^{1/3} - (0.085 \pm 0.001) \tag{3.28}$$

This is then the simplest equation for the estimation of σ_S.

3.3 Interactions of neutral solutes

Once a hard sphere has been placed in the cavity created for it in the solvent, its interactions with the surrounding solvent may be 'switched on'. All solutes, ions as noble gas atoms, have some non-specific interactions with the solvent, which are dealt with in this Section.

At very short ranges there exist repulsion forces. For the hard sphere model these have been assumed to be infinitely large. If U_{hs} is the pair potential energy for hard spheres and r is the distance between the centres of the particles, Figure 3.2a, then $U_{hs} = \infty$ for $r < (\sigma_S + \sigma_X)$ and $U_{hs} = 0$ for $r \geqslant (\sigma_S + \sigma_X)$. A more realistic model for ions and solvents considers the particles as consisting of a 'hard core' surrounded by a more or less 'soft' shell. Various expressions have been proposed for this softening effect on the repulsive pair potential energy

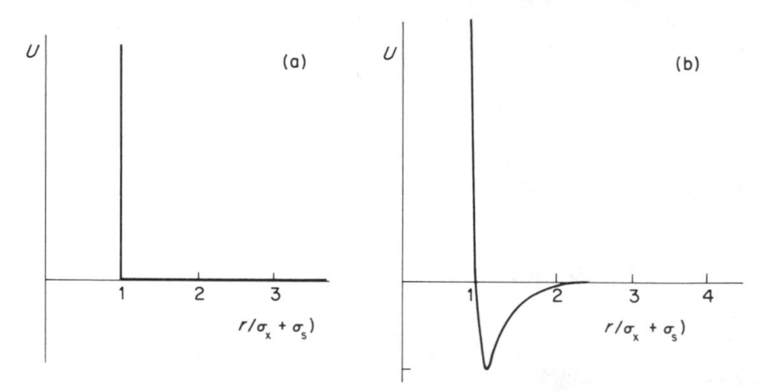

Figure 3.2 (a) The potential energy of a hard-sphere fluid. (b) The potential energy of a 'Lennard–Jones' fluid, $U = 4u[(\sigma/r)^{12} - (\sigma/r)^6]$, where u is the depth of the potential well (the equilibrium attractive potential energy) and $\sigma = \frac{1}{2}[\sigma_X + \sigma_S]$

making it proportional to r^{-n} ($n \geqslant 8$, usually chosen as 9 or 12) or to $\exp(-r/a)$ ($0.30 \leqslant a \leqslant 0.35$ for r in nm). In practice, however, the contribution of this repulsive potential energy term, which is very short range, is neglected. The diameters σ_X and σ_S are implicitly chosen in such a manner, that the effect of repulsion at $r \geqslant (\sigma_S + \sigma_X) \approx 0$. The positive contributions to ΔG^0_{solv} and ΔH^0_{solv} ascribed to the cavity formation thus implicitly involve also the positive contribution to the potential energy arising from the hard core (and soft shell) repulsion.

Beyond the distance $r = (\sigma_S + \sigma_X)$, however, attractive interactions gain the upper hand, Figure 3.2b. These arise mainly from dispersion forces (London forces), which relate to the deformability of the electronic shells of the atoms. The contribution to the molar Gibbs free energy of solvation from the dispersion forces can be written as

$$\Delta G_{disp} = -(8\pi/9)\mathbf{N}^2 C_{disp} V_S^{-1} \sigma_{SX}^{-3} \qquad (3.29)$$

where σ_{SX} is the distance at which the attractive and repulsive interactions between the solvent and solute particles are equal, and C_{disp} is a parameter that depends on the properties of the solvent and the solute specified below. The value of σ_{SX} is generally set equal to $\frac{1}{2}(\sigma_S + \sigma_X)$, with the same values of σ_S and σ_X specified for cavity formation as discussed above, i.e., the hard sphere diameters. The value of C_{disp} can be estimated in various ways, none of which is completely satisfactory.

If the solute is a noble gas or a non-polar volatile compound, the Lennard–Jones intermolecular potential equation, Figure 3.2b, can be applied:

$$C_{disp,LJ} = 4(u_S u_X)^{1/2} \sigma_{SX}^6 \qquad (3.30)$$

where u is the depth of the potential well of the intermolecular interactions in the specified species, and the Berthelot relationship is assumed for the mixture. Values of u are available for only a few solvents, and generally other expressions must be used.

The Kirkwood–Mueller formula has been found[2] to be a more or less successful approximation:

$$C_{disp,KM} = 6\mathbf{e}_0 \alpha_S \alpha_X / \{[\alpha_S/(-\chi_S)] + [\alpha_X/(-\chi_X)]\} \qquad (3.31)$$

where $\mathbf{e}_0$ is the rest energy of the electron, $8.184 \cdot 10^{-14}$ J, α is the polarizability and χ the molecular magnetic susceptibility. It has, however, been shown[7] that α and χ are approximately proportional to each other for a large number of organic compounds: $\alpha/(-\chi) \approx 9.7 \cdot 10^4$, so that equation 3.31 can be approximated by $C_{disp} \approx 5.06 \cdot 10^{-18} \alpha_S \alpha_X$ J m^6.

The values of u, σ, α, and χ required for the calculation of ΔG_{disp} by means of equations 3.29 and 3.30 or 3.31 are given, as far as they could be found, for about 110 solvents in Table 6.4.

If either the solute or the solvent is polar, i.e., has a permanent dipole moment, μ, inductive interactions are added to the dispersive ones. The corresponding contribution to the molar Gibbs free energy of solvation is[2]

$$\Delta G_{\text{ind}} = -(4\pi/3)\mathbf{N}^2(\mu_S^2\alpha_X + \mu_X^2\alpha_S)V_S^{-1}\sigma_{SX}^{-3} \tag{3.32}$$

If both the solute and the solvent are polar, dipole–dipole interactions are added, according to the Keesom orientation expression:

$$\Delta G_{\text{dip}} = -(8\pi/9)\mathbf{N}^2\mu_S^2\mu_X^2(\mathbf{k}T)^{-1}V_S^{-1}\sigma_{SX}^{-3} \tag{3.33}$$

The values of the dipole moments, μ, required for the calculation of ΔG_{ind} and ΔG_{dip} by means of equations 3.32 and 3.33, respectively, are given for about 110 solvents in Table 6.4.

The temperature dependence of ΔG_{disp} and ΔG_{ind} is introduced directly via V_S, equations 3.29 and 3.32, that of ΔG_{dip} via both T and V_S, and in all cases also indirectly via

$$\sigma_{SX} = \tfrac{1}{2}(\sigma_S + \sigma_X)$$

of which σ_S is temperature dependent, see the discussion following equation 3.27. The corresponding ΔH and ΔS terms can, in principle, be calculated from the ΔG terms by means of the Gibbs–Helmholz equation.

The non-specific interactions of ions are expected to be the same as those of atomic or molecular neutral species of similar size and polarizability. It is therefore of interest to explore the extent to which the calculated thermodynamic functions of solvation agree with the experimental ones for such solutes as the noble gases. These are given directly by their solubilities and the temperature coefficients of the solubilities in the various kinds of solvents. The solubilities at 298.15 K are presented in Table 3.1, and can be well represented[9] by the sum of a cavity formation term, ΔG_{cav}, and interaction term, $\Delta G_{\text{disp}} + \Delta G_{\text{ind}}$ (since as $\mu_X = 0$ also $\Delta G_{\text{dip}} = 0$), and a concentration scale term, all together divided by $RT \ln 10$. The concentration scale term depends on the units in which the solubility is expressed. The agreement between the calculated and experimental solubilities depends on which of several sets of u and σ values are employed in equation 3.30.[8,9] The agreement of the experimental solubilities other than at 298.15 K and of the experimentally derived ΔH_{soln} and ΔS_{soln} with the calculated quantities is less impressive.[8,9] It depends also on whether σ_S or both σ_S and σ_X are required to be temperature dependent.[8]

The solubilities of polyatomic gases can also be described well by the above mentioned terms, provided that their interactions with the solvents are limited to dispersion and induction. Dipolar solute gases require also the ΔG_{dip} term, but solutes with quadrupole moments, such as CO, NO, and C_2H_6, and those with many π-electrons, such as CO_2, C_2H_2, and C_6H_6, undergo further interactions,[9] at least with those solvents that are good electron-pair donors and/or acceptors. More about this aspect, as applied to ions, is provided in Section 3.5 below.

The solubilities of still larger solute molecules are of interest with respect to the thermodynamic properties of large model ions, the tetraphenyl-arsonium and -borate ions, discussed in Section 6.3. Table 3.2 presents the solubilities of

Table 3.1 Solubility of the noble gases in various solvents, $-\log(x_X/(P/\text{torr}))$, at 25 °C. Data from ref. a, unless otherwise noted (1 torr = 133.3 Pa)

Solvent	He	Ne	Ar	Kr	Xe	Rn
Water	8.036	7.970	7.480	7.226	6.984	6.668
Heavy water	7.948	7.905	7.449			6.486
n-Hexane	6.466	6.311	5.479	5.041	4.468	4.031
n-Heptane					4.470[m]	
n-Dodecane					4.394[m]	
c-Hexane	6.794	6.615	5.699	5.209	4.561	4.001[c]
Benzene	7.001	6.820	5.935	5.443	4.816	4.277
Toluene	6.885	6.724	5.841	5.356	4.716	4.186
p-Xylene	6.851	6.698	7.786	5.303		
Methanol	7.106	6.974	6.231	5.938		
Ethanol	6.983	6.843	6.084	5.478		4.757
1-Propanol			5.991	5.368	4.624	4.427
2-Propanol						4.589
1-Butanol			5.922	5.516		4.310
1-Pentanol			5.864	5.38[d]		
1-Hexanol			5.824			
1-Octanol	6.799	6.652	5.741	5.304		
c-Hexanol	7.21[f]	7.03[f]	6.16[f]	5.65[f]		
Benzyl alcohol			6.44[g]			
Ethylene glycol			6.952			
Glycerol				6.62[e]	7.36[d]	6.07
Diethyl ether			5.47[g]			
1,4-Dioxane			6.100			
Benzaldehyde			6.20[g]			
Acetone	6.843	6.685	5.925	6.376[e]		4.650
2-Butanone			5.90[g]			
c-Hexanone			6.12[g,h]			
Formic acid						5.674
Acetic acid			6.262	5.835[e]	4.47[i]	4.856
Butyl acetate				5.22[e]		
Chloroform			6.11[e]	5.36[e]	4.835	4.302
Carbon tetrachloride			5.750	5.16[e]	4.760	
Chlorobenzene	7.040	6.890	5.950	5.441	4.751	
Carbon disulphide	7.29	7.11	5.194		4.863	4.201
Dimethyl sulphoxide	7.43[b]	7.32[b]	6.69[b]	6.23[b]	5.65[b]	
Hydrazine	8.194[j]		7.791			
Aniline					5.586	4.754
Nitromethane	7.294	7.148	6.383	5.958	5.578	
Nitrobenzene	7.337	7.242	6.232	5.738	5.114	
N-methyl formamide	7.94[k]	7.77[k]	7.01[k]	6.60[k]		
N-methyl acetamide	7.190		6.243[l]			

References and notes:

[a] Solubility Data Series, Vol. 2 and 4, Pergamon, 1980; [b] C. L. DeLigny and N. G. Vander Veen, *Chem. Eng. Sci.*, 27, 391 (1972); [c] At 291 K; [d] J. A. M. van Liempt and W. van Wijk, *Rec. Trav. Chim.*, **56**, 632 (1937); [e] F. Koeroesy, *Trans. Faraday Soc.*, **33**, 416 (1937); [f] A. Lannung, *J. Am. Chem. Soc.*, **52**, 68 (1930); [g] B. Sisskind and I. Kasarnawsky, *Z. Allgem. Anorg. Chem.*, **200**, 279 (1931); at 273 K; [h] Ref. g, at 298 K; [i] At 301 K; [j] At 288 K; [k] C. L. DeLigny, H. M. Denessen, and M. Alfenaar, *Rec. Trav. Chim.*, **90**, 1265 (1971); [l] At 308 K; [m] G. L. Pollack, *J. Chem. Phys.*, **75**, 5875 (1981); G. L. Pollack and J. F. Himm, *ibid.*, **77**, 3221 (1982); A. Ben-Naim and Y. Marcus, *ibid.*, **80**, 4438 (1984), where data for hydrocarbons from hexane to icosane are presented.

Table 3.2 Solubility of tetraphenyl compounds in various solvents, $-\log(s_X/\text{mol dm}^{-3})$ at 25 °C

Solvent	Ph_4C	Ph_4Si	Ph_4Ge	Ph_4Sn	Ph_4Pb
Water	7.81[k]		7.77[g]		
n-Hexane	2.48[h]				
n-Heptane	2.25[a]	2.72[a]	2.98[a]	3.33[a]	2.81[a]
Benzene	1.60[a]	1.09[a]	1.28[a]	1.67[a]	1.53[a]
p-Xylene	2.03[a]	1.56[a]	1.82[a]	2.15[a]	2.81[a]
Methanol	3.76[e], 3.69[g]	3.26[e]	3.36[c], 3.47[b]	3.6[b]	
Ethanol	3.61[e], 3.63[g]	3.20[e]	3.26[e], 3.35[g]		
Ethylene glycol	3.83[g]		3.57[g]		
Diethyl ether	2.73[a]	2.15[a]	2.33[a]	2.62[a]	2.71[a]
1.4-Dioxane	1.69[a], 2.07[l]	1.39[a]	1.47[a], 1.93[l]	1.91[a]	1.66[a]
Acetone	2.8[d], 2.77[g]		2.18[g]		
Propylene carbonate	3.2[d], 3.27[g]		3.01[g]		
Carbon tetrachloride	1.87[a]	1.42[a]	1.51[a]	1.91[a]	1.81[a]
1,1-Dichloroethane	1.30[h]				
1,2-Dichloroethane	1.47[h]				
o-Dichlorobenzene	2.10[l]		1.62[l]		
m-Dichlorobenzene	2.15[l]		1.77[l]		
Dimethyl sulphoxide	3.1[f], 3.21[g]		2.90[g]		
Sulpholane[j]	2.8[d]				
Pyridine	1.81[a]	1.35[a]	1.46[a]	1.80[a]	1.77[a]
Acetonitrile	3.21[e], 3.20[g]	2.90[e]	2.98[e], 3.01[g]		
Nitromethane	3.4[d]				
Formamide	4.9[d], 3.99[g]		3.74[g]		
N-Methyl formamide	3.65[g]		3.39[g]		
N,N-Dimethylformide	2.5[d], 2.51[g]		2.10[g]	2.0[b]	
N,N-Dimetylacetamide	2.4[d], 2.43[g]		2.12[g]		
N-Methyl pyrrolidinone	2.0[d], 2.06[g]		1.72[g]		
HMPT[i]	2.3[d], 2.32[g]		2.02[g]		

References and Notes:
[a] W. Strohmeier and K. Miltenberger, *Chem. Ber.*, **91**, 1357 (1958); [b] A. J. Parker and R. Alexander, *J. Am. Chem. Soc.*, **90**, 3313 (1968); [c] B. G. Cox and A. J. Parker, *J. Am. Chem. Soc.*, **94**, 3674 (1972); [d] R. Alexander, A. J. Parker, J. H. Sharpe, and W. E. Waghorne, *J. Am. Chem. Soc.*, **94**, 1148 (1972); [e] D. H. Berne and O. Popovych, *Anal. Chem.*, **44**, 817 (1972); [f] B. G. Cox and A. J. Parker, *J. Am. Chem. Soc.*, **95**, 402 (1973); [g] J. I. Kim, *J. Phys. Chem.*, **82**, 191 (1978), mean values; [h] M. H. Abraham and A. Nasehzade, *Can. J. Chem.*, **57**, 2004 (1979); [i] HMPT = hexamethyl phosphoric triamide; [j] Sulpholane = tetramethylene sulphone = tetrahydrothiophene-S,S-dioxide, 30 °C; [k] From ΔG_t^0 (Ph_4C, $H_2O \rightarrow CH_3CN$) of ref. g and $-\log s$ of ref. e; [l] J. I. Kim, *Z. Phys. Chem., Frankfurt*, **113**, 129 (1978).

the tetrahedral tetraphenyl derivatives of carbon, silicon, germanium, tin, and lead at 298.15 K in many solvents. The major contribution to the Gibbs free energy of solution (for the gasified solutes) would be ΔG_{cav}, which increases strongly with increasing σ_X. It is expected to be large for these solutes, which have $\sigma_X > 0.8$ nm, compared with the interaction terms, which decrease with decreasing σ_X and distance from the solvent molecule (although other quantitites that they contain, such as α_X, also increase with σ_X). The measured solubilities shown in Table 3.2 pertain to the solid solutes, and include therefore a contribution from the lattice energy. They do not vary monotonously with the size of the central atom as may be expected from the smoothly varying σ_X, so that the more interesting comparisons are along the vertical columns.

3.4 Electrostatic interactions

If a charged conducting sphere of macroscopic size located in a vacuum is discharged, transfered into a liquid medium, and recharged there to its original charge, the electrostatic work performed is given by the Born equation[10]

$$W_{el} = W_{Born} = (8\pi\varepsilon_0)^{-1}(z\mathbf{e})^2 r^{-1}(1 - 1/\varepsilon) \qquad (3.34)$$

where ε_0 is the permittivity of free space, z is the number of elementary charges $\mathbf{e}$, to which the sphere is charged, r is the radius of the sphere, and ε is the relative permittivity or the bulk dielectric constant of the liquid medium.

As the size of the sphere is permitted to decrease and approach microscopic dimensions, deviations from equation 3.34 become increasingly important. Even if the liquid medium is considered to be continuous, dielectric saturation will occur in the space near the sphere as its electric field $|E| = |z|\mathbf{e}/\varepsilon_0\varepsilon r^2$ increases with decreasing size. The (relative) dielectric constant of the liquid medium may be expressed as (see Figure 3.3):

$$\varepsilon(E) = \varepsilon(0) - bE^2 + \cdots \qquad (3.35)$$

where b is a constant, that depends on the nature of the liquid and on P and T. The modified electrostatic work of transfering the sphere becomes

$$W_{el} = W_{Born} + W_{sat} = W_{Born} + (160\pi^2\varepsilon_0^2)^{-1}(z\mathbf{e})^4 b r^{-5}\varepsilon(0)^{-4} \qquad (3.36)$$

The second term is appreciable only if r is as small as the sum of the radius of an ion and the diameter of a solvent molecule, i.e., 0.3 to 0.5 nm. The value of the parameter b is known for very few solvents only: it is[11] 1.1 for water, 2.5 for methanol and ethanol, 3.0 for 1-propanol, 11 for 1-butanol, 7.5 for 1-pentanol, 5.0 for benzyl alcohol, 1.4 for glycerol, 1.6 for chlorobenzene, nitrobenzene and aniline, all in 10^{-17} m^2 V^{-2}.

The effect of the dielectric saturation at short distances is overshadowed by other effects, which arise when a model of the solvent is employed, that is more realistic than the continuum used in the discussion hitherto. The model that has been widely used is that of Buckingham,[12] which involves Z discrete solvent molecules arranged at a definite geometry around the charged sphere, which is of molecular dimensions and therefore designated henceforth as the ion. The spherical ion is considered to be devoid of other properties than its charge $z\mathbf{e}$ (taken in the algebraic sense) and its radius r_X. The solvent molecules are also spheres, to which the following properties are assigned: a diameter σ_S, a polarizability α_S, a dipole moment μ_S and an electrical quadrupole moment θ_S. Axial symmetry is assumed around the line connecting the ion with the centre of the solvent molecule, and the point dipole of the solvent is aligned with this line,

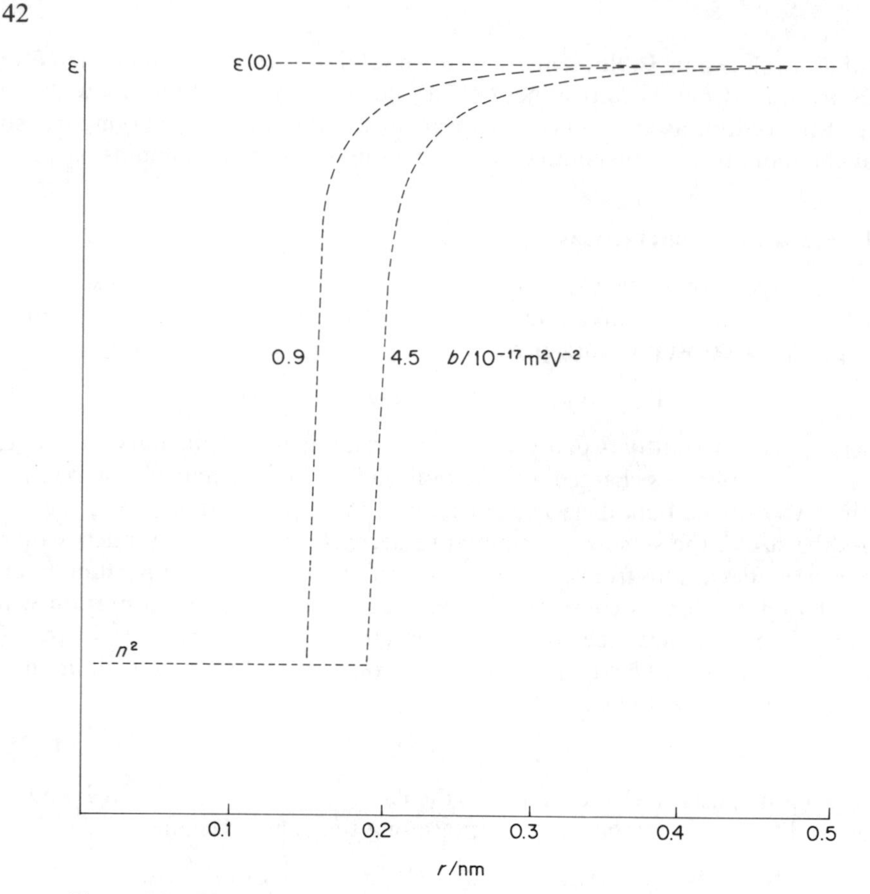

Figure 3.3 The relative permittivity of a hypothetical solvent having a bulk value of $\varepsilon(0)$, a high-field value $n^2 = 0.2\varepsilon(0)$, and a parameter b as indicated, as a function of the distance from a unit point charge

pointed away from a cation and towards an anion, see Fig. 2.1a. This orientation yields the most favourable electrostatic potential energy. The geometries considered are tetrahedral for $Z = 4$ and octahedral for $Z = 6$.

The electrostatic potential energy per solvent molecule for this model is the sum of the ion–dipole, ion–quadrupole, and ion-induced dipole terms, and terms for the mutual interactions of the solvent molecules near the ion among themselves, which depend on the geometry. The result is

$$U_{\text{el}} = -\mathbf{N}|z|\mathbf{e}\mu r^{-2} + \mathbf{N}z\mathbf{e}\theta r^{-3} - \tfrac{1}{2}\mathbf{N}(z\mathbf{e})^2\alpha r^{-4} + C'(Z)\mu^2 r^{-3} - C''(Z)(z/|z|)\mu\theta r^{-4} + O(r^{-5}) \quad (3.37)$$

The terms in r^{-5} and higher negative powers of r are usually neglected, and the value of r to be used in equation 3.37 is generally taken as $r = r_X + \frac{1}{2}\sigma$, where r_X is the crystal ionic radius and σ is equated with σ_S employed in Section 3.2. The values of the coefficients are: $C'(4) = (15/8)(3/2)^{1/2} = 2.296$, $C''(4) = 3.445$, $C'(6) = 7.114$, and $C''(6) = 10.671$.

A deterrent to the general use of equation 3.37 is the unavailability of solvent electrical quadrupole moment data, θ_S. Some estimates for water and ammonia are known, but are considered unreliable.[12] In fact, if equation 3.37 is assumed to account for the electrostatic part of the Gibbs free energy of ion solvation, values of θ_S can be and have been estimated from appropriate experimental data on electrolyte solvation and assumptions concerning the non-electrostatic contributions. The θ_S values obtained in this way[13,14] are 6.7 for water, 9.0 for methanol, 12.0 for ethanol, 4.0 for acetone, and 14.7 for N-methylformamide, all in $10^{-33}\,\mathrm{C\,m^{-2}}$. However, the use of these figures in equation 3.37 will just reproduce the original measured data and not lead to further insight. No independent values of θ_S are known.

Contrary to the dielectric saturation term, equation 3.36, which is of shorter range than the non-negligible terms of equation 3.37, and therefore generally neglected, the Born term is an important contribution. It is considered to apply beyond the shell of the nearest 4 or 6 solvent molecules. The appropriate value of r to be used in equation 3.34 is therefore $r = r_X + \sigma_S$. It is of interest to note that only the terms that involve θ_S also involve z to an odd (the first) power, i.e., depend on the sign of the charge of the ion. If the difference between the electrostatic potential energy of solvation of two ions of equal size r_X but different signs of the charge is considered, Buckingham's model[12] predicts that all the terms will cancel, except those that contain θ_S:

$$U_{\mathrm{el},+}(r_X) - U_{\mathrm{el},-}(r_X) = 2\mathbf{N}|z|e\theta_S r^{-3} + 2C''(Z)\mu_S\theta_S r^{-4} \qquad (3.38)$$

This difference has indeed been stressed by some authors [12-14] as proving that there is a significant difference between the strength of the hydration or the solvation of cations and anions of the same size.

The above considerations lead to the electrostatic contribution to the Gibbs free energy of solvation, and by appropriate differentiation to the entropy, enthalpy, volume change, heat capacity change, etc., for the solvation process. The quantity that is temperature and pressure sensitive is, of course, the bulk relative permittivity ε_S. The derivatives $(\partial\varepsilon_S/\partial T)_P$ and $(\partial\varepsilon_S/\partial P)_T$ are listed in Table 5.5 for water, and the former quantity for many solvents in Table 6.3. The ions are generally considered to be thermally nonexpandible and noncompressible (i.e., r_X is taken as constant), and changes of b in equation 3.36 with T and P are generally neglected. A difficulty arises, however, with a quantity that appears in equation 3.34, namely with the component σ_S of r. The differentiation of $\Delta G_{\mathrm{Born}} = \mathbf{N}W_{\mathrm{Born}}$ with respect to P or T should, therefore, be much more complicated than commonly practiced, because of the dependence of σ_S on these two variables.

The pressure derivative of the electrostatic contribution to the Gibbs free energy of solvation should yield the electrostriction, i.e., the volume diminution formally ascribed to the solvent, due to the presence of the field caused by the charge on the ion. A detailed theory[15] calculates first the effective pressure that is equivalent to the effect of this field E on the solvent in the immediate surroundings of the ion:

$$dP = (\varepsilon_0/4\pi\kappa_{TS})(\partial\varepsilon_S/\partial P)_{E,T}\, dE \qquad (3.39)$$

where κ_{TS} is the isothermal compressibility of the solvent. The theory considers the dependence of the relative permittivity ε_S of the solvent on both P and E, and arrives at an expression that permits the calculation of the expected electrostriction at given distances r from the ion, that in turn determine the value of E. For this calculation the pressure dependencies of the density, the relative permitivity, and the refractive index of the solvent have to be known up to very high pressures, as well as the parameter b of equation 3.35. The implementation to the solvent water is discussed in Section 5.2.

3.5 Short range interactions

The electrostatic model, such as that Buckingham[12] discussed in Section 3.4, is not the only approach to the interactions of an ion with the solvent molecules in its immediate vicinity. An alternative approach recognizes the coordinative bonds that may be formed, i.e., allows for some overlap of the electron orbitals of the ion and the solvent in the solvated ion. Whereas quantum chemical methods can be applied for this in the isolated solvate in the gas phase (Section 2.2), an empirical approach is still used for this purpose for the treatment of ion solvation in the bulk solvent, which employs the concepts of donor–acceptor interactions. Cations and anions are treated somewhat differently in this approach.

Some cations have well defined coordination spheres in solid solvates. These are generally believed to persist in solution. This is particularly true of the ions of the transition metal elements, but also for the ions of some main group elements (Be^{2+}, Mg^{2+}, and Al^{3+}, at least for hydration), and for the ions of some post-transition elements (e.g., Zn^{2+}, Cd^{2+}, and Ag^{+}). The structural, spectroscopic and chemical kinetic evidence for this is discussed in Chapter 4. At this point the idea is accepted, and these cations are considered as being provided with a coordination sphere of definite geometry. This is not the case, however, for many other cations, the solvation spheres of which are diffuse and ill-defined, as for the alkali metal cations and most organic cations.

If solvates of definite geometry and coordination number Z are formed, their formation constants can generally be estimated, i.e. the equilibrium constants for the replacement of some reference solvent S_1 in the solvate $M(S_1)_Z^{z+}$ by the solvent S_2 to form the solvated ion $M(S_2)_Z^{z+}$. This replacement reaction is generally studied (see Section 7.2) in mixed solvents $S_1 + S_2$, but the values may be extrapolated to apply to the pure solvent S_2. The standard Gibbs free energy of the replacement reaction is thus known, and when added to the Gibbs free energy of solvation of the cation M^{z+} with the reference solvent S_1, gives an estimate of the solvation Gibbs free energy with the solvent S_2. Minor corrections have still to be applied for the difference in the interactions of

$M(S_1)_Z^{z+}$ with bulk S_1 and those of $M(S_2)_Z^{z+}$ with bulk S_2, according to the concepts in Sections 3.2 to 3.4.

Cations are generally classified as Lewis acids, and act as acceptors towards the donor atoms of solvents which are Lewis bases. This includes practically all the inorganic cations and most of the polar solvents, although it must be recognized that there exist cations essentially devoid of acceptor properties (e.g. the tetraalkylammonium cations) and polar solvents essentially devoid of Lewis base properties (e.g. fluorosulphonic acid). These still participate in solvation by means of the electrostatic interactions dealt with in Section 3.4.

There are no generally accepted criteria for measuring the electron pair acceptance capability of cations. For cations with rare-gas-like closed electronic shells this quantity is expected to be proportional to their field strengths or their surface charge densities, i.e., $z\mathbf{e}/\varepsilon_0 r_M^2$. A quantity that has been proposed as a measure of the tendency of the ions to undergo covalent bonding is their 'softness'. The softness parameter σ_M is defined[16] as the difference between the ionization potential of the gaseous atom to form the cation and the enthalpy of hydration of the latter, normalized by subtraction of and division by the corresponding difference for the hydrogen ion. A list of cation softness parameters is given in Table 3.3a. It is expected that some combination, $a(z\mathbf{e}/\varepsilon_0 r_M^2) + b\sigma_M$, where a and b are constants, would best express the electron pair acceptance capability of the cations. This expectation is based on the presumption that such a general capability can be defined for each cation, independently of the donor properties of the Lewis base solvents.[17]

The electron pair donation abilities of solvents have been specified much more clearly. It is necessary to distinguish between the donation ability of the solvent molecule in an inert solvent, i.e., that of the isolated solvent molecule, and that measured for the bulk solvent.[4] A further distinction exists between the general ability of a solvent to donate a pair of electrons to a coordinate bond, i.e., its donicity, and the donation ability exhibited towards cations capable of providing a hydrogen atom for the formation of a hydrogen bond, such as trialkylammonium cations. A well established measure of the donation ability of the isolated solvent molecule is its donor number DN, introduced by Gutmann.[18] It is discussed further and compared with other scales of donicity[4] in Section 6.1. A special class of Lewis base donor solvents comprises those which have pronounced hydrogen bond acceptance abilities. These should be of importance in the solvation of cations such as the trialkylammonium ones, as mentioned above, but also in mixed aqueous–organic solvent mixtures, in the solvation of hydrated cations in the second solvation shell. The β-scale has been proposed by Kamlet and Taft[19] for the expression of this kind of solvent donicity. This, again, is discussed further in Section 6.1, where data for some 110 solvents are presented.

Anions generally do not have around them a coordination sphere with a well defined geometry. Since anions are Lewis bases and have lone-pair electrons to donate, they are solvated by solvents according to the Lewis acidity or electron pair acceptance ability of the latter. They also accept hydrogen bonds

Table 3.3a Crystal ionic radii, r_M[a], and softness parameters, σ_M[b], of cations

M^{z+}	r_M/nm	σ_M	M^{z+}	r_M/nm	σ_M
H^+		0.00	Al^{3+}	0.050	−0.25
Li^+	0.060	−0.95	Sc^{3+}	0.075[e]	−0.51
Na^+	0.095	−0.75	Y^{3+}	0.093	−0.68
K^+	0.133	−0.53	La^{3+}	0.115	−0.65
Rb^+	0.148	−0.49	Gd^{3+}	0.094[e]	−0.56
Cs^+	0.169	−0.46	Lu^{3+}	0.086[e]	−0.67
Cu^+	0.096	0.26	Pu^{3+}	0.101[e]	−0.62
Ag^+	0.126	0.18[c]	Cr^{3+}	0.062	−0.06
Au^+		0.45	Fe^{3+}	0.060	0.22
Tl^+	0.144	0.09	Ga^{3+}	0.062	0.29
$NH_4{}^+$	0.148		In^{3+}	0.081	0.44
Be^{2+}	0.031	−0.41	Tl^{3+}	0.095	0.92
Mg^{2+}	0.065	−0.37	Bi^{3+}	0.102[e]	0.61
Ca^{2+}	0.099	−0.65	Ce^{4+}	0.080[e]	−0.54
Sr^{2+}	0.113	−0.59	Th^{4+}	0.099	−0.55
Ba^{2+}	0.135	−0.60	U^{4+}	0.097[e]	−0.38
Mn^{2+}	0.080	−0.11	Pu^{4+}	0.080[e]	−0.21
Fe^{2+}	0.075	−0.06	Zr^{4+}	0.072[e]	
Co^{2+}	0.072	−0.18	$(CH_3)_4N^+$	0.280[f]	0.81[i]
Ni^{2+}	0.070	−0.11	$(C_2H_5)_4N^+$	0.337[f]	
Cu^{2+}	0.070	0.39	$(C_3H_7)_4N^+$	0.372[f]	
Zn^{2+}	0.074	0.37	$(C_4H_9)_4N^+$	0.413[f]	
Cd^{2+}	0.097	0.59	$(C_5H_{11})_4N^+$	0.443[f]	
Hg^{2+}	0.110	1.28	$(C_6H_{13})_4N^+$	0.469[g]	
Sn^{2+}	0.093	0.31	$(C_7H_{15})_4N^+$	0.492[g]	
Pb^{2+}	0.132	0.58	$(C_6H_5)_4P^+$	0.425[h]	
UO_2^{2+}	0.073[g]	−0.38[d]	$(C_6H_5)_4As^+$	0.426[h]	7.32[i]

For annotation see Table 3.3b.

from protic solvents that can donate them. The ability of anions to undergo covalent bonding (coordinative bonding) by the donation of a pair of unshared electrons to an acceptor solvent can be described by their softness, in the same manner as for the cations. The softness parameter σ_X is defined[16] for anions as the difference between the electron affinity of the gaseous atom or radical forming the anion and the enthalpy of hydration of the latter, normalized by subtraction of the corresponding difference for the hydroxide anion and division by the differences between the ionization potential of the hydrogen atom and the enthalpy of hydration of the hydrogen ion. A list of anion softness parameters is given in Table 3.3b. A relation may also exist between the softeness parameters here defined and the polarizability and the sizes of the anions, but this has not been established unambiguously. The hydrogen bond accepting abilities of the monoatomic anions probably decreases according to the decreasing electronegativities of the corresponding atoms: $F > O > Cl > S > \cdots$, but for polyatomic anions (that usually have oxygen atoms in their periphery, as in $ClO_4{}^-$) no clear relationship is known.

Table 3.3b Ionic radii, r_X[a], and softness parameters, σ_X[b] of anions

X^{z-}	r_X/nm	σ_X	X^{z-}	r_X/nm	σ_X
F^-	0.136	−0.71	IO_3^-	0.182	
Cl^-	0.181	−0.16	ClO_4^-	0.236	0.00
Br^-	0.195	0.10	MnO_4^-	0.229[j]	
I^-	0.216	0.40	HCO_2^-	0.158	−0.44
OH^-	0.140	0.00	$CH_3CO_2^-$	0.159	−0.48
SH^-	0.195	0.63	HCO_3^-	0.156[j]	
CN^-	0.182, 0.191[j]	0.48	$B(C_6H_5)_4^-$	0.421[h]	6.86[i]
SCN^-	0.195, 0.213[j]	0.84	S^{2-}	0.184[e]	1.02
N_3^-	0.195[j]	0.78	Se^{2-}	0.209[j]	
BF_4^-	0.232[j]		CO_3^{2-}	0.185	−0.37
NO_2^-	0.155	−0.24	SO_4^{2-}	0.230	−0.31
NO_3^-	0.189	−0.41	SeO_4^{2-}	0.249[j]	
ClO_3^-	0.200	0.14	CrO_4^{2-}	0.256[j]	
BrO_3^-	0.191				

[a] Pauling crystal radii for coordination number 6 for monoatomic ions, thermochemical radii for polyatomic ones, except where otherwise noted; [b] From ref. 16; [c] Ag^+ behaves as if it had a much larger σ_M, 0.4 to 0.7, than that obtained from the enthalpy of hydration and ionization energy; [d] Y. Marcus, *J. Inorg. Nucl. Chem.*, **37**, 493 (1975); [e] R. D. Shanon and C. T. Prewitt, *Acta Cryst.*, **B25**, 925 (1969), **B26**, 1046 (1970); [f] Van der Waals radius, E. J. King, *J. Phys. Chem.*, **74**, 4590 (1970); [g] Scaled to give van der Waals radius by multiplication of structural model radius by 0.837; [h] Y. Marcus, *Rev. Anal. Chem.*, **5**, 53 (1980); [i] S. Glikberg and Y. Marcus, *J. Soln. Chem.*, **12**, 272 (1983); [j] H. D. B. Jenkins and K. P. Thakur, *J. Chem. Educ.*, **56**, 576 (1979).

The electron pair acceptance ability of solvents has been measured on many scales, the Dimroth and Reichardt E_T one[20] having been evaluated for the largest number of solvents, and is discussed further in Section 6.1, where values for some 110 solvents are presented. It is compared there also with other scales, and in particular with a pair of scales that distinguishes between the polarity of a solvent (the π^*-scale) and its hydrogen bond donating ability (the α-scale), as proposed by Taft and co-workers.[21]

If the short range interactions may indeed be described in these terms, then cations should have a contribution $(aze(\varepsilon_0 r_M^2)^{-1} + b\sigma_M)\cdot DN$(normalized) and anions a contribution $c\sigma_X \cdot E_T$(normalized) to their solvation Gibbs free energies, where *a*, *b*, and *c* are suitable constants, independent of the properties of the ions and of the solvents. Marcus and Glikberg[17] have found that this is a viable approach, but that although the anions have practically no electron acceptance abilities, cations do have some electron pair donation abilities beside their obvious acceptance power. A term in E_T is relevant also for the cations, while a term in *DN* is not required for the anions.

3.6 Solvent structural effects

The contributions to the Gibbs free energy of solvation of an ion that have been discussed hitherto have not considered explicitly any structure that the solvent may have and any changes in this structure that the ion may cause. Implicitly, such effects are included in some of the terms, such as cavity formation, which involve bulk properties of the solvent, since these are structure dependent.

The following questions may be asked in this connection:

(i) How is the structure of the solvent defined and measured?
(ii) How much structural change occurs when a solute particle (an ion) is dissolved in the solvent?
(iii) How do such structural changes affect the thermodynamics of the dissolution process?

Since water is the structured solvent par excellence, both the theoretical and the practical considerations have involved this solvent in particular. The concepts can be applied to other structured solvents (e.g., hydrogen peroxide, sulphuric acid, ethylene glycol, or formamide) when suitable modifications are made, provided some essential data are available.

According to Ben-Naim[22] the structural dependence is introduced in the pair potential energy U of the molecules in the pure solvent. This may be divided into a non-structural part U_{nonstr} and a structural one. The former arises from hard-core repulsions, and from attractions due to dispersion forces, dipole–dipole interactions, and higher multipole and induced-multipole interactions. In an ensemble of N solvent molecules (at a given T and P or V), this depends on the configuration (R^N) of the molecules, $U_{\mathrm{nonstr}}(R^N)$, see Section 3.1. The structural part is ascribed to hydrogen bond formation. For the sake of simplicity, the formation of a hydrogen bond is described[22] in terms of one characteristic energy, u_{hb}, which is specific for each solvent, differing, e.g., also for light water, H_2O, and heavy water, D_2O, and a geometrical factor $g(R_i, R_j)$, which can also be called a probability factor. It has a value of ≈ 1, when the mutual configuration (R_i, R_j) of the two partner molecules i and j, i.e., both the coordinates and the orientations, are favourable for hydrogen bond formation (for water an O—O distance of 0.276 nm and an O—H...O bond angle of 180°), and falls more or less sharply to zero when the configuration deviates from the favourable one. The quantity $g(R^N) = \sum_{i \neq j}^{N} g(R_i, R_j)$ is the measure of the extent of hydrogen bonding in a particular configuration (R^N). The pair potential energy in such a configuration is thus

$$U(R^N) = U_{\mathrm{nonstr}}(R^N) + u_{\mathrm{hb}} \cdot g(R^N) \tag{3.40}$$

The amount of structure inherent in the solvent, at a given T and P, is defined as the ensemble average of the geometrical factor or the probability factor, $\langle g \rangle_0$. The quantity $\exp(-U(R^N)/\mathbf{k}T)$ functions as a weighting factor in the statistical mechanical evaluation of this average, made by integration over all the configurations and all the space occupied by the system. The average number of hydrogen bonds formed by a solvent molecule in the pure solvent is $(2/N)\langle g \rangle_0$, a number between 0 and 4 for water. Attempts have been made to estimate the latter quantity for water by spectroscopic means (infrared, Raman and NMR spectroscopy), but no agreed values resulted from these attempts.

If now a solute particle, an ion, is introduced at a fixed point R_0 in the system, the new ensemble average $\langle g \rangle_X$ will differ from that in the pure solvent, since a certain binding energy $B(R^N)$ is added to $U(R^N)$, see equation 3.4, and the weighting factor mentioned above is thereby changed. The difference $\Delta\langle g \rangle_X = \langle g \rangle_X - \langle g \rangle_0$ defines the change in the solvent structure that has occurred on the introduction of the solute particle. This change in $\langle g \rangle$ cannot be measured directly for the general case, but it can be estimated for the particular solvent water.

If it is assumed[22] that the solute particle has the same strength of interaction with light as with heavy water (the same B) and also that U_{nonstr} is the same for these two modifications of water, the only difference remaining is u_{hb}. In fact, the quantity

$$\Delta u_{hb} = u_{O\text{—}D\ldots O} - u_{O\text{—}H\ldots O} = -0.96 \text{ k J mol}^{-1}$$

has been estimated. A comparison of the standard molar Gibbs free energies of solution of the solute X in light and heavy water then gives the desired information[22,23]

$$\exp[-(\Delta G^0_{solution}(\text{X in } D_2O) - \Delta G^0_{solution}(\text{X in } H_2O))/\mathbf{R}T] = \Delta u_{hb} \cdot \Delta\langle g \rangle_X \quad (3.41)$$

In order for the changes in the thermodynamic quantities of the dissolution process, due to these changes in the structure of the solvent, to be evaluated, it is expedient to consider the dissolution as occurring in two steps.[24] In the first, the solute particle is added and all the solute-solvent interactions are permitted to take place, but the solvent keeps the structure it has in the pure state. The solution with this 'frozen-in' structure is designated by ^{fr} in the following. In the second step the solvent structure is permitted to relax to its new equilibrium state. The chemical potential of the solute at equilibrium is

$$\mu_X = \mu_X^{fr} + \Delta\mu_{relax}(\partial\langle g \rangle_X/\partial N_X)_{T,P} \quad (3.42)$$

where $\Delta\mu_{relax}$ is the difference between the chemical potentials of the more structured and the less structured forms of the solvent. For the sake of simplicity, a two-state model of the solvent is employed here. Since, however, the various structural forms of the solvent are at equilibrium after the second step, $\Delta\mu_{relax} = 0$, and whatever structural changes there may be, measured by $(\partial\langle g \rangle_X/\partial N_X)$, these have no effect on the chemical potential of the solute, and $\mu_X = \mu_X^{fr}$.

Not so, however, for other thermodynamic functions: $T(\bar{S}_X - \bar{S}_X^{fr})$ and $\bar{H}_X - \bar{H}_X^{fr}$ are finite (but exactly equal and opposite in sign, so that they mutually compensate) and $\Delta\bar{S}_{relax}$ and $\Delta\bar{H}_{relax}$ are generally also finite quantitites. The quantity μ_X^{fr} does depend on the structure of the (pure) solvent, i.e., on the distribution of its molecules among the various structural states, but not on any structural changes caused by the dissolution process. In order for the quantities introduced here, μ_X^{fr}, $\bar{S}_X^{fr}$, etc. and $\bar{S}_{relax}$, etc., to be calculated, a detailed model of the solvent structure and its energetics is required,[24] but no very realistic and tractable model has so far been presented, not even for the much studied solvent water.

These considerations, although conceptually clear, remain qualitative for the time being, and the appropriate contributions to the thermodynamic functions of solvation due to the structure of the solvent and its changes on dissolution cannot be evaluated. The question of which other contributions to these functions, if any, are covered implicitly by those arising from changes in the solvent structure cannot be answered either. Since this is the case, it may be best to abandon separate considerations of solvent structural effects, expecting them to be covered by other effects, considered under other headings.

The more or less independent terms that must be added in order to obtain the Gibbs free energy, the entropy, the enthalpy, etc. of solvation are then:

(i) A concentration scale term, if the standard state of the solute ion in the gas and the solution phases are not the same.
(ii) A cavity formation term, which has implicit in it a short range structural effect.
(iii) General short range effects of repulsion (taken care of in the cavity term), dispersion forces, and dipole–dipole (and induced dipole) interactions for ions with a non-spherical charge distribution.
(iv) Short range donor–acceptor type interactions with solvent molecules in the first coordinate sphere, or alternatively the formation of definite solvates with a limited number of solvent molecules, as calculated for the gas phase in Chapter 2.
(v) Long range electrostatic interactions, particularly of the Born type, but applying only beyond a certain radius, i.e., that of the first coordination sphere, with eventual corrections for other electrostatic interactions.

Such calculations have been undertaken by several authors in recent years, as seen, for example, in the publications,[25,26] in which the authors, of course, employed their own versions of this general scheme.

References

1a. A. Ben-Naim, *J. Phys. Chem.*, **82**, 792 (1978); b. A. Ben-Naim and Y. Marcus, *J. Chem. Phys.*, **81**, 2016 (1984).
2. R. A. Pierotti, *Chem. Rev.*, **76**, 717 (1976).
3. H. Reiss, H. L. Frisch, and J. L. Lebovitz, *J. Chem. Phys.*, **31**, 369 (1959) and other papers from this group, quoted in ref. 2.
4. Y. Marcus, *J. Soln. Chem.*, **13**, 599 (1984).
5. S. W. Mayer, *J. Phys. Chem.*, **67**, 2160 (1963).
6. J. I. Kim, *Z. Phys. Chem., Frankfurt*, **113**, 129 (1978).
7. T. Nakamura, *Ashahi Garasu Kogyo Gijutsu Shoreikai Kenkyu Hokoku*, **28**, 175 (1976), *Chem. Abstr.*, **87**, 92 780 b (1977; T. Nakamura and H. J. Choy, *Rept. Res. Lab. Eng. Mat. Tokyo Inst. Tech.*, **1**, 97 (1976), *Chem. Abstr.*, **86**, 71 647 c (1977).
8. J. I. Kim, *Z. Phys. Chem., Frankfurt*, **110**, 197 (1978).

9. C. L. DeLigny and N. G. Van der Veen, *Chem. Eng. Sci.*, **27**, 391 (1972).
10. M. Born, *Z. Phys.*, **21**, 45 (1920).
11. J. Malsh, *Phys. Z.*, **29**, 770 (1928); **30**, 837 (1929).
12. A. D. Buckingham, *Disc. Faraday Soc.*, **24**, 151 (1957).
13. C. L. DeLigny, H. J. M. Denessen, and M. Alfenaar, *Rec. Trav. Chim.*, **90**, 1265 (1971); D. Bax, C. L. DeLigny, and M. Alfenaar, *ibid.*, **91**, 453 (1972); D. Bax, C. L. DeLigny, and A. G. Remijnse, *ibid.*, **91**, 965, 1225 (1972).
14. J. I. Kim, *J. Phys. Chem.*, **82**, 191 (1978).
15. J. E. Desnoyers, R. E. Verall, and B. E. Conway, *J. Chem. Phys.*, **43**, 243 (1965).
16. Y. Marcus, *Israel J. Chem.*, **10**, 659 (1972); Y. Marcus, *Introduction to Liquid State Chemistry*, Wiley, Chichester (1977), 265–7.
17. S. Glikberg and Y. Marcus, *J. Soln. Chem.*, **12**, 255 (1983).
18. V. Gutmann and E. Vychera, *Inorg. Nucl. Chem. Letters*, **2**, 257 (1966).
19.a. V. Gutmann, *Chemische Funktionslehre*, Springer, Vienna (1971); b. M. J. Kamlet and R. W. Taft, *J. Am. Chem. Soc.*, **98**, 377 (1976).
20. K. Dimroth, C. Reichardt, T. Siepmann, and F. Bohlmann, *Ann. Chem.*, **661**, 1 (1963); C. Reichardt, *Solvent Effects in Organic Chemistry*, Verlag Chemie, Weinheim (1979).
21. R. W. Taft and M. J. Kamlet, *J. Am. Chem. Soc.*, **98**, 2886 (1976); M. J. Kamlet, J. L. Abboud, and R. W. Taft, *ibid.*, **99**, 6027 (1977).
22. A. Ben-Naim, *J. Phys. Chem.*, **79**, 1268 (1975).
23. Y. Marcus and A. Ben-Naim, to be published (1985).
24. A. Ben-Naim, *J. Phys. Chem.*, **82**, 874 (1978).
25. P. Claverie, J. P. Daudey, J. Langlet, B. Pullman, D. Piazzola, and M. J. Huron, *J. Phys. Chem.*, **82**, 405 (1978).
26. M. H. Abraham and J. Liszy, *J. Chem. Soc. Faraday Trans. 1*, **74**, 1604 (1978).

Chapter 4

Structural and kinetic aspects

4.1 Introduction

The interactions between an ion and solvent molecules discussed in Chapter 3 lead to a characteristic spatial arrangement and orientation of the solvent molecules around the ion. It also leads to differences in the dynamics of these molecules when in the vicinity of the ion and in the bulk solvent. This Chapter deals with the structural and kinetic consequences of these interactions.

For an isolated ion and a single solvent molecule, the relative orientation must minimize the potential energy, as discussed in Section 2.2. When several solvent molecules surround the ion, this favourable orientation is charged to take into account the simultaneous presence of these molecules in the *solvation shells* around the ion, including the mutual interactions of these molecules. These interactions may be repulsive, because of the crowding of their hard cores, but also attractive, through dispersion forces, multipole interactions, and eventually also hydrogen bonding.

Geometric limitations proscribe the presence of more solvent molecules in the immediate vicinity of an ion than its *coordination number*. This number depends on the size of the ion and the size of that atom of the solvent molecule, that is oriented towards the ion and is nearest to it, and in a secondary manner, on the size and bulkiness of the rest of the solvent molecule. However, the tightness of the packing of the solvent molecules around the ion will differ according to the strength of the interactions, and this variation leaves some leeway in the time average of the number of nearest neighbours in the *first solvation shell*. The mere juxtaposition of solvent molecules to the ion in this solvation shell reveals little about the nature and the strength of the interactions, but it is useful to have a clear picture of this number of nearest neighbours and the spatial extent of this solvation shell, as discussed below, Section 4.2.

The *concentric shell model*, originally suggested by Frank and co-workers,[1] can be generalized to take cognizance of the fact that solvent molecules beyond the first solvation shell generally do not behave exactly as those in the bulk solvent. It is expected, for instance, that for a solvent that has a distinct structure (due to hydrogen bonding or dipole—dipole interactions), this structure is disrupted in a certain spherical shell around the ion. This comes about because

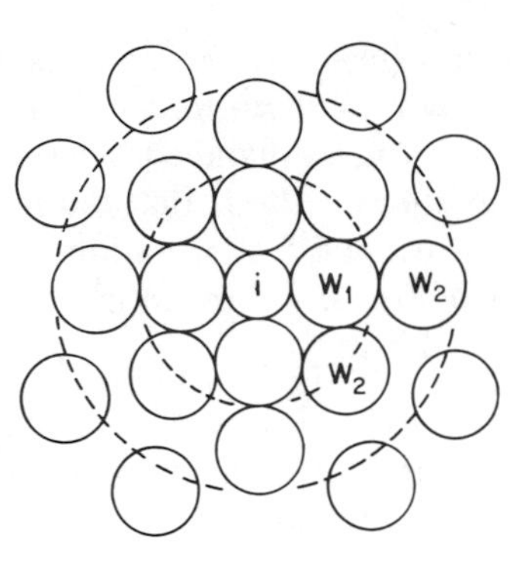

Figure 4.1 The concentric shell model for the solvation of an ion. Note the first coordination shell, which is constituted by the nearest neighbours, and which may, but need not necessarily, be identical with the primary solvation of the ion, and the second and further solvent shells, some of which may constitute the secondary solvation of the ion

of: (i) orientation forces of the electrical field of the ions; and (ii) interactions with solvent molecules belonging to the first solvation shell, that do not have the same spatial arrangement as the bulk solvent. The number of solvent molecules so affected, and the extent of this *second solvation shell* may also be given by the methods described below, Section 4.4, but this region may extend further away from the ion, depending on the properties of the solvent that are concerned. Beyond the second solvation shell the solvent molecules are indistinguishable in their behaviour from those in the pure solvent, hence they constitute the *bulk solvent*. Figure 4.1 presents a schematic drawing of these concentric shells.

The concepts of the concentric shell model, in terms of a first solvation shell, a second solvation shell, and the bulk solvent, should be distinguished from the concepts of primary and secondary solvation used by several authors.[2] The first solvation shell is dictated by the packing of the molecules, and exists whether or not the solvent molecules are bonded or coordinated with the ion, and whether or not their mean residence time near the ion is appreciably longer than near solvent molecules in the pure liquid solvent. Thus large ions of low charge, such as Cs^+ or I^-, have in water a first solvation shell, although they may lack primary solvation (hydration) according to most criteria of this concept. Secondary solvation, again, may not coincide with the second solvation shell, since the former may count some of the solvent molecules in the first solvation shell that do not interact strongly enough with the ion to be counted among the primary solvation, but may not count all the solvent molecules that differ in behaviour significantly from the bulk. The concept of primary and secondary solvation according to various operational definitions is elucidated further in Section 4.4.

4.2 The structure of solvated ions

The actual structure of solvated ions has been studied directly almost exclusively for aqueous electrolyte solutions. However, the techniques should be generally valid also for solutions in other solvents, provided their molecular complexity is not much larger than that of water. This proviso is very difficult to meet (heavy water, D_2O, is an obvious exception), and a compromise must be sought between the amount of information that is desired and the complexity that can be handled. These points will be clarified in the descriptions of the existing methods that follow.

Structural information on the environment of a solvated ion is obtainable from three groups of methods:

(i) diffraction methods, including X-ray, neutron, and electron diffraction, as also EXAFS and similar methods;
(ii) computer simulation methods, including the Monte Carlo and molecular dynamics variants;
(iii) some spectroscopic methods, such as visible–UV or infrared spectroscopy, nuclear magnetic resonance, etc.

The third group of methods yields more indirect information. The former two groups of methods are designed to answer directly the question of how neighbouring solvent molecules are arranged around an ion in the solution. This subject has been reviewed recently by Enderby and Neilson.[3]

For the sake of concreteness consider in the following an ion i placed at the origin of coordinates, and species W (e.g., water molecules or their oxygen atoms) located in its environment. The bulk density of the W particles is N_W^0 particles per volume V. The number of these particles in a spherical shell of thickness dr at a distance r from the ion i is

$$dN_W(r) = 4\pi(N_W^0/V)g_{iW}(r)r^2\,dr \tag{4.1}$$

where $4\pi r^2\,dr$ is the volume of the shell and $g_{iW}(r)$ is a probability function, called the *pair correlation function*. It is the probability of finding a particle of W at the scalar distance r from a particle of i located at the origin, and describes directly the effect of the presence of the particle i there on the local density of the W particles, compared with that in the bulk. Hence $g_{iW}(r)$ is the key quantity for the description of the structure of the solvation shells of the ion. It should be noted that $g_{iW}(r)$ is a spherically symmetrical function, so that it gives, in a sense, only one-dimensional information. Arguments from steric geometry are required in order to form a complete three-dimensional picture of the environment of the ion i.

Figure 4.2 shows the course of $N_W(r) = 4\pi(N_W^0/V)r^2g_{iW}(r)$, which exhibits a series of peaks and valleys, as would also a plot of $g_{iW}(r)$ itself. The curve has essentially zero ordinate values up to $r = r_0$, which is the distance up to which the mutual repulsion of i and W is practically infinite. The curve has a smooth parabolic course, expected for $4\pi(N_W^0/V)r^2$, beyond some distance corresponding to four or five solvent diameters, i.e., there $g_{iW}(r) = 1$. The first peak seen in the curve occurs at r_{max}, which is the mean distance of approach of i and W, whereas the position of the first valley, r_{min}, is the extent of the first solvation shell.

The area under the curve up to a given value of r is the number of W

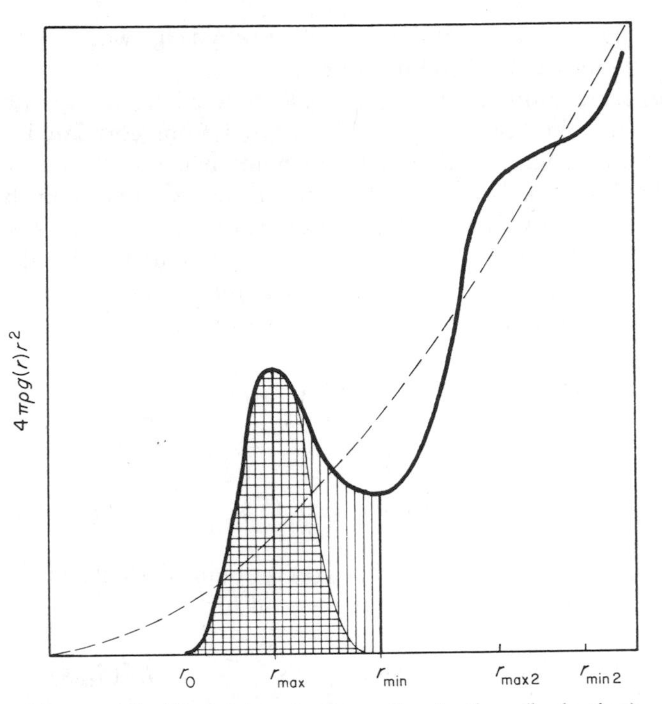

Figure 4.2 The determination of solvation (hydration) numbers h from $g_{iW}(r)$ data. The long-dashed curve is $4\pi\rho r^r$ (where $\rho = N_W^0/V$), the continuous curve is $4\pi\rho g_{iW}(r)r^2$. The cross-hatched area under the curve is symmetric with respect to r_{max}, whereas the vertically hatched area, up to r_{min}, represents h according to equation 4.2. The limits of the second coordination shell around the ion are also shown. (After Y. Marcus, *Introduction to Liquid State Chemistry*, Wiley, Chichester (1977), by permission)

neighbours that i has up to this distance. The integral

$$h = 4\pi(N_W^0/V)\int_0^{r_{min}} g_{iW}(r)r^2\,dr \tag{4.2}$$

is, therefore, the number of solvent molecules in the first solvation shell around the ion i, its (first) solvation number. If the first peak in the curve depicted in Figure 4.2 were symmetrical with respect to r_{max}, it would be better to set h equal to twice the integral up to r_{max}, since this quantity is better defined than r_{min}. However, the mutual repulsion of the W particles as they crowd around the ion i at distances below r_{max} causes the curve to be asymmetrical, and the necessity of using equation 4.2 with the less well defined r_{min}.

The second peak and valley yield the quantities $r_{max\,2}$, $r_{min\,2}$, and h_2, denoting the mean distance between the ion and its next nearest neighbours in the second solvation shell, the mean radial extent of this shell of next nearest neighbours, and the corresponding solvation number. These quantities are less clear-cut than those pertaining to the first solvation shell, when obtained from experimentally measured $g_{iW}(r)$ functions.

The pair correlation function $g_{iW}(r)$ is obtained from experimental data through the interference of radiation scattered from correlated atoms. For radiation of wavelength λ, scattered through an angle θ to the incident beam, a wave-number k is defined as $k = \lambda^{-1}4\pi \sin(\theta/2)$. The partial structure factor $S_{iW}(k)$ is obtained from the diffraction data as detailed further below. It describes the structure of the solution near the ion, i.e., the correlation of the i and W particles, in the so-called k-space. In order to translate this information into ordinary space, the following connection between $S_{iW}(k)$ and $g_{iW}(r)$ is used:

$$S_{iW}(k) = 1 + (4\pi N_W^0/kV)\int_0^\infty (g_{iW}(r) - 1)\sin(kr)r^{-1}\,dr \tag{4.3}$$

or its Fourier-transform

$$g_{iW}(r) = 1 + (2\pi^2 N_W^0 r/V)^{-1}\int_0^\infty (S_{iW}(k) - 1)k\sin(kr)\,dk \tag{4.4}$$

The experimental data from which $S_{iW}(k)$ is obtained are the intensities of the radiations scattered at angles θ to the incident beam

$$I(\theta) = \alpha(\theta)\left[\delta(\theta) + N\sum_j c_j f_j^2 + N\sum_j\sum_{k\neq j} c_j c_k f_j f_k (S_{jk}(k) - 1)\right] \tag{4.5}$$

The calibration parameters $\alpha(\theta)$ and $\delta(\theta)$ depend on the angle θ and on the instrumentation. In the case of $\delta(\theta)$ there is also a dependence on the system studied: this includes the contributions from incoherent and multiple scattering, as well as from inelastic scattering (the Compton and Placzek effects for X-rays and neutrons, respectively). N is the total number of atoms, c_j the concentration of the j-th species (hence $Nc_j = N_j/V$), f_j its coherent scattering amplitude, and the summation extends over all the *atomic* species in the system. The extraction of $S_{iW}(k)$ from $I(\theta)$ data according to equation 4.5 is a very complicated undertaking, even if the calibration parameters $\alpha(\theta)$ and $\delta(\theta)$ are known with good precision.

The simplest aqueous solution of an electrolyte consists of four atomic species: O and H atoms from the solvent, and the monoatomic cations C^{z+} and anions A^{z-}. Among these there are ten two-atom correlations: four homo-correlations of each atomic species with itself and six hetero-correlations of each of these species with the other three. For n different atomic species there are altogether $\frac{1}{2}n(n+1)$ two-atom correlations. This proliferation of terms in the summations in equation 4.5 explains the paucity of diffraction data for solutions of electrolytes with polyatomic ions or in solvents with more than two different kinds of atoms, since these would be very difficult to interpret.

In order to extract the $S_{jk}(k)$ of interest from the experimental data after correction for the calibration factors, i.e., from $[I(\theta) - \alpha(\theta)\delta(\theta)]/\alpha(\theta)$, it is necessary to know the scattering amplitude functions $f_j(k)$. It is then necessary to be able to vary them systematically, in addition to the possible variation of the

c_j, in order to have as many independent linear equations as there are unknown $S_{jk}(k)$. The f factor for X-rays (called the form factor) depends strongly on k (it decreases monotonically with increasing k) and on the atomic number Z of the scattering atoms (it increases linearly with Z). For neutrons f (called the coherent scattering length) is independent of k but depends very irregularly on Z. In fact, it depends on the isotopic composition of the scattering nuclei of the atomic species characterized by Z, and may have negative values (e.g., for ^1H, ^7Li, and ^{62}Ni), though it is positive for the majority of nuclei. This fact permits more flexibility of the neutron scattering method, through proper choice of the isotopic composition, compared with the X-ray scattering method. The latter, in particular, is incapable of yielding information concerning the orientation of the solvent (water) molecules around an ion, since the contribution from the ion–hydrogen correlations to the sums in equation 4.5 is essentially zero. This loss of information reduces also the number of simultaneous equations to be solved to six.

Whereas the scattering amplitude f_j depends (for X-rays) on k, the concentrations c_j do not, and may be varied by the experimenter, subject to the constraint of the electroneutrality of the solution. The upper limit of this variation is the saturation of the solution, the lower one the practical loss of the information concerning ion–solvent correlations compared with the contributions to $I(\theta)$ from solvent–solvent correlations. In view of the sizes of the f factors, the maximal contributions from ion–solvent correlations to $I(\theta)$ that can be achieved in practice are 30% for X-ray and 10% for neutron scattering.

The prescribed f_j factors and the limited range through which the c_j can be varied preclude the collection of $I(\theta, f, c)$ data for X-ray scattering sufficiently precise for the direct evaluation of individual partial structure factors $S_{jk}(k)$. The procedure must be reversed, i.e., models for the solution are assumed, yielding partial pair correlation functions $g_{jk}(r)$, from which the $S_{jk}(k)$ are obtained via equation 4.3, and these are then introduced into equation 4.5 for comparison with the experimental $I(\theta)$ data. The variable parameters of the model, or the model itself, are changed, until agreement within the precision limits of the data is attained. This procedure may not produce a unique solution of the structure, but the precision of the data that can be measured now can rule out many otherwise plausible structures in favour of a particular one, with a very limited range of variability of its parameters.

The ability of the experimenter to vary the isotopic composition of, say, the cation in neutron scattering experiments, permits the employment of an ingenious difference technique.[4] If data are obtained for two solutions differeing only in the isotopic composition of C^{z+}, then the difference in the corrected neutron scattering intensity becomes

$$\Delta_C(k) = A_{CO}(S_{CO}(k) - 1) + A_{CD}(S_{CD}(k) - 1) + A_{CC}(S_{CC}(k) - 1) + A_{CA}(S_{CA}(k) - 1) \quad (4.6)$$

(heavy water is used in the neutron diffraction work, hence H is replaced by D in the species subscripts). In equation 4.6 the A_{Cj} have the form $(f_C - f'_C)f_j c_C c_j$, where f_C and f'_C pertain to the two isotopic compositions. The relative contributions from the ion–ion correlations are small compared with those from ion–solvent correlations, and decrease more rapidly with decreasing salt concentrations. Thus equation 4.6 involves essentially only S_{CO} and S_{CD}, and $\Delta_C(k)$ yields direct information on the distribution of oxygen and hydrogen (deuterium) atoms around the cation C^{z+}.

Whereas the diffraction methods are applied to real, and fairly concentrated solutions, the computer simulation methods deal with model systems, and their relevance to real solutions of electrolytes depends on the nearness of the model employed to reality. This model consists generally of one ion and a large number N of solvent molecules (in practice, 25 to 215), confined in a cubic box having an edge L, so that L^3 is the specified volume, and $(N + 1)/L^3$ is the specified particle density of the system. The solvent molecules interact with the ion and among themselves in a pairwise fashion. The potential energy of the system is the sum of these pairwise potentials $u_{jk}(r)$, which are usually taken to be spherically symmetrical (so that the orientations of the particles are averaged out). The computer simulation consists of taking a random configuration of the system (at the specified volume and temperature or total energy) and changing it in certain steps. Boundary problems are avoided by the use of periodic boundary conditions (Figure 4.3), which assures the constancy of the relative positions of the particles in their confining box, even though the random change of the coordinates would otherwise remove a particle from it. The stepwise changes are continued until the Helmholtz free energy of the model system is minimized, several tens of thousands of steps being generally required.

In the Monte Carlo (MC) method,[5] the initial random configuration is specified by the coordination r_x, r_y, and r_z of each particle, where r_i/L is a random number between zero and unity. The configurational potential energy $U(r^{N+1})$ (where r^{N+1} denotes some configuration of the $N + 1$ particles) is calculated from the pairwise additivity of the postulated pair potentials $u_{jk}(r)$, at the specified temperature T. A so-called Markov chain is now followed, where each step consists of picking at random one of the particles, and giving its coordinates random increments (random fractions of some arbitrary unit increment). The configurational potential energy after the $(t + 1)$th step, $U(t + 1)$, is calculated, and is compared with that obtaining before this step is commenced, i.e., after the tth step, $U(t)$. The step is accepted if $U(t + 1) < U(t)$, and the new configuration is the starting point for the $(t + 2)$th step. If $U(t + 1) \geqslant U(t)$, then the quantity $\exp(-U(t + 1)/U(t))$ is compared with a random number between zero and unity: if the exponent is not smaller, the step is accepted, otherwise the $(t + 1)$th step is rejected, and another step is taken starting from the tth configuration. This procedure is continued till no further substantial decrease (beyond an arbitrary tolerance) of the potential energy is obtained. The configurations are averaged along the process, and the average configuration at the equilibrium state yields the pair correlation function $g_{iW}(r)$ of the system.

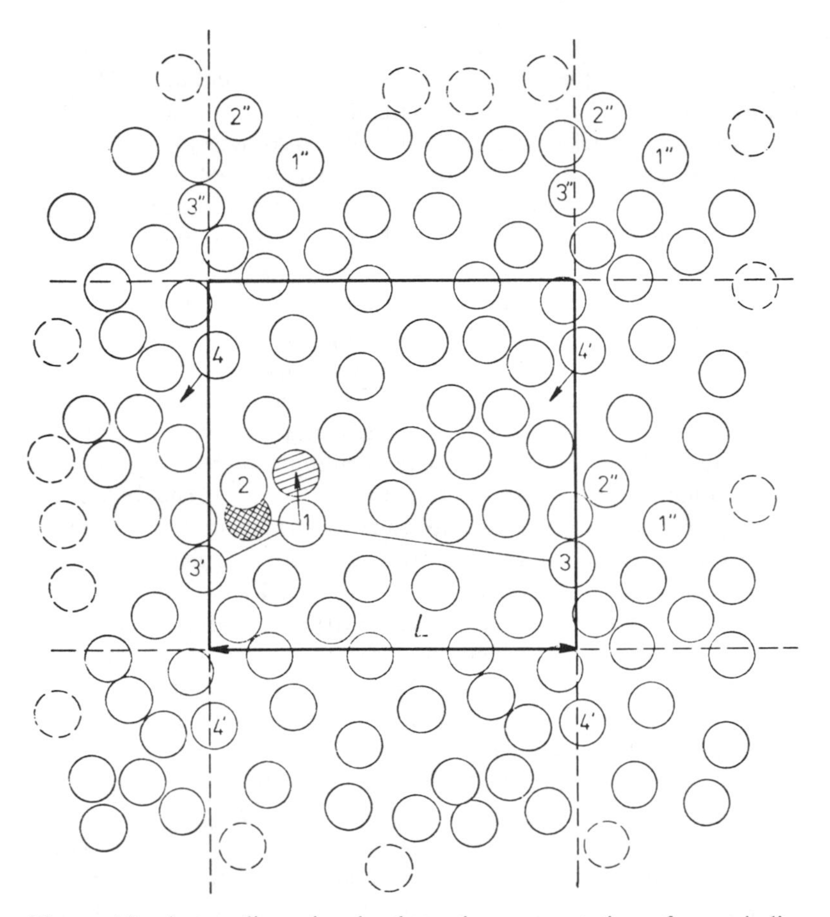

Figure 4.3 A two-dimensional schematic representation of a periodic boundary system of 30 particles of diameter σ (corresponding to $(30)^{3/2} =$ 164 particles in three dimensions). The central square of edge $L = 8\sigma$ is the basic system, the squares around it contain 'ghosts'. The behaviour of the numbered particles and their primed and double-primed ghosts is explained by the following examples. When particle 1 moves towards particle 2, it will cause a large positive increase in the potential energy in the cross-hatched position, but possibly a decrease in it in the diagonally hatched position. As particle 4 moves outside the box through the left-hand wall, particle 4′ enters it from the opposite side. Particle 3′ is used in the calculation instead of particle 3, in order to avoid interactions between particles further apart than $r = L/2$. (After Y. Marcus, *Introduction to Liquid State Chemistry*, Wiley, Chichester (1977), by permission)

In the molecular dynamics (MD) method,[6] the initial configuration specifies in addition to the random coordinates of the particles also random momenta (velocities), subject to the condition of a constant total kinetic energy, (3/2) $(N + 1)\mathbf{k}T$. The step consists of an arbitrary fixed small time interval, during which a sequence of collisions among the particles may occur. The first collision to happen is examined in terms of Newton's laws of motion, in conformation with the conservation of the total energy (kinetic plus potential) and the momentum in this collision. The results of this collision, i.e., the new momenta and coordinates of the particles involved, are taken into account in the next time interval, and all the particles are followed till a further collision occurs. A 'collision' can occur only if the particles come within the range of action of the inter-particle forces, since this causes a change in the potential energy which must be compensated by a corresponding change in the kinetic energy, hence also of the velocities, but at a fixed total momentum of the pair of particles involved. It is therefore very difficult to use complicated pair potentials $u_{ij}(r)$, and in the earlier studies only hard sphere potentials, $u_{ij}(r < \sigma) = \infty$ and $u_{ij}(r \geqslant \sigma) = 0$, where σ is the contact distance of the particles, i.e., the sum of their radii, were used. The Helmholz free energy is averaged over each time interval, and it decreases due to the collisions until equilibrium is attained. Again, configurations in the equilibrium state are averaged to yield $g_{iW}(r)$. The dynamic nature of the procedure yields, however, additional information: the self correlation function of the particles, i.e., their trajectory. From this the self diffusion coefficient and other transport properties of the particles are obtainable.

The key quantity for both the MC and the MD methods is the pair potential $u_{JK}(r)$, which determines the potential energy of the system at each step. The results of the computations bear a close relationship to reality only if this pair potential is chosen judiciously. In addition to the ion–solvent interactions that are discussed in Chapters 3, the mutual interactions among the solvent molecules W must also be known. These differ from those in the bulk solvent, in so far as some of the W particles are already affected by the presence of the electrical field of the ion i at their location. Sometimes, however, acceptable results are obtained even if this possible modification of the solvent molecules is ignored. Many successful calculations have been carried out with the Stillinger and Rahman (ST2) model for water, applicable to a single water molecule in the gas phase. The model is a tetrahedral arrangement of two charges of $+0.2357$ units at a distance of 0.08 nm from the centre and two charges of -0.2357 at a distance of 0.10 nm.[7] A variety of pair potentials has been suggested and tested,[3] recently used versions being a combination of the ST2 model for water with a point charge ion in a Lennard–Jones sphere[8] or a non-empirical quantum mechanical potential.[9] The criteria for the success of the computations is the agreement of the derived structure and thermodynamic properties of the system with experimental data.

Computer simulation methods as well as the diffraction methods involving ionic solutions have so far been confined to aqueous solutions. The results are presented and discussed in Section 5.4. These methods yield the pair correlation function, from which the number of nearest neighbours of kind W around the ion i, h according to equation 4.2, and their mean distance from i, r_{max}, are obtained. The orientation of the solvent molecule relative to the ion, considering now water as a non-spherical particle, has been obtained by the isotope differential mode of the neutron diffraction method, see p. 57, applied directly to the cation–hydrogen atom distances, compared with the cation–oxygen atom distance. However, the isotropic nature of $g_{iW}(r)$ precludes statements concerning the geometrical arrangement of the solvent molecules around the ions, and a symmetrical arrangement, subject to the value of h for the system, is assumed. This can be checked indirectly, provided $g_{WW}(r)$ data show characteristic distances in the solution which are absent in the pure solvent. A comparison of these data with the expectation from models may then decide in favour of a definite geometry.

Several methods for the study of the solutions of ionic compounds tacitly assume the most symmetrical geometries compatible with the coordination numbers in models, e.g., a regular tetrahedron or octahedron and a cube for coordination numbers of 4, 6, and 8, respectively. The consequences of these geometries are then compared with experimental data. Whereas X-ray diffraction is applied in this manner in certain cases, this is the only way in which visible spectroscopy can be used for this purpose. The ligand field caused by the donor atoms of the solvent molecules arranged in these geometries splits the d-orbitals of transition metal ions in a characteristic manner, leading to absorption peaks that can be analysed theoretically. The reverse procedure, of analysing the observed absorption spectrum of a transition metal ion in solution in terms of a definite geometry is not straightforward. It is impossible, if there is some indication that the ligand field lacks the maximal symmetry, e.g., in the case of mixed solvents. In favourable cases, where the absorption spectrum of the solution is practically identical with that of a solid solvate of known structure, it is valid to assume this structure also for the species in the solution. Sometimes it is expedient to use a transition metal ion doped into a colourless compound, consisting of a non-transition metal ion surrounded by the donor atoms of the solvent to be studied (usually oxygen) in an established structure, if the desired solvate cannot be isolated or if its structure is unknown.[10]

For the geometry around non-transition metal ions, and for a decision on whether the nearness of the solvent molecules to an ion indeed constitutes coordination, a different kind of criterion must be set up. An experimental datum that points unambiguously to coordination of the solvent to the ion is the presence of a line in the Raman or far infrared spectrum that must be assigned to the vibration of the coordinative bond. For ions that form coordinative bonds with donor atoms of the solvent molecules in a definite geometrical arrangement, the vibrational spectral lines should be subject to the known rules of group theory with regard to their numbers, activities in the Raman and infrared spectra, dependence on the light polarization in the former, and, in a less predictable manner, also regarding to the actual frequencies and intensities. It is very rare, however, that all the expected lines and no others (except lines due to the bulk solvent) are observed. The case of $Be(H_2O)_4{}^{2+}$ is an example, where new lines are observed in a solution of a beryllium salt with a monoatomic anion in water, not due to the water itself. The observed lines have been ascribed to a regular tetrahedral structure, with $\nu_1(a_1) = 535$, $\nu_2(e) = 82$, $\nu_3(t_2) = 880$, and $\nu_4(t_2) = 355\ cm^{-1}$. However, several additional new lines, due

Table 4.1a Vibrational frequencies, ν/cm^{-1}, for 'cation in cage' species in water and liquid ammonia

Ion	Water	Ammonia	Ion	Water	Ammonia
Li^+	—[a]	241[bc]	Zn^{2+}	360[c]	385[b]
Na^+	—[a]	194[bc]	Cd^{2+}	348[c]	342[b]
K^+	—[a]		Hg^{2+}	360[c]	415[b]
Ag^+		260[b], 296[c]	Pb^{2+}		315[b]
Be^{2+}	535[b], 540[c]	485[b]	Al^{3+}	520[b], 532[c]	
Mg^{2+}	360[bc]	328[bc]	Ga^{3+}	475[b]	477[b]
Ca^{2+}	—[a]	265[bc]	In^{3+}	400[bc]	440[b]
Sr^{2+}	—[a]	243[bc]	Tl^{3+}	450[c]	
Ba^{2+}	—[a]	215[bc]	Cr^{3+}	500[c]	
Mn^{2+}		395[b]	Fe^{3+}	510[c]	
Fe^{2+}		389[b]	Bi^{3+}	390[c]	
Co^{2+}	380[c]		Ce^{4+}	394[c]	
Ni^{2+}	390[c]	405[b]	Th^{4+}	420[c]	
Cu^{2+}		395[b]			

[a] No spectral line found corresponding to such a vibration; [b] from ref. 17; [c] from ref. 11.

perhaps to hydrolysed species, are also observed in the spectrum, and the listed assignments are not above controversy.[11]

Generally, however, the vibrational spectrum shows certain shifts and intensity changes of the lines of the solvent, and at most a single new line, when a salt is dissolved in the solvent. This new line must be ascribed to the interaction of the ions with the solvent, if mutual interactions of the ions can be ruled out (its intensity varies with the first power only of the salt concentration). Excluded from these considerations, of course, are the lines due to internal vibrations of polyatomic ions, such as UO_2^{2+} or NO_3^-. These may be shifted or split, when various solvent environments are compared, and great care must be taken in identifying a new line, uniquely assignable to ion–solvent coordination. This new line is generally assigned to the symmetric stretching of the bonds between the ion and its surrounding polyhedron of donor atoms from the solvent molecules, of unspecified geometry: an 'ion in a cage' model is deemed to describe this situation. Although the number and geometrical arrangement of these coordinated solvent molecules cannot be obtained from the appearance of this new line in the vibrational spectrum, its existence is a direct proof of the coordination of the solvent molecules to the ion and to its having some primary solvation. Table 4.1 presents this qualitative information concerning solvent coordination to ions.

Spectroscopic methods, such as Raman spectroscopy or nuclear magnetic resonance (NMR) spectroscopy, do not yield values of the mean distance apart of i and the nearest W molecules, r_{max}. Hence they are not really structure determining methods for the solution. The NMR method does give information on the mean residence time of a solvent molecule in the first solvation shell, as

Table 4.1b Vibrational frequencies, ν/cm^{-1}, for 'cation in cage' species in various solvents

Solvent	Li^+	Na^+	K^+	Rb^+	Cs^+	NH_4^+	Mg^{2+}
Water	—	—	—				362[h]
Acetone	425[gm]	195[g]	140[g]				
THF[a]	413[g]	192[g]	142[g]				
PC[b]	397[g]	186[g]	144[g]	115[g]	112[g]	184[g]	
Ammonia	241[h]	194[h]					328[h]
Pyridine	385[g]	182[g]	133[k]			196[g]	
4-Me-py[c]	390[g]	178[g]				190[g]	
Formamide	370[n]	210[n]					
DMF[d]	420[l]						
2-Pyrrolidinone	400[n]						
NMP[e]	398[g]	204[g]	140[g]	106[g]		207[g]	
Acetonitrile	402[i]						330[i]
Nitromethane	368[i]						325[i]
DMSO[f]	429[gm]	200[g]	153[g]	122[g]	109[g]	214[g]	
Sulpholane		186[j]	155[j]			197[j]	

[a] THF = tetrahydrofuran; [b] PC = propylene carbonate; [c] 4-Me-py = 4-methyl pyridine; [d] NMP = N-methylpyrrolidinone; [f] DMSO = dimethylsulphoxide; [g] A. I. Popv, *Pure Appl. Chem.*, **41**, 275 (1975); [h] ref. 11; [i] A. Regis and J. Corset, *J. Chim. Phys.*, **69**, 1508 (1972); [j] T. L. Buxton and J. A. Caruso, *J. Phys. Chem.*, **77**, 1882 (1973); [k] M. J. French and J. L. Wood, *J. Chem. Phys.*, **49**, 2358 (1968); [l] C. Lassigne and P. Baine, *J. Phys. Chem.*, **75**, 3188 (1971); [m] A. J. Lees, B. P. Straughan, and D. J. Gardiner, *J. Mol. Struct.*, **54**, 37 (1979); [n] C. N. R. Rao, *J. Mol. Struct.*, **19**, 493 (1973).

discussed in Section 4.3, and may be useful in giving solvation or coordination numbers, as discussed in detail in Section 4.4. In this sense they provide information on the structure of the environment of an ion.

4.3 Solvent dynamics in solvated ions

In addition to being characterized by the mean number and distance of the solvent molecules in the first solvation shell, a solvated ion is characterized also by the mean residence time of the solvent molecules in this shell. This time depends on the kind of bonding that takes place between the ion and the donor atom of the solvent molecule. If a coordinative bond is formed, which may be discernible by the spectroscopic methods discussed in Sections 4.2 and 4.4, a relatively long mean residence time should be expected. If the solvent molecules undergo only electrostatic interactions with the ion, the residence time should not be appreciably longer than in the bulk solvent, and indeed under certain circumstances it is even shorter.

The mean residence time of a solvent molecule S in the first solvation shell of an ion i, τ_i, is determined by the rate of the exchange of these solvent molecules with the bulk solvent. This rate is generally controlled by the rate of the release of one solvent molecule from the solvation shell, since the mechanism is generally a dissociative one. The approach of a solvent molecule from the bulk to the ion is governed by its rate of diffusion, as modified by the field of the ion, and is generally much more rapid then the rate of release, i.e., of a single desolvation step, which is a unimolecular reaction. The rate constant k_r for this release reaction pertains to a single solvent molecule, whereas the mean residence time τ_i is an average for all the h solvent molecules in the first solvation shell. Therefore these quantities are not strictly reciprocals of each other, although they are approximately so.

Three methods are commonly used for obtaining information concerning the rate of desolvation. If this rate is very slow, classical kinetic methods can be applied. Otherwise the relaxation of ultrasound and the relaxation of NMR signals are used. The former method is applicable for τ_i values of several seconds, the latter two methods for τ_i of the order of microseconds. Very specialized techniques have been used to push down the time scale for the solvent exchange to nanoseconds.

The classical kinetic methods involve either the exchange of solvent molecules around the ion with similar solvent molecules that are labeled isotopically (e.g., $H_2{}^{18}O$), or their exchange with some other solvent or ligand. The independence of the rate constant from the nature of the incoming ligand is then a proof of the dissociative mechanism. Such methods have been applied by Taube and co-workers to transition metal ions that form inert complexes,[12] such as Cr^{3+}, Ru^{3+}, or Rh^{3+}. Rates of association reactions of solvated ions with ligands often depend on a solvent release step, hence their bimolecular rate constants are also a measure of the rate of this step. Again, independence from the incoming ligand is a criterion for this being a dissociative step control. Further criteria are the negative and positive signs of the entropy and volume of activation of these reactions, respectively. Even if the reaction is shown to be controlled by the solvent release step, however, there is no direct relationship of the bimolecular rate constant (the dimensions of which are concentration$^{-1}\cdot$time^{-1}) with the unimolecular rate constant k_r (the dimensions of which are time^{-1}).

The ultrasound absorption method deals with a solution of a salt made up of solvated cations CS_n^{z+} and solvated anions AS_p^{z-}. At sufficiently high concentrations these will form outer-sphere ion pairs, which are solvent-separated (see Secion 8.1):

$$CS_n^{z+} + AS_p^{z-} \rightleftharpoons CS_nAS_p^z \quad (z = (z+) - (z-)) \tag{4.7}$$

the solvation shells of both kinds of ions remaining intact. The equilibrium constant and the forward and backward rate constants of reaction 4.7 are K_1, k_1, and k_{-1}, respectively. A solvent molecule S may be reversibly released from the solvation sphere of the cation in this ion pair at a rate compatible with the frequencies of the ultrasonic instrumentation (exchange of the solvent with the solvation shell of the anionic partner of the ion pair is assumed to be much faster and hence irrelevant). The release occurs in a unimolecular step:

$$CS_nAS_p^z \rightleftharpoons CS_{n-1}AS_p^z + S \tag{4.8}$$

If the anion is initially solvated ($p \neq 0$), then it retains its solvation shell in this step, and a solvent-shared ion pair results, whereas if the anion is initially unsolvated ($p = 0$) then a contact ion pair results. In the former case, a further solvent molecule may be eliminated between the partners of the ion pair in a subsequent step, to form a contact ion pair. From a different point of view, the anion may move from the outer coordination sphere of the cation into the inner one in this subsequent step. If a single relaxation of the ultrasonic signal is observed, such a subsequent step is of no concern for the present discussion, since its rate is either much too large or much too small compared with the rate of reaction 4.8. The equilibrium constant and the forward and backward rate constants of the latter reaction are K_2, k_2, and k_{-2}, respectively. The rate of reaction 4.7, being diffusion controlled, is much faster than that of reaction 4.8. The criterion that ensures that the reaction sequence 4.7 and 4.8 indeed controls the release of the solvent from the solvation shell of the cation is the strong dependence of the measured rate on the cation and its virtual independence from the nature of the anion.

The observation of a single relaxation when the absorption of ultrasonic energy is measured as a function of the frequency f is a further criterion, as has been mentioned above, for the applicability of this sequence. The ultrasound absorption coefficient α is given according to Atkinson and co-workers[13] by

$$\alpha f^{-2} = A(1 + f^2 f_r^{-2})^{-1} + B \tag{4.9}$$

The parameter $A = A'(T, c_i)f_r^{-1}$ is a known function of the temperature T and the concentration c_i of the ion pair, and involves the enthalpy and volume change of ion pairing and thermodynamic properties of the solution. The parameter B is the limiting value of αf^{-2} at high frequencies. The quantity f_r is the relaxation frequency, i.e., $1/(2\pi)$ times the reciprocal of the relaxation time of the system. It can be obtained from a rearrangement of equation 4.9, which yields also the parameter A:

$$f^2(\alpha - Bf^2)^{-1} = A^{-1} + (A^{-1}f_r^{-2})f^2 \tag{4.10}$$

The overall rate constant of the solvent release, k_f, is the reciprocal of the relaxation time of the system, hence $k_f = 2\pi f_r$. The rate constant for the actual desolvation step, reaction 4.8, is

$$k_r = k_2 = k_f K_{1+2}^{-1}(1 + k_2 k_{-1}^{-1})(1 + K_2) \tag{4.11}$$

where K_{1+2} is the overall association constant, $K_{1+2} = K_1(1 + K_2)$, which is the quantity directly accessible by experiment. Provided that $k_2 \ll k_{-1}$ and $K_2 \ll 1$, both of which being true as a rule, k_r can be readily evaluated from the sound absorption data.

The data that have been obtained by this method are presented in Table 4.2. They are seen to correspond from very fast exchanges, i.e., mean residence times of the order of nanoseconds, to very slow exchanges, i.e., residence times of many days, that have not been obtained by sound absorption, but by the classical kinetic methods. Figure 4.4 demonstrates this wide spread graphically.

Table 4.2a Unimolecular rate constants for water release from the first hydration shell of cations at 25 °C

Ion	$\log(k_r/s^{-1})$	Ion	$\log(k_r/s^{-1})$	Ion	$\log(k_r/s^{-1})$
Li^+	9.1[a]	Cu^{2+}	9.3[c], 8.3[de]	Ho^{3+}	8.5[f]
Na^+	9.1[a]	Zn^{2+}	7.5[b]	Er^{3+}	8,3[f], 6.7[h]
K^+	9.3[a]	Ru^{2+}	−1[c]	Tm^{3+}	8.2[f], 6.5[h]
Rb^+	9.0[a]	Cd^{2+}	8.2[c]	Yb^{3+}	7.9[f]
Cs^+	9.0[a]	Al^{3+}	−0.8[b]	Lu^{3+}	7.9[f]
Be^{2+}	3.5[b]	Sc^{3+}	5.8[g]	Ti^{3+}	5.0[b], 4.8[c]
Mg^{2+}	5.2[b], 5.7[c]	Y^{3+}	8.3[b]	V^{3+}	3.2[d]
Ca^{2+}	8.5[b]	La^{3+}	8.3[f]	Cr^{3+}	−6.3[b]
Sr^{2+}	8.6[b]	Ce^{3+}	8.5[f]	Fe^{3+}	4.3[b], 3.5[c]
Ba^{2+}	8.9[b]	Pr^{3+}	8.6[f]	Ru^{3+}	−6[c]
V^{2+}	1.9[c]	Nd^{3+}	8.7[f], 7.9[h]	Rh^{3+}	−7.5
Cr^{2+}	9[b]	Sm^{3+}	8.9[f]	Ga^{3+}	3.3[b]
Mn^{2+}	7.4[b], 7.5[d]	Eu^{3+}	8.8[f], 7.9[h]	In^{3+}	4.3[c]
Fe^{2+}	6.5[bd]	Gd^{3+}	8.8[f], 7.3[h]	Tl^{3+}	9.5[c]
Co^{2+}	6.4[b], 6.1[d]	Tb^{3+}	8.7[f], 7.2[h]	Bi^{3+}	4[c]
Ni^{2+}	4.5[b], 4.4[d]	Dy^{3+}	8.6[f], 7.0[h]		

[a] Ref. 13, from sound absorption; [b] R. G. Pearson and P. C. Ellgen, in *Physical Chemistry. An Advanced Treatise*, H. Eyring, ed., Academic Press, New York, Vol. VII (1975), pp. 228–30; [c] Ref. 17, Chapter 11, from NMR; [d] E. v. Goldhammer, in *Modern Aspects of Electrochemistry*, J. O'M. Bockris and B. E. Conway, eds., Plenum Press, New York, Vol. 10 (1975), p. 77, from NMR line widths; [e] for the two axial water molecules in distorted octahedral $Cu(H_2O)_6^{2+}$, the rate for the equatorial water molecules being $2 \cdot 10^4$ times slower; [f] N. Purdie and C. A. Vincent, *Trans. Faraday Soc.*, **63**, 2745 (1967), and D. P. Fay, D. Litchinsky, and N. Purdie, *J. Phys. Chem.*, **73**, 544 (1969), from sound absorption; [g] at −20 °C, J. W. Neely, UCRL-20580 (1971), from NMR line widths; [h] J. Reuben and D. Fiat, *J. Chem. Phys.*, **51**, 4918 (1969), from NMR.

Table 4.2b Unimolecular rate constants[a] for solvent release from the first solvation shell of cations, $\log(k_r/s^{-1})$, at 25 °C

Ion	Water[b]	Methanol	Ethanol	Ammonia	MeCN[c]	DMF[c]	DMSO[c]
Mg^{2+}	5.2, 5.3	3.7	6.4				
Mn^{2+}	7.4, 7.5	5.6			6.9[e]	6.4[d]	6.8[d]
Fe^{2+}	6.5	4.7				5.7[d]	4.0[d]
Co^{2+}	6.4, 6.1	4.3		6.9	5.1, 5.5	5.4, 5.6	5.2, 5.5[d]
Ni^{2+}	4.5, 4.4	3.0	4.0	5.0	3.3, 4.2	3.6, 3.9	3.5, 4.0
Cu^{2+}	9.3, 8.3	8					
V^{3+}	3.2	3.1					
Cr^{3+}	−6.3			−5.1		−7.3	−7.5
Fe^{3+}	4.3, 3.5	3.7, 3.4[d]	4.3		1.6	1.5	1.7
Al^{3+}	−0.8	3.6				−0.8	−1.2
Ga^{3+}	3.3	4.0				0.2	

[a] From ref. 17, pp. 314–317, except for water; when two values are given, they bracket the reported values; [b] From Table 4.2a; [c] MeCN = acetonitrile, DMF = N,N-dimethylformamide, DMSO = dimethylsulphoxide; [d] J. C. Boubel, J. J. Delpuech, and G. Martin, *Mol. Phys.* **33**, 1729 (1977); [e] E. v. Goldhammer and Ch. Barrais, *J. Soln. Chem.*, **9**, 237 (1980).

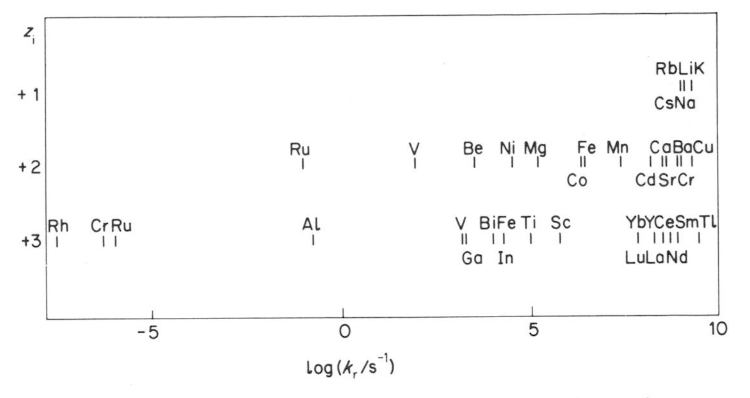

Figure 4.4 The first-order rate constants k_r for the release of a water molecule from the first hydration shell of cations of charge z_i

A major fraction of the results presented in Table 4.2 and in Figure 4.4 have been obtained by NMR line width measurements. The relaxation of those solvent nuclei that give the measured NMR signal in the combined magnetic field and electric field of the neighbouring nuclei is due to several effects acting concurrently. The major relaxation mechanisms that have been noted include orientation relaxation, by rotation of the solvent molecule around the axis from the ion to the centre of the solvent molecule, the exchange of labile atoms, such as hydrogen atoms, between the solvent in the solvation shell and the bulk solvent, and exchange of the entire solvent molecule. If the latter relaxation mode is to be studied, proton NMR should be avoided in favour of nuclei from skeletal atoms of the solvent, preferably those bonded to the ion, since these will be most affected by its near presence. Therefore ^{17}O will be the nucleus of choice for the NMR measurements of the rate of exchange of water or alcohol molecules, and ^{15}N for those of acetonitrile, for example. Sometimes, however, ^{1}H NMR can also be used for the purpose, as in the measurements of the methyl hydrogen atoms of methanol, compared with the hydroxylic ones, if the rates of the release of protons and entire solvent molecules are to be compared.

The longitudinal (spin-lattice) relaxation time T_1 for ^{1}H-nuclei in aqueous solutions of diamagnetic salts is a measure of the rate of reorientation of the water near the ions. In water at 25 °C, $T_1 = 3.4$ s, but in salt solutions it is somewhat larger or smaller, depending on the nature of the ions and their effect on the structure of water,[14] see Section 5.3. Since in such solutions the water exchange rate is much faster than once every few seconds, it is obvious that T_1 is a time average for the orientation of the water molecules in the bulk and in the solvation shells of the ions.

On the other hand, the transverse (spin–spin) relaxation time T_2 (given by the half-width of the NMR peak at half-height) of a ^{17}O nucleus labelling a solvent molecule S is given in the presence of a paramagnetic ion P by

$$T_2^{-1} = T_{2S}^{-1} + \tau_S^{-1}[T_{2P}^{-2} + T_{2P}^{-1}\tau_P^{-1} + \Delta\omega_P^2]/[(T_{2P}^{-1} + \tau_P^{-1})^2 + \Delta\omega_P^2] \quad (4.12)$$

according to Swift and Connick.[15] T_{2S} is the peak half-width in the pure solvent S, T_{2P} is the relaxation time of ^{17}O in the solvation shell of the paramagnetic ion P, τ_S and τ_P are the mean residence times of ^{17}O-labelled solvent molecules in the bulk and the solvation shell, respectively, and $\Delta\omega_P$ is the difference in the Larmor frequencies of ^{17}O in the bulk solvent and in that of the solvation shell. For a molality m_P of the paramagnetic ion the ratio of the number of the solvent molecules in the two environments is $p = hm_PM_S$, where h is the solvation number and M_S the molar mass of the solvent, in kg. Certain limiting cases can be distinguished, depending on the relative magnitudes of T_{2P}^{-2}, $T_{2P}^{-1}\tau_P^{-1}$, and $\Delta\omega_P^2$. In those cases where the rate is controlled by the chemical exchange of the solvent molecules, $\Delta\omega_P^2 \ll T_{2P}^{-2}$ and $\tau_P \gg T_{2P}$, or $\Delta\omega_P^2 \gg T_{2P}^{-2}$ and $\tau_P \simeq T_{2P}$. In both cases

$$\tau_P^{-1} = (T_2^{-1} - T_{2S}^{-1})p^{-1} \tag{4.13}$$

is the resulting rate of release of the solvent molecules from the solvation shell of the paramagnetic ion P.

The method can be applied, of course, also with other nuclei that give suitable NMR signals, such as ^{15}N or ^{13}C. It must also be noted that the bulk solvent in the present context includes any solvent that is in the solvation shells of anions and exchanges much faster than that in the solvation shell of the paramagnetic ion P.

When data are available from both the ultrasound absorption and NMR techniques, e.g., for the rare earth cations in water,[16] the rate constant obtained by the former method, k_r, is somewhat larger than that obtained by the latter, τ_P^{-1}. This is due to the fact that the former quantity pertains to the release of a single solvent molecule, whereas the latter pertains to the average rate of release: it has p in the denominator of equation 4.13, and p is proportional to the solvation number h.

The NMR data have the advantage relative to the sound absorption data in that the former deal directly with the rate of self-exchange of the solvent, and do not require the presence of an anion in the outer sphere of the coordination shell of the cation. Although the independence of the results from the nature of the anion in the latter case attests to the validity of the method, the evidence from the self-exchange is more direct. A disadvantage of the NMR method is that it is applicable only to paramagnetic ions, whereas the sound absorption method can deal with both diamagnetic and paramagnetic ones. Both methods, however, deal only with the solvation shells of cations, and leave those of anions unattended.

4.4 Solvation numbers

The solvation number of an ion is the number of solvent molecules associated with the ion in a specified manner, although this manner is often left unspecified and must be understood from the context. For a given ion, the solvation number depends strongly on the manner of this association and on the method by which this is ascertained. Sometimes there exists a discrepancy between the solvation number operationally defined by the method of measurement and the solvation number that corresponds to the model that conceptually is at the basis of the method. For a given method, the solvation number varies in a systematic way with the properties of the ion (mainly charge and size), but its systematic dependence on the properties of the solvent has not, in general, been clarified.

Two recent monographs[17,18] devote extensive discussions to solvation numbers, but are unable to recommend conceptual or operational definitions of them that are clearly superior to others. It is concluded that solvation numbers may be useful for the discussion of various kinds of problems concerning ions in solution, but that it is impossible to transfer the insight gained by the use of solvation numbers from one problem to the next, in the general case. The classes of problems on which solvation numbers have some bearing, and which can also be formulated in terms of the methods that can be used for the measurement of these numbers, are listed in Table 4.3.

The following concepts are relevant to the discussion of these methods and the results obtained from them. The *first solvation shell* is the set of solvent molecules which are immediate neighbours of the ion. They may or may not interact with the ion strongly: if they, or some of them do interact strongly, those that do are coordinated to the ion, and constitute its *inner coordination sphere* and its *primary solvation*. Criteria for the strength of the interaction that

Table 4.3 Methods dealing with solvation numbers

Method	Kind of solvation
1. Structure determinations:	
1a. Diffraction methods: X-rays, neutrons	1st, eventually also 2nd solvation shell
1b. Computer simulation: Monte Carlo, molecular dynamics	1st, eventually also 2nd solvation shell
2. Methods depending on slow solvent exchange:	
2a. Isotope dilution	Primary solvation
2b. NMR peak area ratio	Primary solvation
2c. Raman or IR peak area ratio	Primary solvation
3. Spectroscopic methods	
3a. Visible–UV spectrometry	Primary solvation
3b. NMR chemical shifts	Primary solvation, but solvation numbers not directly obtained
3c. NMR line widths	Nondefinite solvation
4. Thermodynamic measurements:	
4a. Molar entropies of solvation	Primary and some secondary solvation
4b. Compressibilities	Primary and some secondary solvation
4c. Activity coefficients	Primary solvation(?)
5. Transport methods:	
5a. Transference (Washburn) numbers	Primary, secondary, and further solvation
5b. Ionic mobilities, Stokes radii	Primary and secondary solvation

warrants inclusion in the primary solvation are specific for each method, although the resulting primary solvation numbers disperse only within rather narrow limits.

The *second solvation shell* is the set of solvent molecules which are next-nearest-neighbours of the ion. In general they have some properties that distinguish them from the solvent molecules in the pure solvent. If not, this shell is already a part of the *bulk solvent*, at least with respect to these properties. If they, or some of them do have distinguishing properties, the solvent molecules involved belong to the *secondary solvation* of the ion, together with those solvent molecules from the first solvation shell that have not been included in the primary solvation.

Solvent molecules belonging to the primary and secondary solvation of the ion are oriented by its field in a manner that is incompatible with the normal mutual orientation of the solvent molecules in the bulk solvent. There will, therefore, exist a region around the ion with solvent molecules having neither the central orientation nor the bulk orientation. This disordered region is called the *thawed zone* (a figurative extension of the ice-like structure ascribed to temporarily defined regions in liquid water). In the case where primary solvation is absent, the thawed zone may coincide with the first solvation shell, otherwise it may coincide with the second solvation shell, or, for ions with a very strong field, it may start only beyond the latter shell.

Of the methods listed in Table 4.3, those pertaining to structure determination, 1a and 1b, that have been discussed in Section 4.2, yield unambiguously the solvation number corresponding to the first solvation shell. For the reasons stated in Section 4.2, these methods have been applied so far only to aqueous solutions, and it is unlikely that they may become applicable to solutions in much more complicated solvents. The pertinent solvation (hydration) numbers are shown in Table 4.4.

For the decision on whether these numbers represent also the primary solvation numbers, suitable criteria for the strength of the interaction must be set up. One such criterion, which does not yield solvation numbers but qualitative evidence for the coordination of the solvent, is the appearance of a new line in the vibrational spectrum, as has been discussed in Section 4.2. Another criterion for the existence of a primary solvation of the ion pertains to the slowness of the release of the solvent from the first solvation shell to the bulk solvent, as discussed in Section 4.3. If the rate of solvent exchange is slow with respect to the NMR frequency relevant to suitable nuclei in the solvent molecule, two distinct NMR signals may be observed, one for nuclei in the primary solvation shell of an ion and the other for the relevant nuclei in all the other species: the bulk solvent and in solvation shells of ions (generally anions) where the exchange is more rapid. The distinct signals arise from the different shielding to which these nuclei are exposed in these different environments, hence from the different chemical shifts they suffer. The solvent exchange is sufficiently slow at room temperature only in rare cases (see Section 4.3 and Table 4.2), but it can be slowed down enough at low temperatures, say -60, or -100 °C. In the case

Table 4.4a Number of water molecules in first hydration shell or primary hydration numbers

Ion	*h*	Ion	*h*	Ion	*h*
Li^+	4[a], 6[b]	VO^{2+}	5[l]	Yb^{3+}	8[a], ~4[s]
Na^+	4–8[a], 6[b]	UO_2^{2+}	4[mn]	Lu^{3+}	8[a], 6[s]
K^+	6–8[a], 6[b]	Al^{3+}	6[cdopq]	Ti^{3+}	6[lt]
Cs^+	6–8[a], 7[b]	Sc^{3+}	5[m]	V^{3+}	4[uv]
H_3O^+	4[a]	La^{3+}	8–9[a], 6[s]	Cr^{3+}	6[awxy]
NH_4^+	8[b]	Ce^{3+}	5[r]	Fe^{3+}	6[ay]
Be^{2+}	4[cd]	Pr^{3+}	9[a], ~6[s]	Rh^{3+}	6[z]
Mg^{2+}	6[ac]	Nd^{3+}	8–9[a], ~6[s]	Ga^{3+}	6[cek]
Ca^{2+}	6[a]	Sm^{3+}	9[a]	In^{3+}	6[c]
Mn^{2+}	6[af]	Eu^{3+}	8[a]	Th^{4+}	9[m]
Fe^{2+}	6[ag]	Gd^{3+}	8[a]	F^-	6[ab]
Co^{2+}	6[ah]	Tb^{3+}	8[a]	Cl^-	6[a], 6–8[b]
Ni^{2+}	6[agij]	Dy^{3+}	8[a]	Br^-	6[a]
Cu^{2+}	4 + 2[al]	Ho^{3+}	~3[s]	I^-	6–7[ab]
Zn^{2+}	6[am]	Er^{3+}	8[a], 4–6[s]	NO_3^-	6–9[a]
Cd^{2+}	6[a]	Tm^{3+}	8[a]	SO_4^{2-}	8[a]

[a] From X-ray or neutron diffraction, see Table 5.12, p. 117; [b] from MC or MD computer simulation, see Table 5.12; [c] A. Fratiello, R. E. Lee, V. M. Nishima, and R. E. Schuster, *J. Chem. Phys.*, **48**, 3705 (1968), A. Fratiello, D. D. Davis, S. Peak, and R. E. Schuster, *Inorg. Chem.*, **10**, 1627 (1971), ^{1}H-NMR; [d] R. E. Connick and D. Fiat, *J. Chem. Phys.*, **39**, 1349 (1963), ^{17}O-NMR; [e] D. N. Fiat and R. E. Connick, *J. Am. Chem. Soc.*, **88**, 4754 (1966), ^{17}O-NMR; [f] C. K. Jørgensen, *Acta Chem. Scand.*, **11**, 53 (1957), visible spectrum; [g] A. M. Chmelnik and D. Fiat, *J. Am. Chem. Soc.*, **93**, 2875 (1971), ^{17}O-NMR; [h] N. A. Matwiyoff and D. E. Darley, *J. Phys. Chem.*, **72**, 2659 (1968), ^{1}H-NMR; [i] J. W. Neely and R. E. Connick, *J. Am. Chem. Soc.*, **94**, 3419 (1972), ^{17}O-NMR; [j] T. J. Swift and G. P. Weinberger, *J. Am. Chem. Soc.*, **90**, 2023 (1968), ^{1}H-NMR; [k] S. F. Lincoln, *Aust. J. Chem.*, **25**, 2705 (1972), ^{1}H-NMR; [l] C. K. Jørgensen, *Acta Chem. Scand.*, **11**, 73 (1957), visible spectrum; [m] J. A. Fratiello, R. E. Lee, and R. E. Schuster, *Inorg. Chem.*, **9**, 391 (1970), ^{1}H-NMR; [n] A. Fratiello, V. Kubo, R. E. Lee, and R. E. Schuster, *J. Phys. Chem.*, **74**, 3726 (1970), ^{1}H-NMR; [o] H. W. Baldwin and H. Taube, *J. Chem. Phys.*, **33**, 206 (1960), isotope dilution; [p] N. A. Matwiyoff, P. E. Darley, and W. G. Movius, *Inorg. Chem.*, **7**, 2173 (1968), ^{1}H-NMR; [q] A. Fratiello, R. E. Lee, V. M. Nishida, and R. E. Schuster, *Inorg. Chem.*, **8**, 69 (1969), ^{1}H-NMR; [r] A. Fratiello, V. Kubo, S. Peak, B. Sanchez, and R. E. Schuster, *Inorg. Chem.*, **10**, 2552 (1971), ^{1}H-NMR; [s] A. Fratiello, V. Kubo, and G. A. Vidulich, *Inorg. Chem.*, **12**, 2066 (1973), ^{1}H-NMR; [t] H. Hartmann and H. L. Schläfer, *Z. Phys. Chem.* (*Leipzig*), **197**, 116 (1951), visible spectrum; [u] A. M. Chmelnik and D. Fiat, *J. Magnetic Resonance*, **8**, 325 (1972), ^{17}O-NMR; [v] H. Hartmann and H. L. Schläfer, *Z. Naturf.* **A6**, 754 (1951), visible spectrum; [w] M. Alei, *Inorg. Chem.*, **3**, 44 (1964), ^{1}H-NMR; [x] J. P. Hunt and H. Taube, *J. Chem. Phys.*, **18**, 757 (1950), isotope dilution; [y] H. Hartmann and H. L. Schläfer, *Angew. Chem.*, **66**, 768 (1954), visible spectrum; [z] W. Plumb and G. M. Harris, *Inorg. Chem.*, **3**, 542 (1964), isotope dilution.

where the solvation by water is of interest, such low temperatures are attainable when the water is diluted with acetone, which is deemed a weakly solvating solvent. Provided the ratio of water to metal ions present in the solution is sufficiently large, well in excess of what can be accommodated in the first solvation shell, acetone does not compete successfully with the water; however it does when the water becomes relatively scarce.[19] With many other solvents of interest, such low temperatures can be attained without dilution.

The chemical shift between the two peaks should be independent of the

Table 4.4b Number of solvent[a] molecules in first solvation shell or primary solvation numbers

Ion	MEOH	EtOH	NH_3	MeCN	DMF	DMSO
Li^+				4[b,aa]	4[z]	4[z]
Na^+				4[b]	4[z]	4[z]
Ag^+				4[x], 2[aa]		
Be^{2+}					4[c]	
Mg^{2+}	6[dy]	4[e], 6[f]	6[g]	6[b,aa]	6[z]	6[z]
Ca^{2+}				6[aa]		
Sr^{2+}				6[aa]		
Mn^{2+}				6[h]		
Co^{2+}	6[im]			6[j]	6[k]	6[l]
Ni^{2+}	5[i], 6[m]					6[l]
Zn^{2+}	6[ny]					
Al^{3+}	6[oy]	5.5[e]	6[p]	3[q]	6[r]	6[s]
Cr^{3+}			6[t]			
Fe^{3+}	6[m]					
Co^{3+}			6[w]			
Ga^{3+}	6[o]			6[u]	6[v]	

[a] MeOH = methanol, EtOH = ethanol, MeCN = acetonitrile, DMF = N,N′-dimethylformamide, DMSO = dimethylsulphoxide; [b] I. S. Perelygin and M. A. Klimchuk, *Russ. J. Phys. Chem.*, **47**, 1138, 1402 (1973), IR shifted peak intensity; [c] N. A. Matwiyoff and W. G. Movius, *J. Am. Chem. Soc.*, **89**, 6077 (1967), ^{1}H-NMR; [d] S. Nakamura and S. Meiboom, *J. Am. Chem. Soc.*, **89**, 1765 (1967), ^{1}H-NMR; [e] G. W. Stockton and J. S. Martin, *Can. J. Chem.*, **52**, 744 (1974), ^{13}C-NMR; [f] T. D. Alger, *J. Am. Chem. Soc.*, **91**, 2220 (1969), ^{1}H-NMR; [g] L. W. Harrison and T. J. Swift, *J. Am. Chem. Soc.*, **92**, 1963 (1970), ^{1}H-NMR; [h] L. Burlamacchi, G. Martini, and M. Romanelli, *J. Chem. Phys.*, **59**, 3008 (1973), ESR; [i] Z. Luz and S. Meiboom, *J. Chem. Phys.*, **40**, 1058, 1066 (1964), ^{1}H-NMR; [j] N. A. Matwiyoff and S. V. Hooker, *Inorg. Chem.*, **6**, 1127 (1967), ^{1}H-NMR; [k] N. A. Matwiyoff; *Inorg. Chem.*, **5**, 788 (1966), ^{1}H-NMR; [l] L. S. Frankel, *Chem. Commun.*, **1969**, 1254, ^{1}H-NMR; [m] T. E. Rogers, J. H. Swinehart, and H. Taube, *J. Phys. Chem.*, **69**, 134 (1965), isotope dilution; [n] S. A. Baldwin and T. E. Gough, *Can. J. Chem.*, **47**, 1417 (1969), ^{1}H-NMR; [o] D. Richardson and T. D. Alger, *J. Phys. Chem.*, **79**, 1733 (1975), ^{1}H-NMR; [p] H. H. Glaeser, H. W. Dodgen, and J. P. Hunt, *J. Am. Chem. Soc.*, **89**, 3065 (1967), ^{14}N-NMR; [q] L. D. Supran and N. Sheppard, *Chem. Commun.*, **1967**, 832, ^{1}H-NMR; [r] A. Fratiello and R. E. Schuster, *J. Phys. Chem.*, **71**, 1948 (1967); W. G. Movius and N. A. Matwiyoff, *Inorg. Chem.*, **6**, 847 (1967), ^{1}H-NMR; [s] S. Thomas and W. L. Reynolds, *J. Chem. Phys.*, **44**, 3148 (1966), ^{1}H-NMR; [t] H. H. Glaeser and H. H. Hunt, *Inorg. Chem.*, **3**, 1245 (1964), isotope dilution; [u] A. Fratiello, D. D. Davis, S. Peak, and R. E. Schuster, *Inorg. Chem.*, **10**, 1627 (1971), ^{61}Ga-NMR; [v] W. G. Movius and N. A. Matwiyoff, *Inorg. Chem.* **8**, 925 (1969), ^{1}H-NMR; [w] S. F. Lincoln, *Coord. Chem. Rev.* **6**, 309 (1971), ^{1}H-NMR; [x] T.-C.G. Chang and D. E. Irish, *J. Soln. Chem.*, **3**, 161 (1974), infrared peak area ratio; [y] R. N. Butler and M. C. R. Symons, *Trans. Faraday Soc.*, **65**, 945 (1969), ^{1}H-NMR; [z] I. S. Perelygin and V. S. Osipov, *Russ. J. Phys. Chem.*, **53**, 1036 (1979); I. S. Perelygin and S. Ya. Yamidanov, *ibid.*, **53**, 1354 (1979), IR shifted peak intensities; [aa] P. V. Huong and J. P. Roche, *Mol. Spectr. Dense Phases, Proc. 12th Europ. Congr. Mol. Spectr., Strasbourg, 1975*, p. 613, from IR peak heights of free solvent.

concentration of the salt, for this method to be applicable. The areas under the two peaks are then proportional to the number of solvent molecules belonging to the primary solvation of the ion from which solvent release is slow and to the number of the remaining solvent molecules. The solvation number h is thus obtained in a straightforward manner from the relative peak areas, see Figure 4.5. The precision in h attainable from the ratio of the peak areas is limited, in

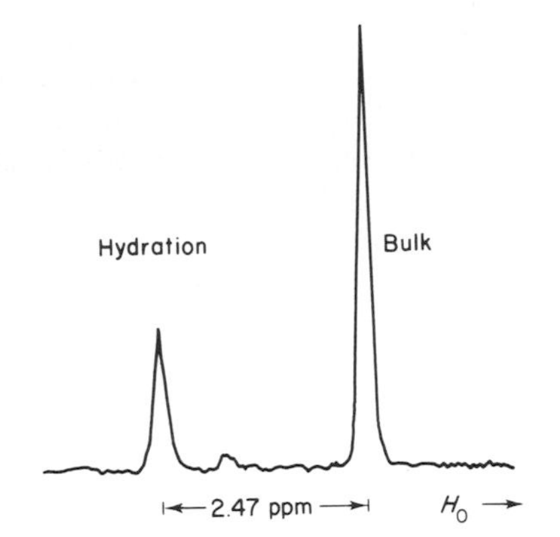

Figure 4.5 The proton NMR spectrum of an aqueous acetone solution of gallium perchlorate at −50 °C, where the exchange of water molecules between the first hydration shell of the cation and the 'bulk' water is sufficiently slow for separate peaks to be observable. Their areas are proportional to the number of water molecules belonging to each kind, i.e. hydration and bulk. No solvation by the acetone occurs under the circumstance chosen for the experiment. (Reprinted with permission from A. Fratiello, R. E. Lee, and R. E. Schuster, *Inorg. Chem.*, **9**, 82 (1970). Copyright 1970 American Chemical Society)

favourable cases, to ±0.1 units. In other cases it may be as bad as ±0.3 units, but generally the figure obtained is sufficiently close to an integer, for the latter to represent the sought primary solvation number. Experience shows that salts of alkali metal cations with anions like chloride or perchlorate do not exhibit split peaks at low temperatures. When a split peak is found for a solution of a multivalent cation with these anions, it is safe to assign the primary solvation number obtained from the peak area ratio, *h*, to the cation.

For diamagnetic multivalent ions the method works best with ^{1}H-NMR, and for the paramagnetic ones with ^{17}O-NMR. The condition of slow solvent exchange is more readily achieved for ^{17}O-NMR than for ^{1}H-NMR, since the rate of exchange of protons is generally a few orders of magnitude faster than that of the entire solvent molecules. However, the ^{17}O signal is lower and broader than the ^{1}H signal. The chemical shift of ^{17}O nuclei in the solvation shell of diamagnetic ions relative to those in the bulk solvent is small with respect to the natural line width of the peak, hence no splitting of the peak is observed, whereas the chemical shifts for ^{1}H nuclei in the two environments differ much more than the line widths. The chemical shifts for nuclei in the primary solvation shells of paramagnetic ions are manyfold larger than for those of diamagnetic ions, for even the ^{17}O signals to be well separated. However, in the case of ^{1}H signals so much line broadening occurs, which causes overlap of the peaks and the unsuitability of these nuclei. Nuclei of the skeletal atoms, ^{17}O and ^{15}N, can still be used also for solutions of diamagnetic ions, if small concentrations of fast exchanging paramagnetic ions, such as Co^{2+} or Eu^{3+} are added. These cause a large shift of the signal from the bulk solvent, so that two well resolved peaks are observed.

The NMR peak area method has been applied to some two dozen ions in water, and for some of them also in nonaqueous solvents, such as liquid ammonia, methanol, ethanol, acetonitrile, dimethylformamide, and dimethylsulphoxide. The resulting solvation numbers are presented in Table 4.4. The primary hydration numbers so obtained generally agree with the number of solvent molecules in the first solvation shell obtained from the structural methods, in the cases where both approaches have been applied to the same ions. Hydration numbers $h = 6$ have been obtained for Mg^{2+}, Fe^{2+}, Co^{2+}, Ni^{2+}, Zn^{2+}, Al^{3+}, and Cr^{3+} (for the last mentioned ion not from NMR, but from isotope dilution). Cases where no agreement between the primary solvation number obtained by the NMR method and the number of solvent molecules in the first solvation shell given by the structural methods are also noted in Table 4.4. These discrepancies are generally 'explained' in an *ad hoc* manner by blaming the intrusion of the anions into the first solvation shell in the concentrated solutions that may contain acetone and have a low dielectric constant. Otherwise, a too low water content is blamed, so that the acetone does compete with the water in the first solvation shell.[19]

A similar method, utilizing the ratio of the areas under spectral peaks, that has, however, been used rather seldom so far, employs the vibrational spectrum of the solvent. In favourable cases, the vibrational frequency of the bond between the coordinated donor atom and the atom next to it in the solvent is shifted sufficiently far away from the corresponding frequency of the bulk solvent, for resolvable lines to be obtained. The peak areas are then proportional to the numbers of solvent molecules belonging to the two categories of 'coordinated' and 'bulk', and a solvation number can be estimated from the composition of the solution and the ratio of these areas. This, for example, is the case for silver ions in acetonitrile, where a primary solvation number of 4 has been established from the infrared spectra of the solutions.[20]

Primary solvation numbers are also obtained from visible spectrometry, albeit in an indirect manner, as described on p. 61 above. In several cases regular octahedral geometry, implying a primary solvation number of 6, is the only one fully compatible with the observed spectral features in solutions of salts of transition metals. Association with the anion or other ligands that may be present in the solution (solvolysis products, for instance) must be definitely ruled out before a primary solvation number of six is accepted for the system.

The primary solvation number, however, does not always equal the number of solvent molecules in the first solvation shell, as has been true in the cases discussed hereto, and listed in Table 4.4. Some of the methods listed in Table 4.3 purport to yield primary solvation numbers, which are definitely smaller than the 4 or 6 nearest neighbours expected in the first solvation shell. Some of these methods, again, yield directly a primary solvation number, others require a primary solvation number to be assumed for a model, the calculated results for which being then compared with the experimental data. The use of the NMR chemical shifts is such a method, in which the primary solvation number features as an important parameter, but cannot be derived directly from the data.

When fast exchange between the solvent molecules in the solvation shells of the ions and the bulk solvent occurs, no separate peaks in the NMR spectrum are observable, only a single peak having a chemical shift that is the weighted average of those that would have been obtained had the exchange been slow. The fast exchange causes the environment of the nucleus from which the NMR signal is obtained to change many times within the time span corresponding to the frequency of this signal. Hence only a mean environment is registered by this nucleus, giving rise to this single peak. The observed chemical shift (δ_{obs}, in ppm) between this peak and the peak for the pure solvent depends on the natures of the cation and the anion of the solute, for a given solvent, and on the concentration of the salt. This latter dependency is contrary to the case of slow exchange, where the chemical shift between the split peaks is independent of the concentration, see p. 71. At relatively low concentrations, indeed, δ_{obs} is proportional to the molal concentration m of the salt, and $\lim(m \to 0)(d\delta_{obs}/dm)$ is, in principle, a well defined quantity for each salt and solvent system. This statement must be applied with due caution, however, because of the practical necessity of working at fairly high concentrations in order to obtained well measurable NMR peaks, which makes a rather long extrapolation to infinite dilution inevitable. This limiting slope should be given by the separate effects of the solvent associated with the cation and that associated with the anion:

$$\lim_{m \to 0} (d\delta_{obs}/dm) = \nu_+ h_+ \delta_+ + \nu_- h_- \delta_- \tag{4.14}$$

where the $\nu_\pm$ are the stoichiometric coefficients in the salt. The problems remain of splitting the observed infinite dilution slope appropriately between the cation and the anion, and the splitting of the contribution of each kind of ion between the primary solvation number $h_\pm$ and the specific (per solvent molecule) solvent shift $\delta_\pm$. Arbitrary choices, such as $\delta_{NH_4^+} = 0$ (on the grounds that it fits into the structure of water used as the solvent), $\delta_{R_4N^+} = 0$ (for large alkyl groups R, due to assumed too weak interactions with the solvent), or $\delta_{K^+} = \delta_{Cl^-}$ (on the ground of their similar mobilities in water) are not very useful.

The first split, to separate the cation and anion shifts, has been made by two methods, that yield essentially the same values, and have been applied to water and methanol as the solvents.

For water, in the method of Davis *et al.*,[21] the preliminary arbitrary choice of $h_{Cl^-} \cdot \delta_{Cl^-} = 0$ has led to two parallel smooth plots of $h_+ \cdot \delta_+$ of univalent and divalent cations against their charge-to-radius ratios. These two lines coalesce for the adjusted $h_{Cl^-} \cdot \delta_{Cl^-} = 0.076$ ppm/(kg. mol^{-1}) at 0 °C. The second method starts from very low temperatures, where at high salt molalities separate NMR signals for bound and free water appear. Here 'bound' is taken to be water that is directly coordinated to the cations, with δ_{bound} independent of the temperature, whereas 'free' water is all the remaining water, i.e., the bulk water, that associated with the anions, and that affected in a secondary manner by the cations, with δ_{free} being temperature dependent. The temperature coefficient for bulk water is now applied to δ_{free}, to obtain the values at the higher temperatures of concern. Since the single line observed at these temperatures is a weighted mean of δ_{bound} and δ_{free}, the value of $h_{Cl^-} \cdot \delta_{Cl^-} = 0.05$ ppm/(kg$\cdot$mol^{-1}) can be calculated.[21] The proximity of these two estimates of $h_{Cl^-} \cdot \delta_{Cl^-}$ lends credibility to the methods, and leads to separate $h_+ \cdot \delta_+$ and $h_- \cdot \delta_-$ values in aqueous solutions.

In methanol, on the other hand, the two separate peaks observed for the hydroxylic ^{1}H signals, with shifts σ_{bound} and σ_{bulk} against the methyl ^{1}H signal as the internal reference, have been studied by Butler and Symons.[22] At low temperatures, a 'primary cation shift' σ'_+ and the corresponding anion shift σ'_- (the so-called 'residual shift' involves σ'_- as well as small contributions from secondary solvation effects of both kinds of ions) are obtained from:

$$\sigma'_+ = (h_+/M_{\mathrm{MeOH}})(\sigma_{\mathrm{bound}} - \sigma_{\mathrm{MeOH}}) \quad (4.15a)$$

$$\sigma'_- = (\sigma_{\mathrm{MeOH}} - \sigma_{\mathrm{bulk}})/\nu_- m \quad (4.15b)$$

where M_{MeOH} is the molar mass of methanol and σ_{MeOH} is its chemical shift. The method has been applied to the perchlorates of Mg^{2+} and Al^{3+} (at -69 °C) and of Zn^{2+} (at -81 °C), with the value $h_+ = 6$ being taken from the peak area ratio. The shifts found are $\sigma'_{\mathrm{Mg}} = -0.30$, $\sigma'_{\mathrm{Zn}} = -0.37$, $\sigma'_{\mathrm{Al}} = -0.82$, and $\sigma'_{\mathrm{ClO_4}} = +0.123$, all in ppm. The shifts of other ions can then be obtained from the splitting of observed salt shifts. The cation shifts, σ'_+, are seen to involve solvation numbers from an independent source, the anion shifts, σ'_-, are seen to be independent of solvation numbers altogether. Both quantitities are of no use for obtaining from chemical shift data separate estimates of the solvation number $h_\pm$ and the specific chemical shift (per solvent molecule) $\sigma_\pm$ (or $\delta_\pm$, if an external reference is used.

Akitt[23] has preferred to relate the observed chemical shifts of ^1H in aqueous solutions to an arbitrary external reference, ethane gas, so that the integral form of equation 4.14 has also a contribution from the chemical shift of the bulk water. When his equations are rewritten for the case of univalent anions ($\nu_+ = 1$), the result is

$$\delta'_{\mathrm{obs}} = (m/55.5)h_+(\delta'_+ - \delta'_{\mathrm{W}}) + (m/55.5)\nu_-(F/2)(\delta'_{\mathrm{F}} - \delta'_{\mathrm{W}}) + \delta'_{\mathrm{W}} \quad (4.16)$$

where primed shift symbols pertain to the ethane reference, $(m/55.5)$ is the molar ratio of salt to water, δ'_{W} is the chemical shift of the bulk water, and ν_- is the number of anions per cation, or its charge number. The effects of cations and anions on the water molecules in their immediate environments are different in kind, a fact that is manifested in the different directions of the corresponding chemical shifts: downfield for cations, upfield for anions. This is also recognized by the symbolization of the effect of the anion, as the number F of hydrogen bonds broken by them, turning the affected water into 'free' molecules, that differ from the bulk water, and have a chemical shift δ'_{F} against the ethane reference. It turns out that to a good approximation non-hydrogen-bonded water (e.g., gaseous water molecules) has a value of $\delta'_{\mathrm{F}} \approx 0$. The shifts δ'_+ of cations of known hydration numbers h_+, such as $Al(H_2O)_6{}^{3+}$ or $Mg(H_2O)_6{}^{2+}$ (and the assumed $Ca(H_2O)_6{}^{2+}$) are obtained from low temperature measurements in the presence of acetone, where separate bulk and hydration water peaks are observed. If it is assumed that these δ'_+ are temperature-independent and that $\delta'_{\mathrm{F}} = 0$, then equation 4.16 can be used to calculate F

$$F = (2/\nu_-)[(55.5/m)(1 - \delta'_{\mathrm{obs}}/\delta'_{\mathrm{W}}) - h_+(1 - \delta'_+/\delta'_{\mathrm{W}})] \quad (4.17)$$

It should be emphasized that F is *not* the hydration number of the anion, but the number of hydrogen bonds broken by it. The parameter F has been found to be proportional to the volume of the anion.[23] Akitt's concepts have been criticized on the grounds that the assumption that δ_+ is temperature independent may not be valid.[24]

A recent attempt has been made by Leyendekkers[25] to analyse the separate estimates of $h_+ \cdot \delta_+$ and $h_- \cdot \delta_-$ in water[21] in terms of the contributions of several terms that yield the internal pressure of electrolyte solutions. The data are shown to be compatible with $h_+ = 6$ and $h_- = 6$ too, except for ClO_4^-, SO_4^{2-}, and CrO_4^{2-}, for which $h_- = 8$. However, this compatibility involves the summation of positive and negative terms, that are much larger than the $h_\pm \cdot \delta_\pm$ data used for the test. Hence these hydration numbers are not uniquely able to account for the data, and cannot be taken as quantities derivable directly from the chemical shifts.

Nuclear magnetic resonance has been used in other attempts to obtain solvation numbers of ions, with as little success as in the utilization of the chemical shifts described above. The use of the mole ratio method, where the chemical shift is plotted against the concentration of a solvating solvent in a non-solvating diluent, and the occurrence of an abrupt break in the curve is noted,[17] is on the whole unreliable.[26] The interpretation of the temperature dependence of the chemical shifts has been suggested as a method for the derivation of solvation numbers,[27] but there have been serious reservations against its use.[17] A similar attitude applies to the use of NMR line widths of the ^1H atoms of water, affected by the presence of diamagnetic ions, after having been broadened by the presence of the paramagnetic Mn^{2+} ion[28]. The method has been calibrated by means of $h_{Be} = 4$, $h_{Al} = 6$, and $h_{Ga} = 6$ (obtained from independent sources), to give the following primary hydration numbers:

Mg^{2+}	Ca^{2+}	Sr^{2+}	Ba^{2+}	Zn^{2+}	Cd^{2+}	Hg^{2+}	Pb^{2+}	Th^{4+}
3.8	4.3	5.0	5.7	3.9	4.6	4.9	5.7	10

These numbers do not agree with the primary hydration numbers obtained for these ions by other methods, as is evident from the very low number given for Mg^{2+}, compared with the accepted value of 6, which is as well established as the numbers used for the calibration of the method.

The thermodynamic methods for the determination of the solvation numbers listed in Table 4.3 relate a bulk property of the solution to the properties of solvent molecules, as these are surmized to be affected by their location in the solvation shells of the ions. The effects are less straightforward than those involved in the spectroscopic methods, at least for those among the latter, where a clear-cut distinction between bound and bulk solvent molecules can be made. The effects used in the thermodynamic methods are attributed to the solvation of the ions, but it is generally impossible to decide on whether primary solvation alone, a combination of primary solvation and some secondary solvation, or these together with some other effects are responsible for the observed behaviour.

The entropy of solvation of an ion is the difference between its molar entropy in the solution standard state and that in the gaseous standard state. It includes the following contributions:[29]

(i) the change of the (free) volume at the disposal of the ion;
(ii) long-range electrostatic effects due to the interaction of the charge of the ion with the permanent and induced dipoles of the solvent, the 'Born term';
(iii) immobilization of solvent molecules (loss of translational and rotational degrees of freedom) in the near environment of the ion;
(iv) other effects, due possibly to the interference of the ion, together with its primary solvation, with the structure of the solvent.

The solvation number relates to item (iii) in this list, so that if the contributions from items (i), (ii), and (iv) are subtracted from the individual ionic entropy of solvation, and the difference is divided by the molar entropy of immobilization per one solvent molecule, the ratio is the solvation number, defined operationally by this procedure.

An application of this concept to ionic hydration has been made several decades ago by Ulich.[30] The individual ionic entropies of hydration have been obtained from data on thermocells (see Section 5.3), the change in the available volume (item (i)) has been set at $\mathbf{R} \ln(\mathbf{R}T/1 \text{ atm} \cdot 1 \text{ dm}^3)$, and the average loss of entropy of water by immobilization has been set at 25 J K^{-1} mol^{-1}. This figure corresponds to the difference in the molar entropy of liquid water at 25 °C and that for ice extrapolated to this temperature. This figure corresponds also to the mean molar entropy per water molecule in crystalline salt hydrates, e.g., for $NaH_2PO_4 \cdot 12H_2O$ compared with $NaH_2PO_4 \cdot 7H_2O$. In spite of these simplifying assumptions, the resulting hydration numbers, for which the claimed accuracy is ± 0.5 units, are quite reasonable. More modern values of the standard molar entropies of hydration (Table 5.12), together with the other quantities used by Ulich[30] lead to hydration numbers with an estimated precision of ± 0.2 units, as presented in Table 4.5. They are in agreement with Ulich's original set within the stated precision in most cases. It is, perhaps, remarkable that reasonable hydration numbers are obtained with this procedure, even though items (ii) and (iv) in the above list have been ignored, and the whole effect on the entropy has been ascribed to solvent immobilization. Perhaps a more interesting application of these concepts is to take explicit cognizance of items (ii) and (iv), and reverse the procedure with independent values of the primary hydration number. This permits then the study of the effects of the ions on the structure of water, item (iv), as is shown in Section 5.4.

This method has scarcely been applied to solvents other than water.[31] However, within the uncertainties due to the imperfect estimates of the quantities involved (in particular the correction of the entropy of fusion from the melting point to 25 °C), it has been so applied by Marcus.[32] Individual ionic

Table 4.5 Ionic hydration numbers, h_i^{∞}, obtained from thermodynamic properties of the solutions

Ion	From[a] ΔS^0_{hydr}	From[b] ${}^{\phi}K^{\infty}$	From[c] $\gamma_{\pm}$
H^+	4.0		3.9
Li^+	4.3	2.9	3.4
Na^+	3.0	3.9	2.0
K^+	1.6	3.1	0.6
Rb^+	1.1	2.8	0
Cs^+	0.9	2.3	0
Cu^+	4.0		
Ag^+	3.1	3.4	
Tl^+	1.2		
NH_4^+	3.4	0.4	0.2
Me_4N^+	4.7	(<0)	
Et_4N^+	4.8		
Be^{2+}	10.2		
Mg^{2+}	12.2	8.0	5.1
Ca^{2+}	7.8	7.5	4.3
Sr^{2+}	7.3	9.9	3.7
Ba^{2+}	5.9	9.4	3.0
Ra^{2+}	4.3		
Mn^{2+}	10.7	7.7	
Fe^{2+}	11.3		
Co^{2+}	10.7	8.8	
Ni^{2+}	11.0	9.4	
Cu^{2+}	9.8	9.2	
Zn^{2+}	10.3	9.3	5.3
Cd^{2+}	9.0	6.0	
Hg^{2+}	7.7		
Hg_2^{2+}	9.0		
Sn^{2+}	7.0		
Pb^{2+}	6.1		
UO_2^{2+}	15.1		7.4
Al^{3+}	18.5		11.9
Sc^{3+}	17.3		
Y^{3+}	16.7		
La^{3+}	21.5	12.7	7.5
Ce^{3+}	21.2		
Pr^{3+}	21.9	12.9	
Nd^{3+}	21.8	12.5	
Sm^{3+}	21.9	11.1	
Eu^{3+}	21.6	10.5	
Gd^{3+}	21.8	11.0	
Tb^{3+}	22.7	11.7	
Dy^{3+}	22.8	12.0	
Ho^{3+}	22.8	12.5	
Er^{3+}	23.5	12.6	
Tm^{3+}	23.3	12.8	
Yb^{3+}	23.5	13.0	
Lu^{3+}	23.3		
Pu^{3+}	21.1		
Fe^{3+}	18.8		
Ga^{3+}	19.5		
In^{3+}	12.9		
Tl^{3+}	14.5		
Bi^{3+}	12.9		
Ce^{4+}	19.0		
Th^{4+}	24.0		
F^-	4.9	5.5	1.8
Cl^-	2.3	2.1	0.9
Br^-	(<0)	0.8	0.9
I^-	(<0)	(<0)	0.9
OH^-	5.9	6.6	4.0
SH^-	2.8		
CN^-	2.1		
OCN^-	2.5		
SCN^-	1.6		0.3
N_3^-	2.2		
NO_2^-	2.6		
NO_3^-	2.0	0.8	0
ClO_3^-	2.1		
BrO_3^-	2.7		
IO_3^-	4.8		
HCO_2^-	3.9		
$CH_3CO_2^-$	5.7	2.6	2.6
ClO_4^-	1.2		0.3
MnO_4	1.5		
BF_4^-	1.6		
HCO_3^-	4.4		
HSO_4^-	4.1		
$H_2PO_4^-$	5.6	6.2	
S^{2-}	4.6		
CO_3^{2-}	8.8		
SO_3^{2-}	8.9		
SO_4^{2-}	6.9	10.6	
CrO_4^{2-}	6.4		
HPO_4^{2-}	9.8	16.4	
PO_4^{3-}	15.8	30.9	
$Fe(CN)_6^{3-}$	4.7	2.7	
$Fe(CN)_6^{4-}$	10.3	12.1	

entropies of transfer ΔS_t^0 from the water to many solvents are known (see Table 6.9), hence also the solvation entropies of these ions. Items (i) and (ii) can be readily estimated for such solvents, and item (iv) may well be negligible for them. Resulting values, on the same basis as for hydration numbers, i.e.,

$$h^\infty = [S^0_{\text{hydr}} + \Delta S_t^0 + \mathbf{R}\ln(\mathbf{R}T/1\ \text{atm}\cdot 1\ \text{dm}^3) - \Delta S^0_{\text{el}}]/\Delta S^F(T) \quad (4.18)$$

where $\Delta S^F(T)$ is the molar entropy of freezing extrapolated to the temperature T (25 °C), are shown in Table 4.6.

The compressibility of an electrolyte solution, compared with that of the solvent, provides an entirely different method for the estimation of the solvation numbers. Passinsky[33] argued that since the primary solvating solvent is strongly compressed by the pressure effect of the electric field of the ion (the electrostriction), a fraction of the electrolyte solution, depending on its concentration, is no longer compressible by an external pressure. Indeed, the specific volume of liquid water levels off beyond an applied pressure of 680 MPa, which is well below the internal pressure exerted by the field of monovalent ions having radii less than 1.5 nm, i.e. practically all ions encountered in aqueous solutions. The incompressible mole fraction of the solution is then[34]

$$(\kappa_{TS} - \kappa_T)/\kappa_{TS} = hmM_S \quad (4.19)$$

where M_S is the molal mass of the solvent, κ_{TS} its isothermal compressibility, and κ_T that of the m-molal electrolyte solution, characterized by the hydration

Table 4.6 Ionic solvation numbers, h_i^∞, from thermodynamic data

Ion	MeOH[ab]	EtOH[ab]	PrOH[ac]	PC[ad]	FA[ae]	DMF[ae]	MeCN[ae]	MeCN[f]	DMSO[ae]
Li^+	4.7	2.9	2.5	5.0	3.4	1.8		1.0	2.4
Na^+	4.1	2.6	2.2	4.4	3.8	1.6	4.0	1.8	2.0
K^+	3.3	2.1	1.9	3.8	2.9	1.4	3.4	1.9	1.9
Rb^+	2.7	1.8	1.8	3.0	2.5	1.2	3.2	1.8	1.8
Cs^+	2.4	1.6	1.5	2.6	2.2	1.1	2.9		1.5
Ag^+	4.2			4.9		1.6	4.3		2.6
Me_4N^+	2.9	1.8	1.8			1.3	3.4		2.0
Et_4N^+	3.4					1.5	3.2	0.2	
Cl^-	1.8	1.2	1.4	2.9	3.3	1.8	3.2	1.6	3.4
Br^-	0.4	0.4	0.7	1.2	1.1	1.1	1.7	1.8	1.1
I^-	(<0)	(<0)	0.2	0.3	(<0)	0.7	0.7	1.7	0.4
N_3^-	2.2			2.8		2.0	3.1		2.3
ClO_4^-	1.4	1.1	1.1		2.6	1.5	2.5	1.5	1.5

[a] From entropies of solvation[32], equation 4.18, with ΔS^0_{hydr} from Table 5.13, ΔS_t^0 from Y. Marcus, *Pure Appl. Chem.*, **57**, in press (1985), and ΔS^F from the stated sources; [b] ΔS^F from R. C. Wilhoit and B. J. Zwolinski, *J. Phys. Chem. Ref. Data*, **2**, Suppl. 1 (1973) and correction to 25 °C via C_p data of G. S. Parks, *J. Am. Chem. Soc.*, **47**, 338 (1925); [c] ΔS^F as in note[b], correction via C_p data of G. S. Parks and H. M. Huffman, *J. Am. Chem. Soc.*, **48**, 2788 (1926); [d] ΔS^F from ΔH^F data of I. A. Vasilev and A. D. Khorkov, *Tr. Khim. Khim. Tekhnol.*, **1974** (1), 103, correction as in note[e]; [e] ΔS^F from Y. Marcus, *Introduction to Liquid State Chemistry*, Wiley, Chichester (1977), pp. 106, 110, correction to 25 °C taken as mean for alcohols, i.e., +13%; [f] from compressibilities, equation 4.21, $^\phi K_S^\infty$ data from I. Davidson, G. Perron, and J. E. Desnoyers, *Can. J. Chem.*, **59**, 2212 (1981) and κ_T from M. J. R. Dack, K. J. Bird, and A. J. Parker, *Aust. J. Chem.*, **28**, 955 (1975).

number h. A restatement of equation 4.19 in terms of the apparent molal isothermal compressibility of the solution

$$^{\phi}K_{\mathrm{T}}^{\infty} = \lim_{m \to 0} [(1000/m\rho_{\mathrm{S}})(\kappa_{\mathrm{T}} - \kappa_{\mathrm{TS}}) + \kappa_{\mathrm{T}}{}^{\phi}V] \tag{4.20}$$

where ρ_{S} is the density of the solvent and $^{\phi}V$ is the apparent molar volume of the electrolyte, leads to

$$h^{\infty} = -{}^{\phi}K_{\mathrm{T}}^{\infty}/\kappa_{\mathrm{TS}}V_{\mathrm{S}} \tag{4.21}$$

where V_{S} is the molar volume of the solvent. It is seen that the solvation number is proportional to the limiting molar apparent (or partial) isothermal compressibility, the proportionality constant being a product of known properties of the solvent. The ionic solvation numbers can be obtained if the values of $^{\phi}K_{\mathrm{T}}^{\infty}$ can be apportioned between the cations and the anions of the electrolyte. This has been done for aqueous electrolytes (see Table 5.11), and with $\kappa_{\mathrm{TS}} = 4.525 \cdot 10^{-5}$ bar^{-1} (1 bar = 0.1 MPa) and $V_{\mathrm{S}} = 18.07$ cm^3 mol^{-1}, the values of h^{∞} shown in Table 4.5 have been obtained. The limiting molar isothermal compressibilities of individual ions are known within ± 1.5 cm^3 mol^{-1} bar^{-1}, hence the hydration numbers should be known to within ± 0.2 units. It is seen that for many ions the h^{∞} values from the entropies of hydration and from the compressibility of solutions are in agreement within the claimed precisions.

The commonly used measure for ion–ion interactions in electrolyte solutions, the mean ionic molal activity coefficient, may also be used for the derivation of solvation numbers, i.e., for learning about ion–solvent interactions. Bjerrum[35] has pointed out that the fact that h moles of solvent are bound to each mole of electrolyte reduces the amount of free solvent in the solution correspondingly, so that an m-molal solution is a mixture of $v \cdot m$ moles of ions and $M_{\mathrm{S}}^{-1} - h \cdot m$ moles of free solvent. The entropy of mixing is affected by this, relative to the case of $v \cdot m$ moles of ions and M_{S}^{-1} moles of solvent if no solvation occurred. This manifests itself in the term $-\ln[1 - (h - v)mM_{\mathrm{S}}]$ that should be added to $\ln \gamma_{\pm}$, where in these expressions v is the number of moles of ions into which a mole of electrolyte dissociates, $v = v_{+} + v_{-}$, and M_{S} is the molal mass of the solvent (in kg mol^{-1}). It has been found, however, that a fit of the experimental $\ln \gamma_{\pm}(m)$ data for aqueous electrolytes to the Debye–Hückel term with this additional term permits the derivation of empirical h values, but they are not additive with respect to individual ionic hydration numbers.

This drawback is remedied when the ratio r of the sizes of the solvated ions and the solvent molecules are taken into account, according to the 'volume statistics' approach of Glueckauf.[36] If an attempt is made to leave h as the only adjustable parameter, then the ratio r and the distance of closest approach a in the Debye–Hückel expression must be expressed in terms of h and independent experimental data. Accordingly[37]

$$a = \sum r_{\mathrm{ci}}[1 + 0.58(1 + \bar{V}_{\mathrm{i}}^{\infty}/h_{\mathrm{i}}V_{\mathrm{s}})^{-1}]^{1/3} \tag{4.22}$$

where r_{ci} is the crystal ionic radius, 0.58 an empirical packing fraction, $\bar{V}_{\mathrm{i}}^{\infty}$ the limiting ionic partial molar volume, V_{S} the molar volume of the solvent, and the summation extends over one cation and one anion. Also

$$r = \sum \nu_i(h_i + \bar{V}_i^{\infty}/V_S) \tag{4.23}$$

where the summation extends over all the ionic species. With the usual A and B parameters of the Debye–Hückel expression, that depend on the nature of the solvent through its dielectric constant, the final equation for the activity coefficient is

$$\begin{aligned}\ln \gamma_{\pm} = & -z_+z_-AI^{1/2}(1 + BaI^{1/2})^{-1} \\ & + \ln\left[\left(1 - \sum \nu_i h_i m M_S\right)(1 + mM_S r)^{-1}\right] \\ & + \left(\sum \nu_i h_i + r\right)\nu^{-1} m M_S\left(\sum \nu_i h_i - \nu + r\right)(1 + mM_S r)^{-1}\end{aligned} \tag{4.24}$$

where I is the ionic strength on the molar ($mol \cdot dm^{-3}$) scale. This equation is seen to be quite a complicated and implicit function of the ionic solvation numbers h_i, and their extrication from experimental $\ln \gamma_{\pm}(m)$ data has been effected only for the solvent water and a limited set of ions.[36] The resulting values are additive, in the sense that each ion has a definite value of h_i that can be used to fit activity coefficients of all the salts containing it that have been tested. This treatment considers h_i to be a fixed parameter, i.e., independent of the concentration, hence equal also to h_i^{∞}. The resulting values are shown in Table 4.5. It is seen that they are considerably lower that those obtained from the entropies of hydration and from the compressibilities. The significance of this discrepancy is obscure.

Hydration numbers have been obtained from the partial molar volumes of ions in a treatment relatable to that of the compressibility of the solutions, but in a modified manner, consisting of the consideration of the incompressible volume fraction, rather than of the incompressible mole fraction leading to equation 4.19. The results of this treatment are discussed further in Section 5.2.

Transport properties of solutions can also be used for the estimation of solvation numbers. In a classical Hittorf transference apparatus, some solvent is co-transported with the ions, as is seen when a presumed non-transportable nonelectrolyte, such as the sugar raffinose for the solvent water, is used as the reference.[38] The amount of this co-transported solvent is called the Washburn number. Subsequent work has shown that some of the nonelectrolyte is also transported in such an experiment, and it is safer to measure the amount of D_2O transported per ion into an H_2O solution, assuming the isotope effect is negligible.[39] Still, the reliability of solvation numbers obtained by this method is questionable.

A more widely used method, that has been applied to aqueous as well as nonaqueous solutions, involves the calculation of Stokes' radii from limiting ionic conductivities, thence the volumes of the solvated ions, and from these the solvation numbers. An advantage of this method is that it yields individual ionic solvation numbers, without the need to apportion those of electrolytes into contributions from the constituent ions. This is possible, since by the measurement of transference numbers (by the Hittorf method, or the more convenient and accurate moving boundary or EMF of a cell with transference methods) and the limiting equivalent conductivity of the electrolyte, the individual ionic limiting conductivity λ_i^∞ is obtained directly from experiment. The Stokes radius is given by

$$r_{iS} = |z_i|\mathbf{F}^2(6\pi\mathbf{N})^{-1}\eta_S^{-1}\lambda_i^{\infty\,-1} = 0.00820|z_i|\eta_S^{-1}\lambda_i^{\infty\,-1} \tag{4.25}$$

where $|z_i|$ is the absolute value of the ionic charge number, **F** the Faraday constant, and η_S the viscosity of the solvent. The numerical coefficient applies for η_S in (Pa · s) and λ_i^∞ in (Ω^{-1} cm^2 equiv^{-1}), to give r_{iS} in nm.

However, this Stokes radius is valid only for large particles moving in a viscous continuum, whereas for ions in a molecular solvent a correction must be applied. The practice is to assume that ions of the type $(C_nH_{2n+1})_4N^+$, with $n \geqslant 2$, are unsolvated, and that their corrected Stokes radii r_{iSc} are equal to their crystal or van der Waals radii (0.400 nm for $n = 2$, 0.452 nm for $n = 3$, 0.494 nm for $n = 4$, 0.530 nm for $n = 5$, 0.560 nm for $n = 6$, and 0.590 nm for $n = 7$). The latter are related to the experimental r_{iS} by some simple function, a linear or preferably a quadratic power series. Corrected Stokes radii are then obtained for the other, solvated, ions, by inter- or extrapolation with the same function.[40] The volume of the solvated ion is then obtained from this corrected radius.

If the volume of the unsolvated ion is subtracted from this volume of the solvated ion, the volume of the solvation shell is obtained. The volume of the unsolvated ion is generally a minor fraction of that of the solvated one for monoatomic cations. Hence no great error is committed if the crystal ionic radius r_{ic} is used for this calculation. The volume of the solvation shell must now be divided by that of the solvating solvent, per molecule, in order to obtain the solvation number. This is the problematical part of the calculation, since this volume is not directly known: the volume per molecule is less than that in the bulk solvent because of electrostriction. Even if electrostriction is ignored, the required volume is not 1/**N** times the molar volume V_S of the solvent since this does not take into account the void volume between the molecules packed in the liquid solvent. The molar volume must be divided by a paking fraction k to obtain the volume per molecule. If an empirical packing factor k is taken to account for both the electrostriction and packing effects, the resulting solvation number is

$$h_i^\infty = (4\pi/3)\mathbf{N}k(r_{iSc}^3 - r_{ic}^3)/V_S \tag{4.26}$$

For the r's expressed in nm and V_S in cm^3 mol^{-1}, k values in multiples of 10^3 are appropriate. A detailed discussion of the k value appropriate for water as the solvent is presented in Section 5.2. For nonaqueous solvents, in ignorance of its true values, it will be set equal to 1000 exactly, to obtain relative solvation numbers presented in Table 4.7. It should be noted that inter-solvent comparisons may not be valid because of possible differences in the applicable k

Table 4.7 Solvation numbers, h_i^∞, derived from corrected Stokes' radii (with the factor k in equation 4.26 fixed at 1000)

Ion	H_2O[a]	MeOH[b]	EtOH[b]	PrOH[d]	Me_2CO[b]	PC[f]	$PhNO_2$[b]	Py[b]	MeCN[c]	HCO_2H[p]
Li^+	7.4	5.0	5.8	4.0[e]	2.7	3.4		2.5	1.2	6.6
Na^+	6.5	4.6	3.8	4.2	2.6	2.9	1.6	2.4	1.4	4.0
K^+	5.1	3.9	3.3	3.0	2.5	2.4	1.4	2.1	1.0	2.9
Rb^+	4.7	3.7	3.1	2.6		1.7				2.1
Cs^+	4.3	3.3	2.9	2.4		1.6			0.5	2.1
Ag^+	5.9									2.0
Tl^+	5.0									
NH_4^+	4.6	3.7	3.7		2.2	1.7	1.4	1.2		2.6
Me_4N^+	1.8					0.5				0.5
Mg^{2+}	11.7	11.0[c]		19.1[e]	18.7[e]				6.0	
Ca^{2+}	10.4	8.0			11.9[e]					
Sr^{2+}	10.4	8.0								
Ba^{2+}	9.6	7.9			11.4[e]					
Zn^{2+}	11.3	10.4[c]							6.0	
F^-	5.5	5.0			2.4					
Cl^-	3.9	4.1	3.0		2.0	1.3	1.0			1.8
Br^-	3.4	3.8	2.9		1.8	1.0	1.1	1.0		1.6
I^-	2.8	3.4	2.6		1.8	0.8	1.2	1.1		
SCN^-		3.4	2.8		1.6			1.0		
NO_3^-	3.3	3.4	2.7			0.7	1.0			
ClO_4^-	2.6	2.8	2.1			0.8	1.1	1.2		

Ion	FA[g]	NMF[b]	DMF[i]	NMA[j]	DMA[k]	TMU[l]	DMSO[f]	TMS[m]	HMPT[n]
Li^+	5.4	6.5[h]	5.0	5.1	4.1	2.8	3.3	1.4	3.4
Na^+	4.0	3.5	3.0	3.5	2.6	2.3	3.1	2.0	1.8
K^+	2.5	3.4	2.6	3.3	2.7	2.4	2.8	1.5	1.6
Rb^+	2.3	3.1[h]	2.5	3.4[h]	2.4	2.2	2.3	1.4	1.4
Cs^+	1.9	3.0	2.3	2.6	2.2	2.0	2.0	1.3	1.2
Ag^+			2.1	2.3[h]	2.4				1.8
Tl^+	1.4		1.8						
NH_4^+	1.4	1.5[h]	1.8	2.7	1.4	1.9		0.9	1.6
Me_4N^+			0.4		0.5	0.4	0.4		
Mg^{2+}				10.3					
Ca^{2+}	6.4[b]		8.6						
Sr^{2+}	7.5[b]		8.6						
Ba^{2+}	6.3[b]		9.0						
Zn^{2+}									
F^-									
Cl^-	1.0	1.3	0.7	2.1	0.9		1.1	0.0	0.3
Br^-	0.8	1.2	0.8	1.7	1.0	0.7	1.0	0.0	0.5[o]
I^-	0.8	1.1	0.7	1.5	0.9		1.0	0.1	
SCN^-	0.8	2.3[h]	0.5	1.3	0.7	0.5	0.7[b]		0.3
NO_3^-	0.9		0.3	1.5	1.0[b]	0.4	0.9		0.5[o]
ClO_4^-	0.9[h]	2.1	0.6	0.9[h]	0.8	0.6	0.8	0.1	0.3

[a] From Table 5.8, with $k = 888$; [b] R. Gopal and H. M. Husain, *J. Indian Chem. Soc.*, **40**, 981 (1963); [c] A. P. Zipp, *J. Phys. Chem.*, **78**, 556 (1974), also values for Li^+(6.5), Ca^{2+}(10.2), and Sr^{2+}(10.4) that deviate from those of Gopal[b]; [d] G. Wikander and U. Isacsson, *Z. Phys. Chem.* (*Frankfurt*), **81**, 57 (1972); [e] from Zipp[c], also a value for Na^+ in propanol (2.9) that deviates from that of Wikander[d], and for Li^+ in acetone (3.4) that deviates from that of Gopal[b]; [f] N. Matsuura, K. Umemoto, and Y. Takeda, *Bull. Chem. Soc. Japan*, **48**, 2253 (1975); [g] J. M. Notley, *J. Phys. Chem.*, **70**, 1502 (1966); [h] R. C. Paul, J. S. Banait, J. P. Singla, and S. P. Narula, *Z. Phys. Chem.* (*Frankfurt*), **88**, 90 (1974), also data for Cl^- (1.9), Br^- (1.6), and SCN^- (1.4) in formamide that deviate from those of Notley[g], and for Cl^- (3.5), Br^- (2.9), and I^- (2.4) in N-methylformamide that deviate from those of Gopal[b]; [i] R. C. Paul, J. P. Singla, D. S. Gill, and S. P. Narula, *J. Inorg. Nucl. Chem.*, **33**, 2953 (1971); [j] R. Gopal and O. N. Bhatnagar, *J. Phys. Chem.*, **69**, 2382 (1965); [k] R. C. Paul, J. S. Banait, and S. P. Narula, *Aust. J. Chem.*, **28**, 321 (1975); [l] R. C. Paul, S. P. Johar, J. S. Banait, and S. P. Narula, *J. Phys. Chem.*, **80**, 351 (1976); [m] M. Della Monica and U. Lamanna, *J. Phys. Chem.*, **72**, 4329 (1968); [n] R. C. Paul, J. S. Banait, and S. P. Narula, *Z. Phys. Chem.* (*Frankfurt*), **94**, 199 (1975); [o] P. Bruno, M. Della Monica, and E. Righetti, *J. Phys. Chem.*, **77**, 1258 (1973); [p] R. C. Paul, R. Sharma, T. Puri, and R. Kapoor, *Aust. J. Chem.*, **30**, 535 (1977).

values. Even comparisons for a given solvent may be inaccurate, if the electrostriction of the solvent molecules is strongly dependent on the sizes of the ions, so that k becomes also somewhat dependent on the sizes.

There are several other methods, not listed in Table 4.3, that deal with the solvation of ions, and from which estimates of the solvation numbers can be obtained. Such estimates, however, are very uncertain, and their use is not recommended for this purpose, although they do yield important qualitative information on the solvation. These methods include measurements of the dielectric constant of electrolyte solutions, their viscosities or diffusion coefficients, and several other properties.

References

1. H. S. Frank and M. W. Evans, *J. Chem. Phys.*, **13**, 507 (1945); H. S. Frank and W.-Y. Wen, *Disc. Faraday Soc.*, **24**, 133 (1957); H. S. Frank, *Z. Phys. Chem.* (*Leipzig*), **228**, 367 (1965).
2. J. O'M. Bockris, *Quart. Rev. Chem. Soc.*, **3**, 173 (1949).
3. J. E. Enderby and G. W. Neilson, *Rep. Progr. Phys.*, **44**, 593 (1981).
4. A. K. Soper, G. W. Neilson, J. E. Enderby, and R. A. Howe, *J. Phys. C: Solid St. Phys.*, **10**, 1793 (1977).
5. M. Metropolis, A. W. Rosenbluth, M. N. Rosenbluth, A. M. Teller, and E. Teller, *J. Chem. Phys.*, **21**, 1087 (1953).
6. B. J. Alder and T. W. Wainwright, *J. Chem. Phys.*, **27**, 1208 (1957); **31**, 459 (1959); W. W. Wood and J. D. Jacobson, *ibid.*, **27**, 1207 (1957).
7. F. H. Stillinger and A. Rahman, *J. Chem. Phys.*, **60**, 1545 (1974).
8. K. Heinzinger and P. G. Vogel, *Z. Naturf.*, **A31** 463 (1976); K. Heinzinger, *ibid.*, **A31**, 1973 (1976).
9. O. Matsuoka, E. Clementi, and M. Joshimine, *J. Chem. Phys.*, **64**, 1351 (1976); H. Kistenmacher, H. Popkie, and E. Clementi, *ibid.*, **59**, 5842 (1973); M. Mazei and D. L. Beveridge, *ibid.*, **74**, 6902 (1981); *idem.*, *ibid.*, **76**, 593 (1982).
10. S. F. Lincoln, *Coord. Chem. Rev.*, **6**, 309 (1971).
11. D. E. Irish and M. H. Brooker, *Adv. Raman Infrared Spectrosc.*, **2**, 212 (1976).
12. J. P. Hunt and H. Taube, *J. Chem. Phys.*, **19**, 602 (1951); H. Taube, *J. Phys. Chem.*, **58**, 523 (1954).
13. T. J. Gilligan and G. Atkinson, *J. Phys. Chem.*, **84**, 208 (1982).
14. H. G. Hertz, *Ber. Bunsenges. Phys. Chem.*, **67**, 311 (1963).
15. T. J. Swift and R. E. Connick, *J. Chem. Phys.*, **37**, 307 (1962).
16. Y. Marcus, in *Gmelin Handbook of Inorganic Chemistry, Rare Earths*, Vol. D3 (1981), p. 1.
17. J. Burgess, *Metal Ions in Solution*, Ellis Horwood, Chichester (1978), Chapters 2, 5, and as specified elsewhere.
18. B. E. Conway, *Ionic Hydration in Chemistry and Biophysics*, Elsevier, Amsterdam (1981), Chapter 29.
19. A. D. Covington and A. K. Covington, *J. Chem. Soc. Faraday Trans. 1*, **71**, 831 (1975).
20. T. C. G. Chang and D. E. Irish, *J. Soln. Chem.* **3**, 161 (1974).
21. J. Davis, S. Omondroyd, and M. C. R. Symons, *Trans. Faraday Soc.*, **67**, 3465 (1971).
22. R. N. Butler and M. C. R. Symons, *Trans. Faraday Soc.*, **65**, 945 (1969).
23. J. W. Akitt, *J. Chem. Soc. Dalton Trans.*, **1973**, 42, 49.
24. R. D. Brown and M. C. R. Symons, *J. Chem. Soc. Dalton Trans.*, **1976**, 426.
25. J. V. Leyendekkers, *J. Chem. Soc. Faraday Trans. 1*, **79**, 1123 (1983).

26. Y. Marcus, *Israel J. Chem.*, **5**, 143 (1967).
27. E. R. Malinowsky and R. S. Knapp, *J. Chem. Phys.*, **48**, 4989 (1968); F. J. Vogrin and E. R. Malinowsky, *J. Am. Chem. Soc.*, **97**, 4876 (1975).
28. T. J. Swift and W. G. Sayre, *J. Chem. Phys.*, **44**, 3567 (1966); **46**, 410 (1967).
29. C. V. Krishnan and H. L. Friedman, in *Water. A Comprehensive Treatise*, F. Franks, ed., Plenum Press, New York, Vol. 3 (1973); Y. Marcus and A. Loewenschuss, in Ann. Reports C, 1984 (publ. 1985), Ch. 6.
30. H. Ulich, *Z. Elektrochemie*, **36**, 497 (1930).
31. C. R. Hedwig, D. A. Owensby, and A. J. Parker, *J. Am. Chem. Soc.*, **97**, 3888 (1975).
32. Y. Marcus, *Proc. IX Int. Conf. Non-Aqueous Solvents*, ICNAS, Pittsburgh (1984), p. 27; full paper to be published.
33. A. G. Passinsky, *Acta Physicochim. USSR*, **8**, 835 (1938); *Zh. Fiz. Khim.*, **11**, 606 (1938); **20**, 981 (1946).
34. D. S. Allam and W. H. Lee, *J. Chem. Soc.*, **1966A**, 5, 426; K. Tanaka and T. Sasaki, *Bull. Chem. Soc. Japan*, **36**, 975 (1963).
35. N. Bjerrum, *Z. Anorg. Allgem. Chem.*, **109**, 275 (1920).
36. E. Glueckauf, *Trans. Faraday Soc.*, **51**, 1235 (1955).
37. Y. Marcus, *Introduction to Liquid State Chemistry*, Wiley, Chichester (1977), pp. 237–8.
38. E. W. Washburn and E. B. Millard, *J. Am. Chem. Soc.*, **31**, 322 (1909); **37**, 694 (1915); M. Fischer and S. Koval, *Bull. Sci. Univ. Kiev.* No. **7**, 137 (1939).
39. A. J. Rutgers and Y. Hendrikx, *Trans. Faraday Soc.*, **58**, 2184 (1962).
40. E. R. Nightingale, *J. Phys. Chem.*, **63**, 1381 (1959); N. Matsuura, K. Umemoto, and Y. Takeda, *Bull. Chem. Soc. Japan*, **48**, 2253 (1975).

Chapter 5

Ion hydration

5.1 The relevant properties of water

Water, as a solvent for ionic solvents, has been studied very extensively, and its properties are well known, both at ambient temperatures and pressures and at elevated ones. Several monographs deal with this topic,[1-3] and international commissions are critically evaluating existing data and presenting them in terms of multiparameter expressions in the temperature and the pressure (or the density).

Liquid water is generally considered to exist between its freezing point, 0 °C, and its normal boiling point, 100 °C, if the pressure is limited to 1 atmosphere (101325 Pa). The bulk mechanical and thermal properties of water in this temperature range are presented in Table 5.1. These comprise the density, *d*, the

Table 5.1 Physical properties of H_2O at 1 atm pressure

t (°C)	d^a ($g\ cm^{-3}$)	V^b ($cm^3\ mol^{-1}$)	α_P^a ($10^{-3}\ K^{-1}$)	κ_T^a ($10^{-10}\ Pa^{-1}$)	γ^c ($10^{-3}\ N\ m^{-1}$)	η^d ($10^{-3}\ Pa \cdot s$)
0	0.999840	18.019	−0.0681	5.089	75.62	1.792
5	0.999964	18.017	+0.0160	4.917	74.90	1.519
10	0.999700	18.021	0.0880	4.781	74.20	1.307
15	0.999100	18.032	0.1509	4.673	73.48	1.138
20	0.998204	18.048	0.2068	4.585	72.75	1.002
25	0.997045	18.069	0.2572	4.525	71.96	0.890
30	0.995647	18.095	0.3032	4.477	71.15	0.798
35	0.994032	18.124	0.3457	4.444	70.35	0.720
40	0.992216	18.157	0.3853	4.424	69.55	0.653
45	0.990213	18.194	0.4225	4.415	68.73	0.596
50	0.988036	18.234	0.4576	4.417	67.90	0.547
55	0.985695	18.277	0.4910	4.429	67.06	0.504
60	0.983199	18.324	0.5231	4.450	66.17	0.467
65	0.980555	18.373	0.5539	4.479	65.32	0.433
70	0.977770	18.425	0.5837	4.516	64.41	0.404
75	0.974849	18.481	0.6128	4.561	63.54	0.378
80	0.971798	18.538	0.6411	4.614	62.60	0.354
85	0.968620	18.600	0.6689	4.675	61.69	0.332
90	0.965320	18.663	0.6962	4.743	60.74	0.313
95	0.961900	18.730	0.7233	4.819	59.80	0.295
100	0.958364	18.799	0.7501	4.902	58.84	0.278

(continued)

Table 5.1—*continued*

t (°C)	p^e (kPa)	ΔH^{Vf} (kJ mol^{-1})	$C_P{}^g$ (J K^{-1} mol^{-1})	$n_D{}^h$	ε^i
0	0.611		75.98	1.33395	87.81
5	0.872	44.88	75.70	1.33385	85.93
10	1.228	44.67	75.52	1.33374	83.99
15	1.705	44.45	75.41	1.33339	82.13
20	2.338	44.24	75.34	1.33298	80.27
25	3.169	44.04	75.29	1.33250	78.46
30	4.245	43.84	75.27	1.33195	76.67
35	5.626	43.65	75.27	1.33134	74.94
40	7.381	43.46	75.28	1.33066	73.22
45	9.590	43.27	75.29	1.32991	71.55
50	12.35	43.09	75.31	1.32909	69.90
55	15.75	42.91	75.34	1.32823	68.31
60	19.92	42.74	75.38	1.32730	66.73
65	25.02	42.57	75.42	1.32630	65.20
70	31.18	42.40	75.47	1.32526	63.70
75	38.56	42.24	75.53	1.32417	62.24
80	47.38	42.09	75.60	1.32302	60.81
85	57.82	41.93	75.67	1.32183	59.41
90	70.12	41.79	75.75	1.32061	58.05
95	84.53	41.64	75.85	1.31934	56.73
100	101.325	41.50	75.95	1.31805	55.41

[a] K. S. Kell, *J. Chem. Eng. Data*, **20**, 97 (1975); [b] V/cm^3 mol^{-1} = (18.015/g mol^{-1})/(d/g cm^{-3}); [c] N. B. Vargaftig, B. N. Volkov, and L. D. Voljak, *J. Phys. Chem. Ref. Data*, **12**, 817 (1983); [d] J. Kestin, M. Sokolov, and W. A. Wakeham, *J. Phys. Chem. Ref. Data*, **7**, 941 (1980); [e] D. Ambrose, J. F. Counsell, and A. J. Davenport, *J. Chem. Thermodyn.*, **2**, 283 (1970), see also D. Eisenberg and W. Kauzmann, *The Structure and Properties of Water*, Clarendon, Oxford (1969), p. 62; [f] calculated numerically from $\Delta H^V = \mathbf{R}T^2(\Delta T)^{-1}\,\Delta \ln p$; [g] Landoldt-Börnstein, **II/4**, 520 (1961); [h] Landoldt-Börnstein, **II/8**, 5–565 (1962); [i] M. Umematsu and E. U. Franck, *J. Phys. Chem. Ref. Data*, **9**, 1291 (1980).

molar volume, V, the (isobaric) thermal expansivity, α_P, the (isothermal) compressibility, κ_T, the surface tension, γ, the viscosity, η, the vapour pressure, p, the molar heat of evaporation, ΔH^V, and the molar heat capacity (at constant pressure), C_p. Some bulk optical and electrical properties: the refractive index (for the yellow sodium D line at 589 nm), n_D, and the dielectric constant (the relative permittivity), ε, are also shown in Table 5.1. For the purpose of interpolation, these quantities have been fitted to power series in the temperature, presented in Table 5.1a.

The liquid range of water, however, extends well beyond the common range of 0 to 100 °C. It extends from the critical point $T_c = 647.3$ K (374.2 °C) down to the glass transition point $T_g = 139$ K actually observed for vitreous water[4] or to $T_g = 159$ to 162 K, predicted for supercooled liquid water.[5] This range is not entirely realizable, because of the homogeneous nucleation,[6] that sets in spontaneously at 232 to 233 K (−40 to −41 °C). However, carefully purified water can be supercooled in the liquid state and studied in capillaries or in emulsions[7] down to 235 K (−38 °C). Some physical data for supercooled water are presented in Table 5.2.

Table 5.1a Interpolation functions for the physical properties of H_2O at 1 atm

$d(t)^a = 0.997045 - 2.564 \cdot 10^{-4}(t-25) - 4.812 \cdot 10^{-6}(t-25)^2 + 2.8736 \cdot 10^{-8}(t-25)^3 - 1.4456 \cdot 10^{-10}(t-25)^4$
for d/g cm^{-3} and $5 \leqslant t/°C \leqslant 100$

$\gamma(t) = 71.93 - 0.1554(t-25) - 2.7 \cdot 10^{-4}(t-25)^2$
for γ/mN m^{-1} and $-10 \leqslant t/°C \leqslant 100$

$C_P(t) = 75.294 - 0.00317(t-25) + 1.60 \cdot 10^{-4}(t-25) + 0.0018 \exp(2.19 - 0.142(t-25))$
for C_P/J K^{-1} mol^{-1} and $0 \leqslant t/°C \leqslant 100$

$\Delta H^V(t) = 44.04 - 0.0401(t-25) + 8.35 \cdot 10^{-5}(t-25)^2$
for ΔH^V/kJ mol^{-1} and $5 \leqslant t/°C \leqslant 100$

$\eta(t)^b = 0.02767 \exp(522.9/((t-25) + 150.64))$
for η/mPa·s and $5 \leqslant t/°C \leqslant 100$

$n_D(t) = 1.33250 - 1.028 \cdot 10^{-4}(t-25) - 1.3394.10^{-6}(t-25)^2 + 2.5 \cdot 10^{-11}(t-25)^4$
for $10 \leqslant t/°C \leqslant 90$

$\varepsilon(t)^c = 78.46 - 0.3595(t-25) + 7.00 \cdot 10^{-4}(t-25)^2$
for $0 \leqslant t/°C \leqslant 100$

$p(t) = \exp(16.3687 - 3878.9/(255 + (t-25)))$
for p/kPa and $0 \leqslant t/°C \leqslant 100$

[a] See G. S. Kell, *J. Phys. Chem. Ref. Data*, **6**, 1109 (1977) for a more accurate 9-parameter equation; [b] see J. T. R. Watson, R. S. Basu, and J. V. Sengers, *J. Phys. Chem. Ref. Data*, **9**, 1255 (1980) for a more accurate 19-parameter equation; [c] see M. Uematsu and E. U. Franck, *J. Phys. Chem. Ref. Data*, **9**, 1291 (1980) for a more accurate 10-parameter equation.

Table 5.2 Physical properties of liquid water below the freezing point at 1 atm

t (°C)	d^a (g cm^{-3})	V^b (cm^3 mol^{-1})	α_P^c (10^{-3} K^{-1})	η^d (mPa·s)	C_P^e (JK^{-1} mol^{-1})	γ^f (mN m^{-1})	κ_T^f (10^{-10} Pa^{-1})
0	0.999840	18.019	−0.068	1.79	75.98	75.62	5.089
−5	0.999256	18.024	−0.169	2.16	76	76.40	5.306
−10	0.998117	18.064	−0.292	2.66	76	77.10	5.583
−15	0.996283	18.079	−0.450	3.34	76	77.7	5.944
−20	0.993547	18.136	−0.661	4.33	77.8		6.425
−25	0.989585	18.205	−0.956		82.8		7.094
−30	0.983854	18.329	−1.400		91.6		8.079
−35	0.9727	18.516			105.8		

[a] B. V. Zhelevskii, *Zh. Fiz. Khim.*, **43**, 2343 (1969), *Russ. J. Phys. Chem.*, **43**, 1311 (1969); [b] interpolated, where necessary, from data in ref. a; [c] from the data of ref. a as $\alpha_P = V^{-1}(\Delta V/\Delta t)$; [d] J. Hallet, *Proc. Phys. Soc.*, **82**, 1046 (1963), via Landoldt-Börnstein **II/5.a**, 128/31 (1969); [e] C. A. Angell, *J. Phys. Chem.*, **77**, 3092 (1973) and Landoldt-Börnstein **II/3**, 421/2 (1956); [f] K. S. Kell, *J. Chem. Eng. Data*, **20**, 97 (1975).

The properties of water at temperatures beyond 100 °C are also of interest, since water can serve as a solvent for ionic substances also at these elevated temperatures, when pressure is applied. Three distinct temperature and pressure regions are relevant: those that apply to the saturation curve, where liquid water is at equilibrium with its vapour; arbitrary pressures above the vapour pressure but temperatures still below the critical one; and temperatures and pressures beyond the critical point, where fluid water no longer exists as a liquid. Some properties of water in these regions of temperature and pressure are presented in Table 5.3. In addition to some of the previously mentioned properties, also the specific electric conductance, κ, and the self-ionization constant (the product of the molalities of the hydrogen and the hydroxide ions), K_W, are presented in Table 5.3. These two related properties pertain to the chemical nature of water as a solvent.

The temperature and pressure derivatives of some of the bulk properties of water are important for some theoretical treatments of ion hydration or of ion–ion interactions. Some of these derivative quantities are themselves known over broad temperature and pressure ranges, whereas others are known less well. Table 5.4 presents the derivative quantities at 25 °C and 1 atm only.

The bulk properties of heavy water are also known for wide temperature and pressure ranges, and should be useful for comparison with those of light water, in connection with ion hydration and ion transfer from light to heavy water. These are presented in Table 5.5 up to the normal boiling point and beyond it along the saturation line.

The properties of light and heavy water, including molecular properties, are compared in Table 5.6.

The structure of water is of interest in relation to its properties as a solvent of ions, since it is perturbed by them to a greater or lesser extent. The structure of a

Table 5.3a Physical properties of H_2O on the saturation line (liquid–vapour equilibrium) at various temperatures

t (°C)	p[a] (M Pa)	d[b] (g cm^{-3})	η[c] (mPa · s)	ε[d]	γ[e] (10^{-3} N m^{-1})	$-\log K_W$[f] (mol^2 kg^{-2})	κ[g] (10^{-6} Ω^{-1} cm^{-1})
25	0.00317	0.99705	0.890	78.46	71.96	13.995	0.055
50	0.01235	0.98800	0.547	69.90	67.93	13.275	0.167
75	0.03856	0.97482	0.374	62.24	63.54	12.712	0.394
100	0.10133	0.95836	0.282	55.41	58.78	12.265	0.765
125	0.23203	0.93905	0.218	49.36	53.79	11.912	1.270
150	0.47574	0.91704	0.182	43.94	48.70	11.638	1.69
175	0.8926	0.8945	0.143	39.16	43.28	11.432	2.42
200	1.5551	0.86486	0.134	34.74	37.81	11.289	2.98
225	2.5505	0.8343	0.118	30.80	32.03	11.208	3.28
250	3.9777	0.79914	0.1062	27.08	26.19	11.191	3.28
275	5.949	0.7410	0.096_7	23.63	20.17	11.251	2.86
300	8.592	0.71242	0.085_9	20.26	14.39	11.406	2.29
325	12.058	0.6490	0.082_0	16.84	7.26		
350	16.537	0.57504	0.065_9	13.16	3.79	12.30	0.630
374.2	22.093	0.31776	0.043	8.63	0		

[a] E. Schmidt, *Properties of Water and Steam in SI Units*, Springer, New York (1969); [b] as ref. [a]; [c] J. T. R. Watson, R. S. Basu, and J. F. Sengers, *J. Phys. Chem. Ref. Data*, **9**, 1255 (1980); [d] M. Uematsu and E. U. Franck, *J. Phys. Chem. Ref. Data*, **9**, 1291 (1980); [e] as ref. [a]; [f] W. L. Marshall and E. U. Franck, *J. Phys. Chem. Ref. Data*, **10**, 295 (1981); [g] from $\kappa/\Omega^{-1}\ \text{cm}^{-1} = (K_W/\text{mol}^2\ \text{kg}^{-2})^{1/2} \cdot (d/\text{kg dm}^{-3}) \cdot (\Lambda^\infty/\Omega^{-1}\ \text{cm}^2 \cdot 10^{-3})$, and $\Lambda^\infty(\text{H}^+ + \text{OH}^-)/\Omega^{-1}\ \text{cm}^2 = 300 + 9.5(t/°\text{C}) - 0.017(t/°\text{C})^2$, from D. P. Pearsson, *Diss. Abstr.*, **20**, 3972 (1960).

Table 5.3b Physical properties of H_2O at elevated pressures and temperatures

	10 MPa[a]			30 MPa[a]			100 MPa[a]			300 MPa[b]		
t (°C)	d ($g\ cm^{-3}$)	η ($mPa \cdot s$)	ε	d ($g\ cm^{-3}$)	η ($mPa \cdot s$)	ε	d ($g\ cm^{-3}$)	η ($mPa \cdot s$)	ε	d ($g\ cm^{-3}$)	η	ε
25	1.0015	0.888	78.85	1.0101	0.888	78.85	1.0379	0.890	82.08	1.0993		87.34
50	0.9923	0.549	70.27	1.0007	0.553	70.98	1.0275	0.571	73.22	1.0862		78.17
75	0.9782	0.380	62.59	0.9869	0.386	63.28	1.0132	0.405	65.42	1.0735		70.14
100	0.9629	0.284	55.76	0.9718	0.289	56.44	0.9998	0.309	58.55	1.0609	0.359	63.10
125	0.9401	0.236	49.70	0.9527	0.242	50.39	0.9829		52.49	1.0471		56.94
150	0.9223	0.185	44.30	0.9329	0.190	45.01	0.9650	0.206	47.14	1.0321	0.249	51.55
175	0.8969	0.160	39.47	0.9092	0.164	39.87	0.9448		42.39	1.0164		46.82
200	0.8760	0.137	35.11	0.8882	0.141	35.91	0.9213	0.156	38.17	0.9962	0.199	42.65
225	0.8389	0.118	31.13	0.8557	0.123	32.01	0.9004		34.40			38.97
250	0.8061	0.108	27.43	0.8256	0.113	28.43	0.8766	0.128	31.01	0.976	0.171	35.69
275	0.7668	0.097	23.90	0.8046	0.103	25.11	0.8496		27.95			32.77
300	0.7157	0.086	20.39	0.7509	0.093	21.95	0.8227	0.109	25.17	0.933	0.148	30.15
325	—	—	—	0.6980	0.085		0.7916					
350				0.6443	0.075	15.66	0.7606	0.096	20.29	0.895	0.130	25.65
375					0.060		0.7266					
400				0.353	0.044	5.91	0.691	0.085	16.05	0.856	0.116	21.94
450				0.148_5	0.031	2.07	0.615	0.075	12.39		0.105	18.85
500				0.115_2	0.032	1.68	0.529	0.066	9.29	0.778	0.096	16.25
550				0.098_3	0.033	1.51	0.445	0.059	6.88		0.091	14.07
600				—	—	—	0.375	0.056		0.697		

[a] The data for d and η are from E. Schmidt, *Properties of Water and Steam in SI Units*, Springer, New York (1969), those for ε are from M. Umematsu and E. U. Franck, *J. Phys. Chem. Ref. Data*, **9**, 1291 (1980), dashes signify that the temperature is beyond the liquid range for the given peessure; [b] from G. S. Kell, in *Water: A Comprehensive Treatise*, F. Franks, ed., Plenum, New York, Vol. 1, pp. 385, 484 (1972).

Table 5.4 Relative temperature and pressure derivatives of some bulk properties of H_2O at 25 °C and 1 atm

Property	$(\partial \ln(property)/\partial T)_P$ (10^{-3} K^{-1})	$(\partial \ln(property)/\partial P)_T$ (10^{-10} Pa^{-1})
Density, d	$-\alpha_P = -0.2547$[a]	$\kappa_T = 4.525$[a]
Molar volume, V	$\alpha_P = 0.2547$[a]	$-\kappa_T = -4.525$[a]
Vapour pressure, p	$\Delta H^V/RT^2 = 59.56$[b]	
Surface tension, γ	-2.140[c]	
Viscosity, η	-22.81[d]	-2.319d
Refractive index, n_D	-0.07715[e]	1.102[f]
Dielectric constant, ε	-4.588[g]	4.71[g]

[a] G. S. Kell, *J. Phys. Chem. Ref. Data*, **6**, 1109 (1977); [b] Landoldt-Börnstein **II/2c**, p. 61 (1960); [c] Landoldt-Börnstein **II/3**, p. 421 (1956); [d] J. Kestin, H. E. Halifa, H. Sookiazian, and W. A. Wakeham, *Ber. Bunsengesell. Phys. Chem.*, **82**, 181 (1978); [e] from Table 5.2; [f] H. Eisenberg, *J. Chem. Phys.*, **43**, 3887 (1965); [g] B. B. Owen, R. C. Miller, C. E. Miller, and H. L. Cogan, *J. Phys. Chem.*, **65**, 2065 (1961).

liquid is generally described by its radial distribution function (if monatomic) or by the partial radial distribution functions of its constituent atoms (if molecular) (see Section 3.2). In the case of water, the additional feature of hydrogen bonding between adjacent molecules must also be taken into account. The structure of water has been studied experimentally by Narten and co-workers, using X-ray, neutron (for D_2O, which is assumed to have the same structure), and electron diffraction measurements.[8] Computer simulation by Monte Carlo[9] and molecular dynamics[10] methods has been able to reproduce successfully the results of the diffraction work and to predict some of the thermodynamic properties, but not all of them, adequately. The simulation results are very sensitive to the input potential function, and this must still be refined in order to attain better agreement for the thermodynamic functions, especially of the derivative ones. The main features of the structure of water, expressed by the partial distribution functions $g_{O\text{-}O}$, $g_{O\text{-}H}$, and $g_{H\text{-}H}$, are shown in Table 5.7, the former function also in Figure 5.1. The coordination number Z for a given water molecule is 4.8, i.e., it has this number of nearest neighbours. Not to all of them is it hydrogen-bonded, however, there being on the average only about three hydrogen bonds per water molecule.[9] This number is expected to vary with the temperature and the density (i.e., the pressure), and its accurate definition, estimation, and determination, has been accomplished only recently. Marcus and Ben-Naim[11] used statistical thermodynamic arguments and data on the Gibbs free energy of solvation (i.e., the condensation from the vapour) of H_2O and D_2O molecules in their respective liquids for the evaluation of the extent of hydrogen bonding in liquid water, see Figure 5.2. Shown is the average number of hydrogen bonds per water molecule, the quantity $(2/N)\langle g\rangle_O$, defined in Section 3.6, at varying temperatures.

Table 5.5 Bulk properties of heavy water, D_2O, at 1 atm ($\leqslant T_b$) or at saturation

t (°C)	d^a (g cm^{-3})	V^b (cm^{-3} mol^{-1})	η^c (mPa·s)	p^d (MPa)	ΔH^{Ve} (kJ mol^{-1})	C_P^f (JK^{-1} mol^{-1})	$\kappa_T{}^g$ (10^{-10} Pa^{-1})	γ^h (mN m^{-1})	$n_D{}^i$	ε^j
10	1.10596	18.109	1.685	0.00103	46.23		4.977		1.32922	83.68
15	1.10585	18.111	1.405	0.00144	45.97	84.52	4.861	73.35	1.32906	81.76
20	1.10531	18.120	1.277	0.00200	45.72	84.72	4.755	72.60	1.32878	79.89
25	1.10442	18.134	1.103	0.00274	45.46	84.52	4.678	71.85	1.32841	78.06
30	1.10320	18.154	1.033	0.00370	45.20	84.12	4.615	71.10	1.32793	76.27
35	1.10167	18.179	0.864	0.00495	44.95	84.12	4.543	70.30	1.32737	74.52
40	1.09989	18.209	0.807	0.00655	44.69	84.12	4.513	69.4	1.3267	72.82
50	1.09560	18.280	0.671	0.01112	44.17	83.92	4.491	67.7	1.3251	69.47
60	1.09051	18.365	0.569	0.01824	43.65		4.518	66.0	1.3231	66.36
70	1.08466	18.464	0.488	0.02885	43.13		4.604	64.2	1.3207	63.39
80	1.07815	18.576	0.424	0.04426	42.60		4.771	62.5		60.56
90	1.07107	18.699	0.374	0.06609	42.09		5.052	60.8		57.86
100	1.06319	18.838	0.334	0.09625	41.55	83.92	5.496	58.6		55.28
150	1.01712	19.691	0.216	0.4656	49.70	84.92				
200	0.95804	20.905	0.156	1.5474	35.29	87.72				
250	0.88339	22.671	0.125	3.9994	31.00	94.59				
300	0.78456	25.527		8.701	25.11					
350	0.62329	32.133		16.845						
370.7	0.328	55.0		21.66	0					

[a,b] P. G. Hill and R. D. ChrisMacMillan, *J. Phys. Chem. Ref. Data*, **9**, 735 (1980), G. S. Kell, *ibid.*, **6**, 1109 (1977); [c] N. Matsunaga and A. Nagashima, *J. Phys. Chem. Ref. Data*, **12**, 933 (1983); [d] P. G. Hill and R. D. ChrisMacMillan, *J. Phys. Chem. Ref. Data*, **9**, 735 (1980), P. G. Hill, R. D. ChristMacmillan and V. Lee, *ibid.*, **11**, 1 (1982); [e] calculated numerically from $\Delta H^V = \mathbf{R}T^2(\Delta T)^{-1}\Delta \ln p$; [f] G. S. Kell, in *Water: A Comprehensive Treatise*, F. Franks, ed., Plenum, New York, Vol. 1, pp. 363–412 (1972); [g] see ref. [f]; [h] J. R. Heiks, M. K. Barrett, L. V. Jones and E. Orban, *J. Phys. Chem.*, **58**, 488 (1954), Landoldt-Börnstein, **II/3**, 421 (1956); [i] Landoldt-Börnstein, **II/8**, 5-565 (1962); [j] see ref [f].

Table 5.6 A comparison of some physical properties of H_2O and D_2O

Property	Unit	Value for H_2O	Value for D_2O
General properties			
Molar mass, M	$kg\ mol^{-1}$	0.018015	0.020028
Freezing point, T_m	K (°C)	273.15 (0.00)	276.96 (3.81)
Triple point, T_{tr}	K (°C)	273.16 (0.01)	276.97 (3.82)
Temp. of max density	K (°C)	277.13 (3.98)	284.38 (11.23)
Normal boiling point, T_b	K (°C)	373.15 (100.00)	374.55 (101.42)
Critical temperature, T_c	K (°C)	647.35[b] (374.20)	643.89[b] (370.74)
Critical pressure, P_c	MPa(atm)	22.093[b] (218.3)	21.66[b] (219.5)
Critical density, d_c	$kg\ m^{-3}$	317.76[c]	364.1
Critical molar volume, V_c	$cm^3\ mol^{-1}$	55.3	55.0
Values at T_m			
Heat of fusion, ΔH^f	$kJ\ mol^{-1}$	6.008	6.339
Entropy of fusion, ΔS^f	$J\ K^{-1}\ mol^{-1}$	22.01	22.68
Volume change on fusion, ΔV^f	$cm^3\ mol^{-1}$	−1.632	−1.561
Molar vol. of solid, $V(s)$	$cm^3\ mol^{-1}$	19.65	19.679
Molar vol. of liquid, $V(l)$	$cm^3\ mol^{-1}$	18.018	18.118
Therm. expans., solid, $\alpha_P(s)$	K^{-1}	$1.39 \cdot 10^{-4}$	$1.39 \cdot 10^{-4}$
Therm. expans., liquid, $\alpha_P(l)$	K^{-1}	$-0.59 \cdot 10^{-4}$	$-0.32 \cdot 10^{-4}$
Values for liquid at 298.15 K			
Density, d	$kg\ m^{-3}$	997.047[d]	1104.48[a]
Molar volume, V	$cm^3\ mol^{-1}$	18.069	18.133
Therm. expans., α_P	K^{-1}	$2.55 \cdot 10^{-4}$[d]	$2.18 \cdot 10^{-4}$
Isotherm. compress., κ_T	Pa^{-1}	$4.525 \cdot 10^{-10}$[e]	$4.553 \cdot 10^{-10}$
Surface tension, γ	$N\ m^{-1}$	0.07196	0.07193
Viscosity, η	$Pa \cdot s$	$8.903 \cdot 10^{-4}$[d]	$11.03 \cdot 10^{-4}$[f]
Self diffusion coefficient, D	$m^2 s^{-1}$	$2.5 \cdot 10^{-9}$[i]	$2.0 \cdot 10^{-9}$[i]
Vapour pressure, p	Pa	3166[d]	2737[b]
Heat of vapourization, ΔH^V	$kJ\ mol^{-1}$	44.04[d]	45.46[f]
Entropy of vapour, ΔS^V	$JK^{-1}\ mol^{-1}$	118.78	122.26
Molar entropy, S^0	$JK^{-1}\ mol^{-1}$	70.08	76.11
Heat capacity			
at constant pressure, C_P	$JK^{-1}\ mol^{-1}$	75.29[d]	84.52[f]
at constant volume, C_V	$JK^{-1}\ mol^{-1}$	74.48	83.7
Refractive index, n_D		1.33250[d]	1.32841
Dielectric constant, ε		78.46[d]	77.94
Dipole orient. parameter, g		2.7[e]	
specific conductance, κ	$\Omega^{-1}\ m^{-1}$	$1.38 \cdot 10^{-6}$[e]	
Ionization constant K_W	$mol\ L^{-1}$	$1.81 \cdot 10^{-16}$	$3.54 \cdot 10^{-17}$
Heat of ionization, $\Delta H^0_{K_W}$	$kJ\ mol^{-1}$	56.27	60.33
Values for gas at 298.15 K			
Heat of formation, ΔH^0_f	$kJ\ mol^{-1}$	−241.84	−249.20
Entropy, S^0	$JK^{-1}\ mol^{-1}$	188.87	198.36

Table 5.6—*continued*

Property	Unit	Value of H_2O	Value for D_2O
Heat capacity at constant			
pressure, C_P	$JK^{-1}\,mol^{-1}$	33.58[m]	34.26[m]
constant volume, C_V	$JK^{-1}\,mol^{-1}$	25.19	
$H^0_{298} - H^0_0$	$kJ\,mol^{-1}$	9.908	9.954
Molecular properties			
Bond length, O—H(D)	nm	0.09572[g]	0.09575[g]
Bond angle	deg	104.52[g]	104.47[g]
Moment of inertia, I_A	$kg \cdot m^2$	$1.0220 \cdot 10^{-31}$[g]	$1.8384 \cdot 10^{-31}$[g]
I_B	$kg \cdot m^2$	$1.9187 \cdot 10^{-31}$[g]	$3.8340 \cdot 10^{-31}$[g]
Length of hydrogen bond	nm	0.2765	0.2766
Vibrational fundamentals			
ν_1	cm^{-1}	3657.05	2671.69
ν_2	cm^{-1}	1594.59	1178.33
ν_3	cm^{-1}	3755.79	2788.02
Dipole moment, μ	D (1 D = 3.33564 $\cdot 10^{-30}$ C·m)	1.834[e]	1.84[j]
Electrical quadrupole			
moment, θ	$C\,m^2$	$1.87 \cdot 10^{-39}$[h]	
Polarizability, α	m^3	$1.456 \cdot 10^{-30}$[e]	$1.536 \cdot 10^{-30}$[j]
Ionization potential, I	V	12.62[e]	
Diamagnetic susceptibility, χ_m		$-7.149 \cdot 10^{-7}$	
Collision diameter, σ	nm	0.274[e]	
Intermolecular potential energy			
minimum, $u/\mathbf{k}$	K	732[e]	
Other properties			
Solubility parameter, δ	$J^{1/2}\,cm^{-3/2}$	49.4[k]	48.7[b]
Donor number, DN	$kJ\,mol^{-1}$	139.5[k]	
Acceptance index, E_T	$kJ\,mol^{-1}$	264.0[k]	
Taft polarity index, π^*		1.090[k]	
Taft H-bond donor index, α_{av}		1.017[k]	

Data from: E. M. Arnett and D. R. McKelvey, in *Solute–Solvent Interactions*, J. F. Coetzee and C. D. Ritchie, eds, Dekker, New York, NY, Vol. 1, pp. 392–4 (1969) and G. Nemethy and H. A. Scheraga, *J. Chem. Phys.*, **41**, 680 (1964); except for:

[a] G. S. Kell, *J. Phys. Chem. Ref. Data*, **6**, 1109 (1977); [b] P. G. Hill and R. D. Chris Macmillan, *J. Phys. Chem. Ref. Data*, **9**, 735 (1981); [c] J. T. R. Watson, R. S. Basu, and J. V. Sengers, *J. Phys. Chem. Ref. Data*, **9**, 1255 (1981); [d] Table 5.1; [e] Y. Marcus, *Introduction to Liquid State Chemistry*, Wiley, Chichester (1977); [f] Table 5.6; [g] W. S. Benedict, N. Gailar, and E. K. Plyler, *J. Chem. Phys.*, **24**, 1139, 1165 (1956); [h] R. M. Glaeser and C. A. Coulson, *Trans. Faraday Soc.*, **61**, 389 (1965); [i] G. S. Kell, in *Water: A Comprehensive Treatise*, F. Franks, ed., Plenum, New York, Vol. 1, 411 (1972); [j] L. G. Groves and S. Sugden, *J. Chem. Soc.*, **1935**, 971; [k] Table 6.5; [l] from $((\Delta H^v - \mathbf{R}T)/V)^{1/2}$; [m] A. S. Friedman and L. Haar, *J. Chem. Phys.*, **22**, 2051 (1954).

Table 5.7 The main features of the radial distribution function of water at 25 °C[8,9]

Extremum	$g_{O\text{–}O}$	$g_{O\text{–}H}$	$g_{H\text{–}H}$
First maximum, r/nm	0.283	0.190	0.235
g (Z)	2.31 (4.8)	0.80	1.04
First minimum, r/nm	0.345	0.245	0.300
g	0.85	0.50	0.47
Second maximum, r/nm	0.453	0.335	0.400
g	1.12	1.70	1.08
Second minimum, r/nm	0.56	Not included in model	
g	0.86		
Third maximum, r/nm	0.65	Not included in model	
g	1.09		

5.2 The size of hydrated ions

The size of hydrated ions may be discussed in terms of three concepts: the limiting partial molar volumes of the ions, $\bar{V}_i^\infty$, their molar 'hydrated volumes', $\bar{V}_{ih}^\infty$ (i.e., the molar volume of the hydrated ions) and their hydration numbers, h_i^∞. All pertain to infinite dilution, where only ion–water but no ion–ion interactions take place, hence additivity rules concerning the contributions of the cations and the anions to the molar volume of the electrolytes must be obeyed. The first concept is straightforward, as far as thermodynamic components are concerned, for which $\bar{V}^\infty$ is simple to measure via the apparent molar volume, ${}^\phi V$:

$$\bar{V}^\infty = {}^\phi V^\infty = \lim_{m \to 0} {}^\phi V = \lim_{m \to 0} [(1000/d_W)(M - (d - d_W)(1 + mM)/(md)] \tag{5.1}$$

where m is the molality of the solute (moles per kg of solvent, here water), M its molar mass (in kg/mol), d the density of the solution, and d_W that of water. Accurate density measurements (to $\pm < 10^{-5}$ g cm^{-3}) down to low molalities $m \leqslant 0.01$ mol/kg are generally required in order to obtain $\bar{V}^\infty$ values accurate to better than ± 0.1 cm^3/mol. This quantity should then be separable into the contributions of the cations and the anions, which should be constant from one electrolyte to the other.

Since thermodynamics provides no clues as to how this separation ought to be performed, the *conventional* way to do so is to assign a value to one ion, and rely on the additivity principle. The value conventionally assigned is $\bar{V}_{conv}^\infty(H^+) = 0$ at all temperatures and pressures. The conventional limiting partial molar ionic volumes, $\bar{V}_{i\,conv}^\infty$ at 25 °C and 1 atm are shown in Table 5.8. This conventional assignment leads to unexpected results. For example, the

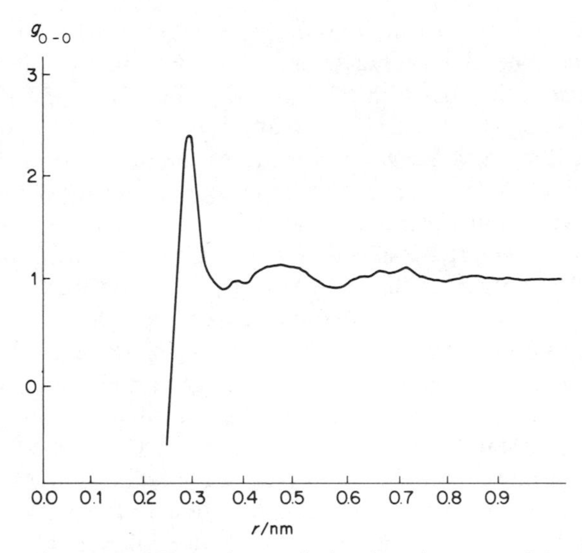

Figure 5.1 The pair correlation function $g_{O\text{-}O}$ of the water molecules in liquid water, as obtained from X-ray and neutron diffraction measurements[8] (Reproduced by permission of American Institute of Physics)

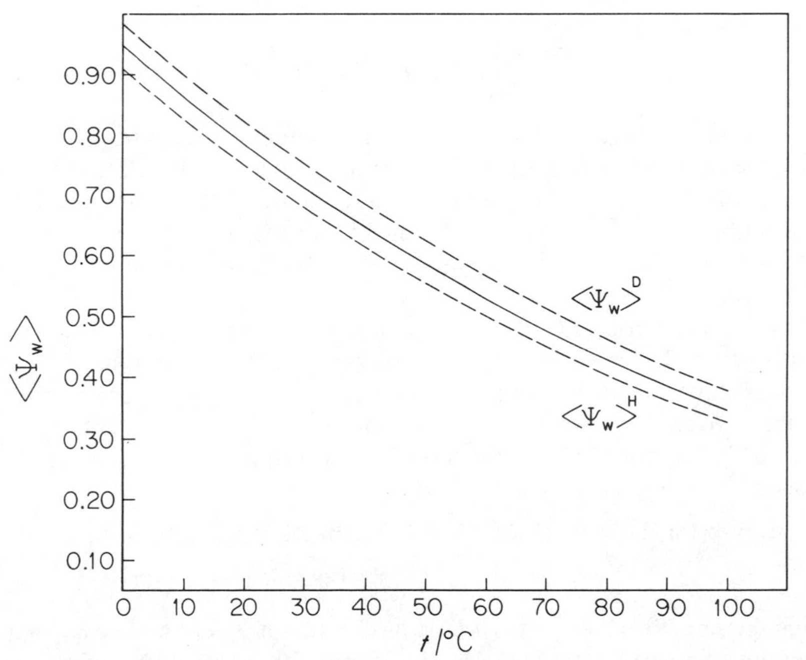

Figure 5.2 The mean number of hydrogen bonds per water molecule $\langle\Psi_W\rangle$ in liquid water as a function of the temperature. The full-drawn line pertains to 'mean' water, i.e., the mean between light and heavy water, the dashed ones on either side to D_2O and H_2O[11]

limiting partial molar expansibilities $\bar{E}_{i\,conv}^{\infty} = (\partial \bar{V}_{i\,conv}^{\infty}/\partial T)_P$ are found to be nonmonotonous functions of the temperature. For many ions they have a maximum in the region between 50 and 75 °C, but this is partly an artifact, due to the assignment of $\bar{V}_{conv}^{\infty}(H^+) = 0$ at all temperatures.

The work of Zana and Yeager[12] constitutes the only successful attempt to measure individual ionic partial molar volumes in water experimentally, using the ultrasonic vibration potentials. The values obtained for a given ion are independent of the counter-ion within an error of ± 2 cm^3/mol, with a few exceptions (OH^-, SO_4^{2-}, Ba^{2+}, where it is substantially higher). This imprecision is unfortunate, in view of the much more accurate conventional volumes, since it forces the investigators to resort to extrathermodynamic assumptions in order to obtain 'absolute' limiting partial ionic molar volumes. Within this imprecision, however, the experimental values of Zana and Yeager[12] (which can be summarized by the single value $\bar{V}^{\infty}(H^+) = -5.4$ cm^3/mol at 22 °C) agree with those obtained by the use of the better of these assumptions.

Two extrathermodynamic assumptions recommend themselves most for the evaluation of individual ionic limiting partial molar volumes. One of them is the extrapolation method, applied to the tetraalkylammonium salts. This may be done in either the Conway mode, equation 5.2a, or the Jolicoeur mode, equation 5.2b:[13a]

$$\bar{V}^{\infty}(X^-) = \lim_{M_{R_4N^+} \to 0} (\bar{V}^{\infty}(R_4NX) - b \cdot M_{R_4N^+}) \tag{5.2a}$$

$$= \lim_{n_C \to 0} (\bar{V}^{\infty}(R_4NX) - b' \cdot n_C) \tag{5.2b}$$

Since the molar mass of the tetraalkylammonium cation, $M_{R_4N^+}$, and the number of carbon atoms in their alkyl chain, n_C, are linearly related ($M_{R_4N^+} = 18.039 + 14.027 \cdot n_C$), as discussed by Krumgalz,[13b] there is a difference between the results: $\bar{V}_i^{\infty}$ of anions obtained by extrapolation to $n_C = 0$ is larger by 20.09 cm^3/mol at 25 °C than those obtained by extrapolation to $M_{R_4N^+} = 0$. Conway's modification, equation 5.2a, leads to $\bar{V}^{\infty}(H^+) = -6.2 \pm 0.8$ cm^3/mol at 25 °C, and is consistent with the experimental determination[12] and with the results of the other recommended extrathermodynamic assumption.

This assumption consists of the division of the limiting partial molar volume of the reference electrolyte, tetraphenylarsonium tetraphenylborate (Ph_4AsBPh_4), between its constituting ions according to their van der Waals volumes:[14]

$$\bar{V}^{\infty}(Ph_4As^+)/\bar{V}^{\infty}(BPh_4^-) = V_{vdW}(Ph_4As^+)/V_{vdW}(BPh_4^-)$$

$$= 195.3/188.3 = 1.0337 \pm 0.0034 \tag{5.3}$$

In this division, $\bar{V}^{\infty}(Ph_4As^+)$ is 1.66% larger than it would have been, had the $\bar{V}^{\infty}$ of the reference electrolyte been divided equally between the cation and the anion (a procedure favoured for other thermodynamic functions, see Section 6.3). The values resulting from the application of equation 5.3 and the additivity principle lead to $\bar{V}^{\infty}(H^+) = -6.7 \pm 0.7$ cm^3/mol.

So-called absolute values of $\bar{V}_i^\infty$ of the ions, calculated from the conventional ones by adding $z_i\bar{V}^\infty(H^+) = -6.4 \cdot z_i$ cm^3/mol (at 25 °C), i.e., z_i times the mean from the two assumptions, are shown in Table 5.8. They are expected to be within ± 0.7 cm^3/mol of the "true" values.

The limiting partial molar ionic volume, $\bar{V}_i^\infty$, of an ion is related to its hydrated volume, V_{ih}^∞ by

$$V_{ih}^\infty = \bar{V}_i^\infty + h_i^\infty \cdot V_{H_2O} \quad (5.4)$$

where h_i^∞ is the (limiting) hydration number and V_{H_2O} is the molar volume of pure water. This is an operational definition, from which either V_{ih}^∞ or h_i^∞ is obtained if the other is known. The former quantity can be estimated by two methods, both not entirely satisfactory. One involves the use of the 'hydrated radius', r_{ih}^∞, and is based on the equation

$$V_{ih}^\infty = \mathbf{N}\, k\, (4\pi/3)(r_{ih}^\infty)^3 \quad (5.5)$$

where k is a packing coefficient, which for close-packed spheres is 1.725 (i.e., they occupy 100/1.725 = 58.0% of the volume), but other assignments have been made. The 'hydrated radius', in turn, is related to the Stokes radius, obtained from the limiting ionic equivalent conductivity, λ_i^∞, and the viscosity of water, η_{H_2O}:

$$r_S^\infty/\text{pm} = 8.20|z_i|(\eta_{H_2O}/\text{Pa}\cdot\text{s})^{-1}(\lambda_i^\infty/\Omega^{-1}\,\text{cm}^2\,\text{equiv}^{-1})^{-1} \quad (5.6)$$

It has been found, however, that the Stokes radii obtained from applying equation 5.6 must be 'corrected'[15] in order to give acceptable hydrated radii. The Stokes radii of the larger tetraalkylammonium ions, Et_4N^+, Pr_4N^+, Bu_4N^+, and Pn_4N^+, are related to their van der Waals ionic sizes r_{vdW} according to

$$r_{vdW}/\text{pm} = 323.7 + 0.1337(r_S^\infty/\text{pm}) + 4.857 \cdot 10^{-4}(r_S^\infty/\text{pm})^2 \quad (5.7)$$

These ions have been considered to be unhydrated, but the smaller, hydrated ions, are assumed to obey the same relationship, but with r_{ih}^∞ replacing r_{vdW}. Even the minimal hydrated radii obtained in this manner, $r_{ih}^\infty = 346$ pm for Rb^+, Cs^+, Br^-, and I^-, however, yield with equation 5.5 and $k = 1.725$ excessively large hydrated volumes, and these with equation 5.4 and the limiting partial molar volumes excessively large hydration numbers. The packing coefficient employed seems to be at fault. If this calculation is applied to the larger tetraalkylammonium ions, the resulting V_{ih}^∞ values are uniformly 1.942 ± 0.010 times larger than the $\bar{V}_i^\infty$ values of these four ions. Since they are considered as unhydrated, according to equation 5.4 with $h_i^\infty = 0$, these two quantities should be equal. A reduction of k_{pack} by this factor, i.e. setting $k = 0.888$, leads then to more reasonable values of V_{ih}^∞ of the ions. These and the h_i^∞ values derived from them by equation 5.4 are shown in Table 5.8.

The values of V_{ih}^∞ obtained by the application of equations 5.5 and 5.6 to conductivity data, λ_i^∞, have the advantage that the latter are individual ionic quantities that can be obtained accurately by experiment. Hence no division of

Table 5.8 Limiting partial molar volumes (conventional, $\bar{V}^{\infty}_{\mathrm{i\,conv}}$, and absolute, $\bar{V}^{\infty}_{\mathrm{i}}$), hydrated volumes, V^{∞}_{ih}, and hydration numbers, h^{∞}_{i}, of ions in aqueous solutions at 25 °C

Ion	$\bar{V}^{\infty}_{\mathrm{i\,conv}}$ [a] ($cm^3\ mol^{-1}$)	$\bar{V}^{\infty}_{\mathrm{i}}$ ($cm^3\ mol^{-1}$)	V^{∞}_{ih}(Stokes) [e] ($cm^3\ mol^{-1}$)	h^{∞}_{i} [f]	V^{∞}_{ih}(compres.) [g] ($cm^3\ mol^{-1}$)	h^{∞}_{i} [h]
H^+	0.00	−6.4				
Li^+	−0.88	−7.3	125.9	7.4	34.6	2.3
Na^+	−1.21	−7.6	109.0	6.5	52.1	3.3
K^+	9.02	2.6	94.4	5.1	44.2	2.3
Rb^+	14.07	7.7	92.8	4.7	49.4	2.3
Cs^+	21.34	14.9	92.8	4.3	38.7	1.3
Ag^+	−0.7	−7.1	99.4	5.9	34.2	2.3
Tl^+	10.6	4.2	93.6	5.0	36.3	1.8
H_3O^+	18.0[b]	11.6				
NH_4^+	17.86	11.5	94.4	4.6	17.4	0.3
Me_4N^+	89.57	83.2	115.4	1.8		
Et_4N^+	149.12	142.7	143.4	0.0		
Pr_4N^+	214.44	208.0	206.9	0.0		
Bu_4N^+	275.66	269.3	270.1	0.0		
Pn_4N^+	339.2	332.8	331.7	0.0		
Ph_4P^+	292.2	285.8				
Ph_4As^+	300.65	294.3				
Be^{2+}	−12.0	−24.8	218.1	13.5		
Mg^{2+}	−21.17	−34.0	176.9	11.7	126.0	8.9
Ca^{2+}	−17.85	−30.7	156.7	10.4	128.6	8.9
Sr^{2+}	−18.16	−31.0	156.7	10.4		
Ba^{2+}	−12.47	−25.3	146.7	9.6	134.4	9.2
Mn^{2+}	−17.7	−30.5	189.6	12.2		
Fe^{2+}	−24.7	−37.5	174.5	11.8		
Co^{2+}	−24.0	−36.8	169.6	11.5		
Ni^{2+}	−24.0	−36.8	147.8	10.3		
Cu^{2+}	−27.76	−36.8	147.8	10.3		
Zn^{2+}	−21.6	−34.4	178.2	11.3		
Cd^{2+}	−20.0	−32.8	173.2	11.4	80.0	6.3
Hg^{2+}	−19.3	−32.1				
Pb^{2+}	−15.5	−28.3	143.4	9.5		
Al^{3+}	−42.2	−61.4	241.7	16.8		
La^{3+}	−39.10	−58.3	208.3	14.8	158.8	12.1
Ce^{3+}	−41.5[d]	−59.7	208.3	14.9		
Pr^{3+}	−42.53	−61.5	i	i		
Nd^{3+}	−43.31	−62.5	i	i	172.5	13.1
Pm^{3+}	−42.6[d]	−61.8	i	i		
Sm^{3+}	−42.33	−61.5	i	i		
Eu^{3+}	−41.0[b]	−60.2	i	i		
Gd^{3+}	−40.41	−59.6	i	i		
Tb^{3+}	−40.24	−59.4	i	i		
Dy^{3+}	−40.83	−60.0	i	i		
Ho^{3+}	−41.76	−61.0	i	i		
Er^{3+}	−42.86	−62.1	i	i	172.4	13.1
Tm^{3+}	−43.8[b]	−63.0	228.2	16.2		
Yb^{3+}	−44.22	−63.4	i	i	189.1	14.0
Lu^{3+}	−45.3[b]	−64.5	i	i		
Cr^{3+}	−39.5	−58.7	219.5	15.5		
Fe^{3+}	−43.7	−62.9	215.3	15.5		
In^{3+}	−25.6[b]	−44.8				
Th^{4+}	−53.5	−79.1				
F^-	−1.16	5.2	103.7	5.5	125.5	6.7
Cl^-	17.83	24.2	93.6	3.9	72.9	2.7
Br^-	24.40[e]	30.8	92.8	3.4	61.9	1.7
I^-	36.22	42.6	92.8	2.8	37.1	(−0.3)
OH^-	−4.04	2.4			139.5	7.6

Table 5.8—*continued*

Ion	$\bar{V}^{\infty a}_{i\,conv}$ ($cm^3\ mol^{-1}$)	$\bar{V}^{\infty}_{i}$ ($cm^3\ mol^{-1}$)	V^{∞}_{ih}(Stokes)[e] ($cm^3\ mol^{-1}$)	$h^{\infty f}_{i}$	V^{∞}_{ih}(compres.)[g] ($cm^3\ mol^{-1}$)	$h^{\infty h}_{i}$
CN^-	24.2[b]	30.6				
OCN^-	26.12	32.5				
SCN^-	35.7	42.1			41.3	0.0
$SeCN^-$	49.68	56.1				
NO_2^-	26.2	32.6				
NO_3^-	29.00	35.4	95.3	3.3	48.1	0.7
ClO_3^-	36.66	43.1	97.7	3.0	73.2	1.7
BrO_3^-	35.3	41.7	103.7	3.4	88.2	2.6
IO_3^-	25.3	31.7	120.1	4.9	152.8	6.7
ClO_4^-	44.12	50.5	96.9	2.6		
MnO_4^-	42.5	48.9	100.3	2.9		
ReO_4^-	48.18	54.6	103.7	2.7		
BF_4^-	44.18	50.6				
SO_3F^-	47.93	54.3				
HCO_3^-	23.4	29.8			60.5	1.7
HSO_4^-	35.67	42.1				
$HCrO_4^-$	44.1[b]	50.5				
$HSeO_4^-$	31.1	37.5				
$H_2PO_4^-$	29.1	35.5				
$H_2AsO_4^-$	35.2	41.6				
$SO_3NH_2^-$	41.49	47.9				
BPh_4^-	277.62	284.0				
$HCOO^-$	26.27	32.7				
CH_3COO^-	40.46	46.9			113.3	4.0
$C_2H_5COO^-$	54.0	60.4				
$C_3H_7COO^-$	70.4[b]	76.8				
CH_2ClCOO^-	50.2[b]	56.6				
$C_6H_5O^-$	68.7	74.4				
$C_6H_5SO_3^-$	108.9	115.3				
$CH_3C_6H_5SO_3^-$	119.6	126.0				
S^{2-}	−8.2	4.6				
CO_3^{-2}	−4.3	8.5	137.1	7.1	267.2	14.4
SO_3^{2-}	8.9	21.7				
SO_4^{2-}	13.98	26.8	123.0	5.3	214.0	10.4
SeO_4^{2-}	21.0	33.8	127.9	5.2		
CrO_4^{2-}	19.7	32.5	120.1	4.9	254.8	12.4
MoO_4^{2-}	28.9	41.7	128.9	4.8		
WO_4^{2-}	25.7	38.5	136.0	5.4		
HPO_4^{2-}	7.7	20.5				
$C_2O_4^{2-}$	16.0	28.8				
$S_2O_3^{2-}$	34.0	46.8				
$Cr_2O_7^{2-}$	73.0	85.8				
AsO_4^{3-}	−15.6	3.6				
$P_3O_9^{3-}$	70.1[b]	89.3				
$Fe(CN)_6^{3-}$	120.8	140.0				
$P_4O_{12}^{4-}$	78.0[b]	103.6				
$Fe(CN)_6^{4-}$	74.0	99.6	164.4	3.8		

[a] F. J. Millero, in *Water and Aqueous Solutions*, R. A. Horne, ed., Wiley-Interscience, New York (1972); [b] J. W. Akitt, *J. Chem. Soc. Faraday Trans. 1*, **76**, 2259 (1980); [c] F. H. Spedding, P. F. Cullen, and A. Habenschuss, *J. Phys. Chem.*, **78**, 1106 (1974); [d] Y. Marcus and N. Soffer, *J. Soln. Chem.*, **10**, 549 (1982); [e] from Stokes radii listed in Nightingale,[15] corrected according to equation 5.7 and converted to V^{∞}_{ih} by equation 5.5 with $k_{pack} = 0.888$; [f] from $V^{\infty e}_{ih}$ and equation 5.4; [g] from equation 5.8 and the data of Padova;[17] [h] from $V^{\infty g}_{ih}$ and equation 5.4; [i] Stokes radii given by R. Lundkvist, E. K. Hulet, and P. A. Baisden, *Acta Chem. Scand.*, **A35**, 653 (1981) yield V^{∞}_{ih} values some seven times larger and h^{∞}_{i} values 2 units larger than those of La^{3+}, Ce^{3+}.

the hydrated volumes of the electrolytes into their cationic and anionic contributions according to extrathermodynamic assumptions needs to be made. They have the disadvantage, however, that the "correction" to the Stokes radius and the choice of the packing coefficient, k, based on the larger tetra-alkylammonium ions are rather arbitrary when applied to the smaller ions. Furthermore, the hydrated volumes and radii and the hydration numbers obtained in this manner describe just one aspect of the hydration, i.e., the amount of water that accompanies the ion when it migrates under the influence of an electric field.

A different aspect of hydration, that leads to the hydrated volumes of electrolytes and to their hydration numbers, is based on considerations of compressibility. The compressibility of electrolyte solutions is lower than that of pure water. If it is accepted that the ions are incompressible as are also the electrostricted water molecules, i.e., those already compressed by the very large pressure excerted by the fields of the ions, then a fraction of the electrolyte solution, that depends on its concentration, is incompressible.[16-18] If this fraction is construed on a volume basis,[17] then the incompressible volume fraction is

$$(\kappa_{\mathrm{TH_2O}} - \kappa_{\mathrm{T}})/\kappa_{\mathrm{TH_2O}} = c_{\mathrm{s}} V_{\mathrm{sh}}/1000$$

where $\kappa_{\mathrm{T}} = -(1/V)(\partial V/\partial P)_T$ is the isothermal compressibility of the c_{s} molar (mol dm^{-3}) solution, $\kappa_{\mathrm{TH_2O}}$ that of water, and V_{sh} the hydrated volume of the salt. This relationship leads to the limiting hydrated molar volume of the solute:

$$V_{\mathrm{sh}}^{\infty}/\mathrm{cm^3\ mol^{-1}} = \lim_{c_{\mathrm{s}} \to 0} (1000/c_{\mathrm{s}})(1 - \kappa_{\mathrm{T}}/\kappa_{\mathrm{TH_2O}}) \tag{5.8}$$

If, however, the fraction is understood on a number basis[18] the incompressible mole fraction of the solution is

$$(\kappa_{\mathrm{TH_2O}} - \kappa_{\mathrm{T}})/\kappa_{\mathrm{TH_2O}} = h_{\mathrm{s}} m_{\mathrm{s}} M_{\mathrm{H_2O}}$$

where m_{s} is the molality of the solute in mol (kg water)$^{-1}$, h_{s} is the hydration number, and $M_{\mathrm{H_2O}}$ is the molar mass of water in kg mol^{-1}, which equals also the reciprocal of the number of moles of water per m_{s} moles of solute. This leads to

$$h_{\mathrm{s}}^{\infty} = \lim_{m_{\mathrm{s}} \to 0} (1/m_{\mathrm{s}} M_{\mathrm{H_2O}})(1 - \kappa_{\mathrm{T}}/\kappa_{\mathrm{TH_2O}}) \tag{5.9}$$

The combination of equations 5.8 and 5.9 leads to the relationship

$$V_{\mathrm{sh}}^{\infty} = h_{\mathrm{s}}^{\infty} V_{\mathrm{H_2O}} \tag{5.10}$$

but an apparent discrepancy exists between the V_{sh}^{∞} of equation 5.8, which includes the volume of the solute itself, and the V_{sh}^{∞} of equation 5.10, which is the volume of only the water of the hydration shells of the ions constituting the solute. No theoretical guidance to the preference of the one or the other expressions is available.

There exists no obviously valid way to divide these limiting hydrated molar volumes into their individual ionic contributions, although the additivity rule does operate. The results of Padova[17] (from equation 5.8), in which this division has been made in an unclear manner, are shown in Table 5.8, as are also values of h_i^∞ obtained from these V_{ih}^∞ and the $\bar{V}_i^\infty$ of this Table by means of equation 5.4. They are seen to be considerably smaller than those obtained from the Stokes radii. For other estimates of h_i^∞ see Section 4.4.

A further concept that is relevant to this discussion is that of the intrinsic volume of the ions. The intrinsic molar volume of an electrolyte is the molar volume it would have in the molten state if it could be supercooled to the temperature at which the electrolyte solution is studied, e.g., to 25 °C. A point to be remembered, however, is that thus defined the intrinsic molar volume of an electrolyte is not additive in terms of the volumes of its ions. Because in this state the ions strongly interact with each other, they mutually affect their volumes, and furthermore, packing effects come into play if the sizes of the ions of the opposite signs are not similar. For instance, the (supercooled) molar volumes of the larger tetraalkylammonium bromides, iodides and tetrafluoroborates can be extrapolated according to equation 5.2a to yield the molar volumes of the anions:[19] -10, -16, and 22 $cm^3\,mol^{-1}$, respectively. This shows that the anions occupy interstitial spaces between the large cations. When the volumes are estimated for salts with commensurate cations, the values become $\sim$32 and $\sim$44 $cm^3\,mol^{-1}$ for bromide and iodide, respectively. The intrinsic volumes of the tetraalkylammonium cations themselves, from the tetrapropyl to the tetraheptyl, are obtained from those of the salts by difference, their mean values from data for the three kinds of salts are[19] (accurate to $\pm 3\,cm^3\,mol^{-1}$):

$$V_{intr}(R_4N^+)/cm^3\,mol^{-1} = -33 + 1.42(M_{R_4N^+}/g\,mol^{-1}) \tag{5.11}$$

Intrinsic volumes of smaller ions can be obtained as follows. The limiting partial molar volume of an ion, $\bar{V}_i^\infty$, can be expressed as the difference between its intrinsic molar volume and the molar electrostriction, $V_{i\,els}$, that it causes in the surrounding water:

$$\bar{V}_i^\infty = V_{i\,intr} - V_{i\,els} \tag{5.12}$$

As has been shown,[20,21] the molar electrostriction $V_{i\,els}$, is proportional to the absolute value of the ionic charge, $|z_i|$, in a series of ions of similar sizes. The constant of proportionality is[21] 25.4 $cm^3\,mol^{-1}$. The intrinsic volume $V_{i\,intr}$, in $cm^3\,mol^{-1}$, calculated from this relationship and equation 5.12, is found for monatomic cations and anions to equal 10^3 times the cube of the sum of the crystal radius of the ion (r_{ic} in nm) and $0.174_6 \pm 0.010$ nm. For small ions, with $r_{ic} < r_{H_2O} \sim 0.14$ nm, the value of $V_{i\,intr}$ is approximately (within $\pm 30\%$) the same as the molar volume of water, 18.07 $cm^3\,mol^{-1}$ at 25 °C. It has been suggested by Akitt[21] that such an ion is capable of replacing just one molecule of water in the hydrogen-bonded structure of water. Larger monatomic ions and polyatomic ones replace more than one water molecule, oxyanions replacing as many as there are oxygen atoms in the anion. The question remains, whether the intrinsic volume defined operationally by equation 5.12 and the simple expression $V_{i\,els} = 25.4|z_i|$ has the same physical significance as that defined as the molar volume of the (supercooled) liquid electrolyte.

Other expressions have been suggested for the molar electrostriction an ion causes, so that other values of $V_{i\,intr}$ result from the application of equation 5.12. In particular, the electrostriction caused per mole of water in the hydration shells can be estimated, either from the dependencies of the apparent molar compressibilities and volumes on the concentration[17] or from the pressure effects of the field of the ion.[22] The molar volume of electrostricted water is then $V_{H_2O\,el} = 16.0\ cm^3\ mol^{-1}$ according to the first method or 13.8 to 16.7 $cm^3\ mol^{-1}$, for monovalent ions of crystal radii from 0.060 to 0.350 nm, according to the second. From equations 5.4 and 5.12 then results

$$V^{\infty}_{ih} = \bar{V}^{\infty}_{i} + h^{\infty}_{i} V_{H_2O} = V_{i\,intr} + h^{\infty}_{i} V_{H_2O\,els} \tag{5.13}$$

which permits the calculation of V^{∞}_{ih}, h^{∞}_{i}, or $V_{i\,intr}$, if the other two of these quantities are known.

The pressure derivative of the electrostatic contribution to the Gibbs free energy of solvation should yield the electrostriction, i.e. the volume change formally ascribed to the solvent due to the presence of the charge on the ion. A detailed theory[22] calculates the effective pressure that is equivalent to the effect of the field E of the ion on the solvent in its immediate surroundings:

$$dP = (\varepsilon_0/4\pi\kappa_{TS})(\partial\varepsilon_S/\partial P)_{E,T}\, dE \tag{5.14}$$

where κ_{TS} is the isothermal compressibility of the solvent. The theory considers the dependence of ε_S on both P and E, and arrives at an expression that permits the calculation of $\Delta V_{els} = h^{\infty}_{i}(V_{H_2O} - V_{H_2D\,els})$ as a function of E, provided that

Table 5.9 The electrostriction in water at 25 °C (according to ref. 22a)

$r/\lvert z\rvert^{1/2}$ nm	E $10^{10}\ V\ m^{-1}$	$\Delta V_{H_2O\,els}$ $cm^3\ mol^{-1}$
1.18	0.013	0.04
0.74	0.033	0.20
0.50	0.073	0.54
0.31	0.190	1.00
0.23	0.345	1.36
0.177	0.576	1.82
0.146	0.854	2.36
0.119	1.284	3.17
0.084	2.58	5.06
0.076	3.16	5.71

According to ref. 22b, higher values of the electrostriction of water are obtained from a modification of the theory. It yields the following pairs of values of ($E/10^{10}\ V\ m^{-1}$, $\Delta V_{H_2O\,els}/cm^3\ mol^{-1}$): (0.042, 0.30) (0.099, 0.90) (0.189; 1.55) (0.270, 2.27) (1.145, 4.89).

the pressure dependencies of the density, the dielectric constant and the refractive index of the solvent are known up to high pressures, and the value of b of equation (3.35) is known. The calculation has been made for water as the solvent,[22] yielding the results in Table 5.9. The average r for which the value of ΔV_{els} applies is obtained from the dependence of E on r, i.e., $E = z\mathbf{e}/\varepsilon_0 \varepsilon r^2$.

5.3 Thermodynamics of ion hydration

The standard molar Gibbs free energy, enthalpy and entropy changes are represented henceforth collectively with the symbol ΔY^0. For the following process

$$I^z(g) \rightarrow I^z(aq) \tag{5.15}$$

they are called the standard molar Gibbs free energy, enthalpy and entropy of hydration, $\Delta Y^0_{i\,hydr}$ of the ion I^z, the charge number of which is to be taken algebraically. The standard states are the hypothetical ideal gas at 1 atm (101325 Pa) pressure and the hypothetical ideal 1 m aqueous solution.

This process cannot be carried out experimentally, and even if it could, it would be encumbered with spurious space charges and with surface potentials that are of no direct interest for the present purpose. The process can, however be thought of as consisting of the difference between two processes of formation from the elements in their standard state: of the gaseous ion,

$$I(ss) \rightarrow I^z + z\,e^- \tag{5.16}$$

and of the aqueous ion,

$$I(ss) \rightarrow I^z(aq) + z\,e^- \tag{5.17}$$

The state of the electron produced in process 5.17 is questionable, but for the sake of consistency it must be the same state as of the one produced in process 5.16, i.e., the gaseous state.

Process 5.16 can be realized experimentally for monoatomic ions and for a very few polyatomic ones. The element (or substance) is first converted from its standard state at the temperature under discussion to the gaseous monoatomic state (or to the gaseous radical) by sublimation, vaporization and dissociation as required. The ΔY^0 for these processes are generally known for atoms producing monoatomic ions and for some radicals producing polyatomic ones.[23] The gaseous atom (or radical) is then converted to the ion by ionization for cations and by electron capture for anions. The respective ionization potentials or electron affinities are again known for monoatomic ions,[23] but very few data for polyatomic ones are available. These energy changes are converted to enthalpies by taking into account the standard molar enthalpy of the gaseous electrons (i.e., their kinetic energy, $(5/2)\mathbf{R}T = 6196.5\ \mathrm{J\ mol^{-1}}$ at 298.15 K), and to Gibbs free energies by taking into account their standard molar entropies ($S^0(e^-(g)) = 20.895\ \mathrm{J\ K^{-1}\ mol^{-1}}$). The quantities $\Delta Y^0_f(I^z(g))$ are thus known and tabulated for most monoatomic and some polyatomic ions.[23] The standard molar entropy of the gaseous ions, based on the third law of thermodynamics, are calculated for monoatomic ions having no electronic entropy from the Sackur–Tetrode equation

$$S^0(\mathrm{I}^z(\mathrm{g})) = (3/2)\mathbf{R}\ln(M/M^0) + (5/2)\mathbf{R}\ln(T/T^0) + \text{const.} \tag{5.18}$$

where M is the molar mass of the ion, M^0 is the unit molar mass, and T^0 is the unit temperature, the constant being -1.165 with the usual units. For polyatomic ions $S^0(\mathrm{I}^z(\mathrm{g}))$ is calculated similarly, with the internal degrees of freedom (rotation and vibration) also taken into account.[24]

Process 5.17, on the contrary, cannot be carried out experimentally. It is replaced by the process

$$\mathrm{I(ss)} + z\mathrm{H}^+(\mathrm{aq}) \rightarrow \mathrm{I}^z(\mathrm{aq}) + (z/2)\mathrm{H}_2(\mathrm{g}) \tag{5.19}$$

which can be carried out, either directly in electrochemical cells or calorimeters, or indirectly, via various secondary processes. By convention, process 5.17 applied to hydrogen ions, i.e., the formation reaction of aqueous hydrogen ions, $\frac{1}{2}\mathrm{H}_2(\mathrm{g}) \rightarrow \mathrm{H}^+(\mathrm{aq}) + \mathrm{e}^-(\mathrm{g})$, is assigned the value zero for both the Gibbs free energy and the enthalpy changes at all temperatures. Hence the measurable ΔY^0 for reaction 5.19 for other ions are also the conventional ΔY^0_{if} for the formation of the aqueous ions, process 5.17. The conventional Gibbs free energy of formation and enthalpy of formation of many ions have been tabulated.[23]

The standard partial molar entropy of aqueous hydrogen ions is conventionally assigned the value of zero at all temperatures: $\bar{S}^0_{\mathrm{conv}}(\mathrm{H}^+(\mathrm{aq})) = 0$. This convention, again, makes the measured standard molar entropy change of reaction 5.19 correspond to the conventional standard partial molar entropies $\bar{S}^0_{\mathrm{i\,conv}}$ of the ions, and these, again, are tabulated.[23]

The conventional standard molar Gibbs free energies, enthalpies, and entropies of hydration, $\Delta Y^0_{\mathrm{i\,hydr\,conv}}$ for process 5.15, obtained as the difference between $\Delta Y^0_{\mathrm{if\,conv}}(\mathrm{I}^z(\mathrm{aq}))$ (process 5.19) and $\Delta Y^0_{\mathrm{if}}(\mathrm{I}^z(\mathrm{g}))$ (process 5.16), as well as the conventional standard partial molar entropies of the ions, $\bar{S}^0_{\mathrm{i\,con}}$, are presented in Table 5.10.

A consistency test that can be applied is that the calculated $\Delta G^0_{\mathrm{i\,hydr\,conv}}$ and $\Delta H^0_{\mathrm{i\,hydr\,conv}}$ should differ by exactly $T\Delta S^0_{\mathrm{i\,hydr\,conv}}$, as for any other process. A further consistency test is the application of the additivity rule for ionic properties at infinite dilution. For two electrolytes, designated as 1 and 2, with, e.g., a common anion and cations having the same charge,

$$\Delta Y^0_{+(1)} - \Delta Y^0_{+(2)} = \Delta Y^0_{\pm(1)} - \Delta Y^0_{\pm(2)} \tag{5.20}$$

where subscript + pertains to the cations and $\pm$ to the electrolytes. Also, the ΔY^0 of solution of an electrolyte in water should equal the difference between its ΔY^0_{f} (or its S^0) and the corresponding sum of the $\Delta Y^0_{\mathrm{if\,con}}$ (or ($\bar{S}^0_{\mathrm{i\,conv}}$) of its constituent aqueous ions.

As for the limiting partial molar volumes, dealt with on p. 98, attempts have been made to obtain absolute values of the standard partial molar entropies, $\bar{S}^0_{\mathrm{i}}$, of the ions. These, when the respective entropies of the gaseous ions are

Table 5.10 Conventional[a] standard molar Gibbs free energies and enthalpies of hydration and partial entropies of aqueous ions and the derived absolute values,[b] all at 298.15 K

Ion	Conventional			Absolute		
	$\Delta G^0_{i\,hydr}$ (kJ mol^{-1})	$\Delta H^0_{i\,hydr}$ (kJ mol^{-1})	$\bar{S}^0_{i\,aq}$ (JK^{-1} mol^{-1})	$\Delta G^0_{i\,hydr}$ (kJ mol^{-1})	$\Delta H^0_{i\,hydr}$ (kJ mol^{-1})	$\bar{S}^0_{i\,aq}$ (JK^{-1} mol^{-1})
H^+	0	0	0	−1056	−1094	−22.2
Li^+	575.2	571.9	13.4	−481	−522	−8.8
Na^+	680.7	687.1[p]	59.0	−375	−407	36.8
K^+	752.5	769.5	102.5	−304	−324	80.3
Rb^+	775.1	794.9	121.5	−281	−299	99.3
Cs^+	798.4	820.0	133.1	−258	−274	110.9
NH_4^+	764[d,q]	774[d]	113.4	−292	−320	91.2
Me_4N^+		639[t]	210		−455	188
Et_4N^+		710[t]	282[s]		−384	261
Pr_4N^+			336[s]			314
Cu^+	521.0	517.9	40.6	−535	−576	18.4
Ag^+	616.4	620.1	72.7	−440	−474	50.5
Au^+		520[k]			−574	
In^+	760.0	680[k]		−296	−414	
Tl^+	746.1	763.8	125.5	−310	−330	103.3
VO_2^+			−42.3			−64.5
Be^{2+}	−292.2	−303.6	−129.7	−2404	−2492	−174.1
Mg^{2+}	274.1	257.0	−138.1	−1838	−1931	−182.5
Ca^{2+}	597.2	603.7	−53.1	−1515	−1584	−97.5
Sr^{2+}	726.3	736.1	−32.6	−1386	−1452	−77.0
Ba^{2+}	853.7	874.4	9.6	−1258	−1314	−34.8
Ra^{2+}	853.0	885.0	54	−1259	−1303	10
V^{2+}		520[k]			−1668	
Cr^{2+}	252[i]	273.2	18.0[i]	−1860	−1915	−26.4
Mn^{2+}	341.8	331.9	−73.6	−1770	−1856	−118.0
Fe^{2+}	263.7	233.6	−137.7	−1848	−1944	−182.1
Co^{2+}	189.9	170.0	−113.0	−1922	−2018	−157.4
Ni^{2+}	114.0	87.0	−128.9	−1998	−2101	−173.3
Cu^{2+}	95.7	83.1	−99.6	−2016	−2105	−144.0
Zn^{2+}	148.6	135.7	−112.1	−1963	−2052	−156.5
Cd^{2+}	376.3	373.0	−73.2	−1736	−1815	−117.6
Hg^{2+}	346.3	353.0	−32.2	−1766	−1835	−76.6
Hg_2^{2+}			84.5			40.1
Sn^{2+}	615.7	628.8	−17	−1496	−1559	−61
Pb^{2+}	678.1	697.4	10.5	−1434	−1491	−33.9
Pd^{2+}	191.7	152.0	−184	−1920	−2036	−228
Pt^{2+}	146.2	140[k]		−1966	−2048	
Sm^{2+}		742			−1446	
Eu^{2+}	720.9	725	−8	−1391	−1463	−52
Yb^{2+}		604			−1584	
VO^{2+}			−133.9			−178.3
UO_2^{2+}	783[n]	821[n]	−97.5	−1329	−1367	−141.9
Al^{3+}	−1363	−1406	−321.7	−4531	−4688	−388.3
Sc^{3+}	−632.8	−657.9	−255	−3801	−3940	−322
Y^{3+}	−289.1	−285.1	−251	−3457	−3567	−318
La^{3+}	13.1	−3.4	−217.6	−3155	−3285	−284.2
Ce^{3+}	−40.9	−58.2	−205	−3209	−3340	−272
Pr^{3+}	−83.1	−101.8	−209	−3251	−3384	−276
Nd^{3+}	−119	−138	−206.7	−3287	−3420	−273.3
Pm^{3+}	−89[g]			−3257	−3445	
Sm^{3+}	−164	−183	−211.7	−3332	−3465	−278.3

(continued)

Table 5.10—*continued*

Ion	Conventional			Absolute		
	$\Delta G^0_{\mathrm{i\,hydr}}$ (kJ mol^{-1})	$\Delta H^0_{\mathrm{i\,hydr}}$ (kJ mol^{-1})	$\bar{S}^0_{\mathrm{iaq}}$ (JK^{-1} mol^{-1})	$\Delta G^0_{\mathrm{i\,hydr}}$ (kJ mol^{-1})	$\Delta H^0_{\mathrm{i\,hydr}}$ (kJ mol^{-1})	$\bar{S}^0_{\mathrm{iaq}}$ (JK^{-1} mol^{-1})
Eu^{3+}	−202	−226	−222	−3370	−3508	−289
Gd^{3+}	−217	−240	−205.9	−3385	−3522	−272.5
Tb^{3+}	−241	−271	−226	−3409	−3553	−293
Dy^{3+}	−265	−295	−231.0	−3433	−3577	−297.6
Ho^{3+}	−313	−330	−226.3	−3480	−3621	−293.4
Er^{3+}	−333	−365	−244.3	−3501	−3647	−310.9
Tm^{3+}	−357	−386	−243	−3525	−3668	−310
Yb^{3+}	−411.5	−433.2	−238	−3579	−3715	−305
Lu^{3+}	−354	−386	−264	−3522	−3668	−331
Ac^{3+}		−90[k]			−3372	
U^{3+}	−47[g]	−54[g]	−192	−3215	−3336	−259
Np^{3+}		−250[k]			−3532	
Pu^{3+}	−77[g]	−128[g]		−3245	−3410	
Am^{3+}		−310[k]			−3592	
Ti^{3+}		−1030[k]			−4312	
V^{3+}		−1140[k]			−4422	
Cr^{3+}		−1360[k]			−4642	
Mn^{3+}		−1330[k]			−4612	
Fe^{3+}	−1103.1	−1152.7	−315.9	−4271	−4435	−382.5
Co^{3+}	−1335.1[j]	−1382[j]	−305	−4503	−4664	−372
Ga^{3+}	−1353.5	−1400.2	−331	−4521	−4682	−398
In^{3+}	−821.1	−818	−151	−3989	−4100	−218
Tl^{3+}	−807.5	−816.0	−192	−3976	−4098	−259
Sb^{3+}		−530[k]			−3812	
Bi^{3+}	−318	−314[h]	−118.8	−3486	−3596	−185.4
$Co(NH_3)_6{}^{3+}$			146			79
Ce^{4+}	−1905	−1915	−301	−6129	−6292	−390
Zr^{4+}			−509[u]	−6799[u]	−6991[u]	−598
Hf^{4+}			−466[u]	−6975[u]	−7159[u]	−555
Th^{4+}	−1599	−1645	−422.6	−5823	−6021	−511.4
Pa^{4+}		−2240[k]			−6616	
U^{4+}		−2270[k]	−410		−6646	−499
F^-	−1527.7	−1513.4	−13.8	−472	−519	8.4
Cl^-	−1403.5	−1470.3	56.5	−347	−376	78.7
Br^-	−1376.8	−1438.7	82.4	−321	−345	104.6
I^-	−1339.4	−1394.4	111.3	−283	−300	133.5
OH^-	−1545.0[e]	−1622.7	−10.8	−439	−529	11.4
SH^-	−1359[c,r]	−1427[c]	62.8	−303	−333	85.0
SeH^-		−1463[m]	79		−369	101
CN^-	−1361[f,l]	−1420[l]	94.1	−305	−326	116.3
$N_3{}^-$	−1343[e,q]	−1442[l]	107.9	−287	−348	130.1
$I_3{}^-$		−1259[d]	239.3		−165	261.5
CNO^-		−1498[f]	106.7		−404	128.9
SCN^-	−1343[k,q]	−1401[k]	144.3	−287	−307	166.5
$BF_4{}^-$	−1256[d,q]	−1314[d]	180.0	−200	−220	202.2
$ClO_3{}^-$	−1343[k,q]	−1406[k]	162.3	−287	−312	184.5
$BrO_3{}^-$	−1396[e,q]	−1470[e]	161.7	−340	−376	183.9
$IO_3{}^-$		−1569[e]	118.4		−475	140.6
$ClO_4{}^-$	−1270[d,f]	−1326[d]	182.0	−214	−232	204.2
$NO_2{}^-$	−1395[e,q]	−1460[e]	123.0	−339	−366	145.2
$NO_3{}^-$	−1362[e,f]	−1423[e]	146.4	−306	−329	168.6

(*continued*)

Table 5.10—*continued*

Ion	Conventional			Absolute		
	$\Delta G^0_{\text{i hydr}}$ (kJ mol^{-1})	$\Delta H^0_{\text{i hydr}}$ (kJ mol^{-1})	$\bar{S}^0_{\text{iaq}}$ (JK^{-1} mol^{-1})	$\Delta G^0_{\text{i hydr}}$ (kJ mol^{-1})	$\Delta H^0_{\text{i hydr}}$ (kJ mol^{-1})	$\bar{S}^0_{\text{iaq}}$ (JK^{-1} mol^{-1})
MnO_4^-	-1301^f	-1356^f	191.2	-245	-262	213.4
ReO_4^-			201.3			223.5
$HCOO^-$		-1517^e	92		-423	114.2
CH_3COO^-		-1517^e	86.6		-423	108.8
HCO_3^-		-1487^e	91.2		-393	113.4
HSO_4^-		-1383^e	131.8		-289	154.0
$H_2PO_4^-$			90.4			112.6
S^{2-}	-3436^c	-3554^c	-14.6	-1324	-1366	29.8
Se^{2-}			0^f			44.4
CO_3^{2-}	-2609^f	-2749^f	-56.9	-479	-561	-12.5
$C_2O_4^{2-}$			45.6			90.0
SO_3^{2-}			-29			15.4
SO_4^{2-}	-3202^f	-3326^f	20.1	-1090	-1138	64.5
SeO_4^{2-}			54.0			98.4
CrO_4^{2-}			50.2			94.6
MoO_4^{2-}			27.2			71.6
HPO_4^{2-}			-33.5			10.9
PO_4^{3-}			-222			-155
AsO_4^{3-}			-162.8			-96.2
$Fe(CN)_6^{3-}$			270.3			336.9
$Fe(CN)_6^{4-}$			95.0			183.8

[a] Unless otherwise noted, the data are from the NBS compilation, ref. 23; [b] $\Delta G^0_{\text{i hydr}} = \Delta G^0_{\text{i hydr conv}} - 1056 \cdot z$ kJ mol^{-1}, $\Delta H^0_{\text{i hydr}} = \Delta H^0_{\text{i hydr conv}} - 1094 \cdot z$ kJ mol^{-1}, $\Delta S^0_{\text{i hydr}} = (\Delta H^0_{\text{i hydr}} - \Delta G^0_{\text{i hydr}})/298.15$ J K^{-1} mol^{-1} (not tabulated), $\bar{S}^0_{\text{iaq}} = \bar{S}^0_{\text{iaq conv}} - 22.2 \cdot z$ J K^{-1} mol^{-1}; [c] from D. R. Rosseinsky, *Chem. Rev.*, **65**, 467 (1965); [d] from D. F. Halliwell and S. C. Nyburg, *Trans. Faraday Soc.*, **59**, 1126 (1963); [e] from Y. Marcus, *Introduction to Liquid State Chemistry*, Wiley, Chichester (1977), p. 250; [f] from V. P. Vasil'ev, E. K. Zolotarev, A. F. Kapustinskii, K. P. Mishchenko, E. A. Podgornaya, and K. B. Yatsimirskii, *Zh. Fiz. Khim.*, **34**, 1763 (1960), *Russ. J. Phys. Chem.*, **34**, 840 (1960); [g] from S. Goldman and L. R. Morss, *Canad. J. Chem.*, **53**, 2695 (1975); [h] from A. F. M. Barton and G. A. Wright, *Proc. 1st Austr. Conf. Electrochem., Sidney 1963*, 124 (1964); [i] from W. M. Latimer, *Oxidation States of the Elements and their Oxidation Potentials*, Prentice-Hall, Englewood Cliffs, 2nd edn. (1952); [j] S. I. Drakin and V. A. Mikhailov, *Zh. Fiz. Khim.*, **36**, 1698 (1962), L. Brewer, L. A. Bromley, P. W. Gilles and N. L. Loefgren, in L. L. Brewer, ed., *Chemistry and Metallurgy of Miscellaneous Materials, Thermodynamics*, McGraw-Hill, New York (1950), pp. 165–8; [k] from Y. Marcus, *Israel J. Chem.*, **10**, 659 (1972) and references therein; [l] from H. D. B. Jenkins, K. F. Pratt and T. C. Waddington, *J. Inorg. Nucl. Chem.*, **39**, 213 (1977), H. D. B. Jenkins and K. F. Pratt, *J. Chem. Phys. Solids*, **38**, 573 (1977); [m] Y. Marcus, unpublished calculations (1971); [n] from Y. Marcus, *J. Inorg. Nucl. Chem.*, **37**, 493 (1975); [p] possibly erroneous, see p. 111; [q] A. P. Altschuler, *J. Chem. Phys.*, **24**, 642 (1956), **28**, 1254 (1958); [r] G. A. Krestov, *Zh. Fiz. Khim.*, **42**, 866 (1968); [s] D. A. Johnson and J. F. Martin, *J. Chem. Soc. Dalton Trans.*, **1973**, 1585; [t] S. I. Nwankwo, *Thermochim. Acta*, **47**, 157 (1981); [u] V. P. Vasil'ev and A. I. Lytkin, *Zh. Neorg. Khim.*, **21**, 2610, 3037 (1976).

subtracted, yield the absolute entropies of hydration of the ions. The attempts to obtain the absolute standard partial molar entropies of ions have been based mainly, but not exclusively, on the evaluation of experimental results from thermocells, as summarized by Rosseinsky[25] and Conway.[26] The EMF of a cell consisting of the two similar electrodes $Hg/Hg_2Cl_2/KCl(aq)$, m (with m equal in both half cells) at different temperatures is determined by the entropy change of the reaction $\frac{1}{2}Hg_2Cl_2(s) + e^-(Hg) \rightarrow Hg(l) + Cl^-(aq)$, corrected for the entropy of transfer of hydrated chloride ions across the junction between the two solutions at the different temperatures.[27] From such measurements and estimates of the entropy of transfer, the value of $\bar{S}^0(Cl^-(aq))$ is obtained, hence that of $\bar{S}^0(H^+(aq)) = -22.2 \pm 1.4\,J\,K^{-1}\,mol^{-1}$ at 25 °C. From this other standard partial molar entropies of aqueous ions are calculated by the additivity rule from data on electrolytes. The standard molar entropy of gaseous H^+ is obtained accurately from the Sackur–Tetrode equation as 108.84 $J\,K^{-1}\,mol^{-1}$, hence its standard molar entropy of hydration is $-131 \pm 1\,J\,K^{-1}\,mol^{-1}$. The absolute standard partial molar entropies of aqueous ions are presented in Table 5.10 and the absolute standard molar entropies of hydration in Table 5.13, p. 125.

Attempts have also been made to estimate the absolute Gibbs free energy and enthalpy of hydration from the conventional ones, to which they are related by

$$\Delta y^0_{i\,hydr}(I^z) = \Delta Y^0_{i\,hydr\,conv}(I^z) + z \cdot \Delta Y^0_{i\,hydr}(H^+) \quad (5.21)$$

The absolute value of $\Delta Y^0_{i\,hyd}$ for the hydrogen ion therefore is the key quantity that must be determined.

Of the methods that have been used for the estimation of the absolute enthalpies of hydration of ions, that of Halliwell and Nyburg[28] seems to be the most promising. The method involves differences between the conventional enthalpies of hydration (see Table 5.10) of ions of the same sizes, applied at first to monovalent ions and later also to large, complex divalent ones, with consistent results. For monovalent ions:

$$\Delta H^0_{+\,hydr\,conv}(r) - \Delta H^0_{-\,hydr\,conv}(r) = \Delta H^0_{+\,hydr}(r) - \Delta H^0_{-\,hydr}(r) - 2\,\Delta H^0_{i\,hydr}(H^+) \quad (5.22)$$

For cations and anions of the same r, the difference in the first two terms on the right-hand side of equation 5.22 will reflect only the relatively small difference in energy inherent in the different orientation of the water molecules around cations and anions, due to their electrical quadrupole moments[29] (Section 3.4, equation 3.38). The main components of the hydration enthalpies will have been cancelled out. If the difference in equation 5.22 is plotted against the reciprocal of the cube of the effective radius r, i.e., the crystal ionic radius plus 0.138 nm, which is approximately the radius of a molecule of water, extrapolation to zero yields the value of $-2\,\Delta H^0_{i\,hydr}(H^+)$ as the intercept (corresponding to infinitely large ions). For the relatively small monovalent ions, the long extrapolation needed contributed most to the uncertainty of the resulting value, $\Delta H^0_{i\,hydr}(H^+) = -1090.8 \pm 10.5\,kJ\,mol^{-1}$. For the divalent ions, which were larger, the main uncertainties arose from the lattice energies (although these partly cancelled in the difference), but the value obtained, $\Delta H^0_{i\,hydr}(H^+) = -1094.5 \pm 20\,kJ\,mol^{-1}$ agrees well with the former value. The extrapolation is somewhat sensitive to the ionic radii chosen, but not as much in the present case, where extrapolation is made against $(r_c/nm + 0.138)^{-3}$, as in several other methods where r_c^{-1} or r_c^{-2} have been used.[26]

Somewhat different values of this quantity have been recommended in recent publications. The values $\Delta H^0_{\mathrm{i\,hydr}}(\mathrm{H^+})/\mathrm{kJ\ mol^{-1}} = -1100 \pm 6$ (in the text)[37a] and -1094 ± 6 (derivable from the Table)[37b] and -1104 ± 17[26] have been obtained by substantially Halliwell and Nyburg's procedure[28] and older $\Delta H^0_{\mathrm{i\,hydr\,conv}}$ data for the alkali metal cations but various values for the ionic radii. However, when the $\Delta H^0_{\mathrm{i\,hydr\,conv}}$ data that are derivable from the more recent compilation[23] are used instead, the considerably less negative value -1058 ± 5 is obtained, the data for Na^+ being clearly outside of the correlation, hence not included. A considerably more negative value, -1136 ± 3, has been derived[30] from the differences between the enthalpies of gaseous hydrate cluster formation of cations and anions (see Section 2.3). The value $\Delta H^0_{\mathrm{i\,hydr}}(\mathrm{H^+}) = -1094\ \mathrm{kJ\ mol^{-1}}$ is used here to convert the $\Delta H^0_{\mathrm{i\,hydr\,conv}}$ values recorded in Table 5.10 to the 'absolute' values, also recorded there.

The standard molar Gibbs free energies of hydration of individual ions have been estimated from direct experimental data on the Volta potential between a flowing aqueous salt solution and a jet of mercury, obtained by Randles:[31] Hg jet/N_2/KCl(aq), m/Hg_2Cl_2/Hg and the EMF of the cell Hg/Hg_2Cl_2/KCl(aq), m/K(Hg). The value $\Delta G^0_{\mathrm{i\,hyd}}(\mathrm{K^+}) = -336.9 \pm 2.0\ \mathrm{kJ\ mol^{-1}}$ has been derived from these data, and yields with the conventional value (Table 5.10) and equation 5.21 the value $\Delta G^0_{\mathrm{i\,hyd}}(\mathrm{H^+}) = -1089.7\ \mathrm{kJ\ mol^{-1}}$. A value of $-1098 \pm 1\ \mathrm{kJ\ mol^{-1}}$ has been obtained from the differences in the Gibbs free energies of gaseous hydrate cluster formation of cations and anions.[30] These values, however, are inconsistent with the $\Delta H^0_{\mathrm{i\,hydr}}(\mathrm{H^+})$ value chosen, $-1094\ \mathrm{kJ\ mol^{-1}}$, and the $\Delta S^0_{\mathrm{i\,hydr}}(\mathrm{H^+})$ value chosen, $-131\ \mathrm{J\ K^{-1}\ mol^{-1}}$, which in turn yield the value $\Delta G^0_{\mathrm{i\,hydr}}(\mathrm{H^+}) = -1056\ \mathrm{kJ\ mol^{-1}}$. This latter value has been used to convert the $\Delta G^0_{\mathrm{i\,hydr\,conv}}$ data recorded in Table 5.10 to the 'absolute' values, also recorded there.

The situation concerning the derivative thermodynamic functions, such as the limiting partial molar heat capacities, compressibilities and thermal expansibilities of individual ions, is much less satisfactory than that concerning the volumes (p. 98) and the entropies (p. 110). Some of the methods used for the latter quantities have also been used for the former, but the uncertainty is larger, because of the more obscure relationship of these second derivative functions (i.e., of the chemical potential with respect to pressure and temperature) to the sizes of the ions.

The R_4NX-extrapolation method, as used, e.g. in equation 5.2 for the volumes, has also been employed for the limiting partial molar heat capacities of the ions in water,[32] with the result that $\bar{C}^\infty_p(\mathrm{Br^-}) = -268 \pm 60\ \mathrm{J\ K^{-1}}$ at 25 °C. However, the relationship between the limiting partial molar heat capacities of the larger tetraalkylammonium ions and their sizes (i.e., molar masses) is not as straightforward as for the corresponding volumes. The latter are expected to vary linearly with the number of the $-CH_2-$ groups that they contain, which contribute directly to the intrinsic volume. The contribution of a $-CH_2-$ group to the heat capacity is more subtle. The conformational mobility of a $-CH_2-$ group does contribute to the intrinsic heat capacity of the tetraalkylammonium cation, but a large contribution to the partial heat capacity comes also from structural changes induced in the surrounding solvent. These changes have also an electrostatic contribution, obtainable from the second derivative of the electrostatic contribution to the Gibbs free energy of hydration (see Section 3.4), with respect to the temperature. This has been estimated[33] as $-12\ \mathrm{J\,K^{-1}\,mol^{-1}}$ for the cation $(HOC_2H_4)_4P^+$, which otherwise would have the same $\bar{C}_p^\infty$ as the neutral analogue $(HOC_2H_4)_4C$, the hydrogen bonding of the four terminal -OH groups and the hydrophobic interactions of the $-CH_2CH_2-$ links, in addition to their intrinsic contributions to the heat capacity, being the more important ones.[34] The value of $\bar{C}_p^\infty(Cl^-) = -142 \pm 4\ \mathrm{J\,K^{-1}\,mol^{-1}}$, that has been estimated in this manner,[26] is more sharply defined than the value obtained from the R_4NX-extrapolation dealt with above, but consistent with it within the wide limits of error of the latter. The individual limiting partial molar heat capacities of aqueous ions obtained on this basis from data on electrolytes[32] are presented in Table 5.11.

Well defined values of the individual ionic limiting partial molar heat capacities of the aqueous ions would be of great value, since this property is very sensitive to changes in the interactions of the ions with the surrounding water molecules and to structural changes in this solvent. So far, values as those recorded in Table 5.11 are uncertain to a degree that precludes their use for the estimation of such structural effects. It is, perhaps, of significance that according to the scale used, all the $\bar{C}_p^\infty$ values of the cations are positive and all those for the anions are negative, irrespective of whether the ions enhance the structure of the water or destroy it.

In contrast with the Gibbs free energy, enthalpy and entropy of hydration, which are commonly used concepts, the heat capacity of hydration is rarely used. It can be readily obtained from the limiting partial molar heat capacity, at least for a monoatomic ion, by the subtraction of the standard molar heat capacity of the gaseous ion, $C_p^0(I^z(g)) = (5/2)\mathbf{R}$.

The adiabatic compressibilities of electrolyte solutions are readily obtained from the velocity u of sound in them: $\kappa_S = u^{-2}\rho^{-1}$. The apparent molal compressibility is obtained from this in the usual manner, and from this the limiting partial molal compressibility:

$$\bar{K}_S^\infty = {}^\phi K_S^\infty = \lim_{m\to 0} [(1000/md_{H_2O})(\kappa_S - \kappa_{SH_2O}) + \kappa_S{}^\phi V] \tag{5.23}$$

The latter quantity can be divided into the individual ionic contributions by taking the pressure derivatives of the corresponding partial molar volumes. On this basis[36] the limiting partial molal adiabatic compressibility of the chloride ion is $-(17.0 \pm 1.0)\cdot 10^{-9}\ \mathrm{cm^3\,mol^{-1}\,Pa^{-1}}$, and the values for the other ions are obtained on the additivity principle, see Table 5.11. The corresponding isothermal compressibilities are less readily obtained, since they are related to the adiabatic ones by[36]

$$\begin{aligned} 10^9\bar{K}_T^\infty = {} & 10^9\bar{K}_S^\infty + 36.79({}^\phi E^\infty/\mathrm{K^{-1}})(\bar{V}^\infty/\mathrm{cm^3\,mol^{-1}}) \\ & - 1.135\cdot 10^{-3}(\bar{C}_p^\infty/\mathrm{JK^{-1}\,mol^{-1}}) \end{aligned} \tag{5.24}$$

but the limiting apparent molar expansibilities and heat capacities are not so well known. The data for complete electrolytes are better known than the individual ionic values, and the former can be split into the ionic contributions by noting that $\bar{K}_{\mathrm{Ti}}^{\infty} - \bar{K}_{\mathrm{Si}}^{\infty}$ is linear with the size of the ions (it is negative for small ions and positive for large ones). Thus $\bar{K}_{\mathrm{T}}^{\infty}(\mathrm{Cl}^-) = -(16.0 \pm 1.5) \cdot 10^{-9}\ \mathrm{cm}^3\ \mathrm{mol}^{-1}\ \mathrm{Pa}^{-1}$ gives a consistent set of values, on employing the additivity principle. This value for the chloride ion is somewhat less negative than that derived from considerations of dielectric saturation.[37] The values of $\bar{K}_{\mathrm{Ti}}^{\infty}$ obtained on the former basis are shown in Table 5.11.

Limiting partial molar expansibilities of individual ions are obtained when the methods applicable to the division of the limiting partial molal volumes into their ionic contributions are applied at several temperatures, and the temperature dependence of the resulting absolute $\bar{V}_{\mathrm{i}}^{\infty}$ values are calculated. This way produces data of larger experimental uncertainties, but of greater general reliability, than attempts to relate conventional limiting partial molal expansibilities to the sizes of the ions or similar methods. The method consists, thus, of employing equation 5.2a in the modified form:

$$\bar{V}^{\infty}(\mathrm{X}^-, t/^{\circ}\mathrm{C})/\mathrm{cm}^3\ \mathrm{mol}^{-1} = \lim_{M_{\mathrm{R_4N}^+} \to 0} [\bar{V}^{\infty}(\mathrm{R_4NX}, t)/\mathrm{cm}^3\ \mathrm{mol}^{-1} - (1.0837 + 0.00115(t/^{\circ}\mathrm{C})) \cdot M_{\mathrm{R_4N}^+}/\mathrm{g} \cdot \mathrm{mol}^{-1}] \quad (5.25)$$

where the numerical values inside the square brackets correspond to the values of the parameter b in equation 5.2a at temperatures between 5 and 45 °C.[13b] These limiting absolute partial molar volumes of the anions X^- at the different temperatures (e.g., the values of $\bar{V}^{\infty}(\mathrm{Br}^-)/\mathrm{cm}^3\ \mathrm{mol}^{-1} = 30.8 \pm 0.1$ (5 to 35 °C) and 30.0 (45 °C)) can then be used with the $\bar{V}^{\infty}$ data of electrolytes at these temperatures to give the $\bar{V}_{\mathrm{i}}^{\infty}(t)$ data for the individual ions, from which the limiting partial molal expansibilities can be calculated. The values obtained in this manner are shown in Table 5.11.

5.4 Structural aspects of hydration

Our knowledge of the structure of aqueous electrolyte solutions has advanced considerably in recent years. On the one hand, computer simulation studies involving ions imbedded in *ca.* 50 to 200 times as many water molecules have produced realistic information concerning the average distances between the ions and the nearest and next-nearest water molecules, as well as concerning the numbers of these neighbours. On the other hand diffraction studies have produced similar evidence concerning actual solutions, consisting of *ca.* 5 to 100 moles of water per mole of salt. These methods are presented in some detail in Section 4.2. There is no direct evidence concerning what happens to the structure of the water beyond the first two hydration shells of an ion, but there are certain inferences from the properties of dilute electrolyte solutions. These are discussed in this Section in terms of the structure-màking and -breaking effects of the ions.

Table 5.11 Conventional limiting partial molar heat capacities, expansivities, and compressibilities of aqueous ions at 298.15 K

Ion	$\bar{C}^{\infty}_{pi\ conv}$ (J K^{-1} mol^{-1})	$\bar{E}^{\infty}_{i\ conv}$ [k] (10^2 cm^3 K^{-1} mol^{-1})	$\bar{\kappa}^{\infty}_{i\ conv} \cdot \bar{V}^{\infty}_{i\ conv}$ [m] (10^9 cm^3 Pa^{-1} mol^{-1})
H^+	0[a]	6.4	0
Li^+	69[a], 62[c]	5.0	−34
Na^+	46[a], 43[c]	12.8	−42
K^+	22[a], 13[c]	11.2	−37
Rb^+	17[b], −8.5[c]	9.2	
Cs^+	0[b], −23[c]	11.4	−27
NH_4^+	80[a], 70[c]	9.8	
$(CH_3)_4N^+$	238[h], 237[c]	11.3[e]	−27[f]
$(C_2H_5)_4N^+$	513[c]	13.4[e]	−60[f]
$(C_3H_7)_4N^+$	922[c]	17.5[e]	−129[f]
$(C_4H_9)_4N^+$	1351,[g] 1339[c]	25.7[e]	−358[f]
$(C_6H_5)_4P^+$	1222,[g] 1212[c]		
$(C_6H_5)_4As^+$	1243,[g] 1239[c]	44.1[e]	
Ag^+	22[a]		
Tl^+	8[b]		
Be^{2+}	100[b]		−23
Mg^{2+}	42,[b] −15[c]	14.6	−83, −80[l]
Ca^{2+}	8,[b] −28[c]	16.0	−71, −77[l]
Sr^{2+}	0,[b] −35[c]	18.8	−93[l]
Ba^{2+}	−33,[b] −46[c]	21.0	−99, −93[l]
Ra^{2+}	−50[b]		
Mn^{2+}	50[a]		−78[l]
Co^{2+}			−87[l]
Ni^{2+}			−91[l]
Cu^{2+}	29[b]		−62, −90[l]
Zn^{2+}	46[a]		−70, −91[l]
Cd^{2+}	8[b]		−57, −65[l]
Pb^{2+}	33[b]		
Al^{3+}	46[b]		
Sc^{3+}	4[b]		
La^{3+}	−13[a]		−139[j]
Ce^{3+}			−152
Pr^{3+}	−29[a]		
Nd^{3+}	−21[a]		
Sm^{3+}	−21[a]		
Eu^{3+}	8[a]		
Gd^{3+}	0[a]		
Tb^{3+}	17[a]		
Dy^{3+}	21[a]		
Ho^{3+}	17[a]		
Er^{3+}	21[a]		
Tm^{3+}	25[a]		
Yb^{3+}	25[a]		
Lu^{3+}	25[a]		
Cr^{3+}	42[b]		
Fe^{3+}	33[b]		

(continued)

Table 5.11—*continued*

Ion	$\bar{C}^{\infty}_{pi\,conv}$ (J K^{-1} mol^{-1})	$\bar{E}^{\infty}_{i\,conv}$ [k] (10^2 cm^3 K^{-1} mol^{-1})	$\bar{\kappa}^{\infty}_{i\,conv} \cdot \bar{V}^{\infty}_{i\,conv}$ [m] (10^9 cm^3 Pa^{-1} mol^{-1})
Zr^{4+}	33[b]		
U^{4+}	13[b]		
UO_2^{2+}	17[b]		
F^-	−107,[a] −117[c]	−4.8	−33[j]
Cl^-	−136,[a] −127[c]	−3.4	−8
Br^-	−142,[a] −131[c]	−1.6	2
I^-	−142,[a] −122[c]	1.4	18
OH^-	−149,[a] −138[c]	−1.4	
SH^-	−142[b]		
SCN^-	−40[a]		
ClO_3^-	−75[i]		
BrO_3^-	−100[i]		
IO_3^-	−121[i]		
ClO_4^-	−59[b]	4.4	
MnO_4^-	−71[i]		
ReO_4^-	−17[b]		
NO_2^-	−97[a]		
NO_3^-	−87[a]	1.0	
HCO_2^-	−88[a]		
$CH_3CO_2^-$	−6,[a] 25[c]		
HCO_3^-	−86,[a] −50[c]		
HSO_3^-	−65[a]		
HSO_4^-	−31[a]		
$H_2PO_4^-$	−115[a]		
$B(C_6H_5)_4^-$	1079,[g] 1070[c]	20.0	
CO_3^{2-}	−414,[b] −266[c]		
SO_3^{2-}	−364[b]		
SO_4^{2-}	293,[a] −279[c]	−3.0	
HPO_4^{2-}	−322[a]		
PO_4^{3-}	−619[b]		

[a] From NBS compilation, ref. 23; [b] from C. M. Criss and J. W. Cobble, *J. Am. Chem. Soc.*, **86**, 5390 (1964), adjusted by addition of $2z$ J K^{-1} mol^{-1}, to conform with the NBS values;[a] [c] J. E. Desnoyers, C. DeVisser, G. Perron and P. Acker, *J. Soln. Chem.*, **5**, 605 (1976); [d] R. M. Noyes, *J. Am. Chem. Soc.*, **86**, 971 (1964); [e] F. J. Millero and W. Drost-Hansen, *J. Phys. Chem.*, **72**, 1758 (1968); [f] H. L. Friedman and C. V. Krishnan, in *Water: A Comprehensive Treatise*, F. Franks, ed., Plenum, New York, Vol. 3, 74 (1973); [g] C. Jolicoeur, P. R. Philip, G. Perron, P. A. Leduc and J. E. Desnoyers, *Can. J. Chem.*, **50**, 3167 (1972); [h] M. Mastroiani and C. M. Criss, *J. Chem. Eng. Data*, **17**, 222 (1972); [i] V. B. Parker, *Thermal Properties of Aqueous Uni-Univalent Electrolytes*, NSRDS-NBS 2 (1965); [j] J. V. Leyendekkers, *J. Phys. Chem.*, **85**, 305 (1981); [k] from the dependence of $\bar{V}^{\infty}_{i\,con}$ on T in the range $273 = T/K = 323$, estimated accuracy $\pm 1.5 \cdot 10^{-2}$ cm^3 K^{-1} mol^{-1}; [l] A. LoSurdo and F. J. Millero, *J. Phys. Chem.*, **84**, 710 (1980); [m] R. M. Noyes, *J. Am. Chem. Soc.*, **86**, 971 (1964).

The structural data obtained by X-ray diffraction pertain only to the relative location of the electron clouds of the ion and the water molecules, and disregards the location of the hydrogen atoms of the latter. For this reason, the ammonium and hydrogen (hydronium) ions, NH_4^+ and H_3O^+, are not perceived by the X-rays, since they are isoelectronic with water molecules. Hence electrolytes containing these cations give X-ray diffraction patterns characteristic of the structure of the surroundings of the anions only. Neutron diffraction, on the contrary, is sensitive to the location of the hydrogen atoms (or rather deuterium atoms, that must be used to avoid neutron loss by absorption). Difference methods, utilizing different isotopes of the atoms of interest, can provide direct information on the partial pair correlation functions, pertaining to specific pairs of species of atoms. In particular, the distance of the hydrogen (deuterium) atoms of the water from the ion and of the oxygen atoms in the solvation shell from one another provide information on the spatial geometry in the hydration shell. Similar information can be obtained by the computer simulation method if the interaction potentials used are non-spherically symmetric.

A summary of the structure of the surroundings of many ions in aqueous solutions is presented in Table 5.12. In spite of a certain spread of the data, a large degree of consistency is seen. This is demonstrated in Figure 5.3, as a plot of the distances d between the centres of the ions and the nearest water molecules against the Pauling ionic crystal radii r_c. The data for the 23 ions for which d values are available are seen to fall on a single straight line with a correlation coefficient of 0.9929: $d = (0.1393 \pm 0.0056) + 1.000\, r_c$. The intercept corresponds to the radius of a water molecule, and has an uncertainty of only $\pm 4.0\%$. This shows that the Pauling crystal radius is a good measure of the radius of the bare ion inside its hydration shells in aqueous solutions.[38]

The information concerning a possible second hydration shell is much more scant. Positive evidence for its existance involves mainly ions exhibiting a high electrical field, i.e., those of small size and a large charge. The mean distance from the centre of the ion to that of a water molecule in this second hydration shell is 0.414 ± 0.004 nm, and the number of these next-nearest neighbours is 12. No individuality of the ions can be discerned with these values.

As the diffraction methods are applied at present, fairly concentrated solutions must be used, typically in the range from 10 to 60 moles of water per mole of salt, see Table 5.12. Some more concentrated solutions have been studied too (3.6 H_2O/NH_4F and 4.0 $H_2O/LiCl$), but in only very special cases have more dilute solutions been studied (up to 96 $H_2O/H_3O^+Cl^-$ and up to 99 $H_2O/NdCl_3$ for X-ray diffraction, and up to 645 $H_2O/NiCl_2$ for neutron diffraction). In solutions within the typical concentration range ion–ion interactions superimpose on ion–water interactions. When the number of total water molecules present per ion falls below the number h of nearest neighbours in the first hydration shell, some of the water molecules must be shared between the ions. The structures in the environments of the cations and the anions should then differ from those in more dilute solutions, where the hydration shells are

Table 5.12 The structure around the ions of aqueous electrolytes

Salt	$R_W = \frac{H_2O}{Salt}$	Method	Cations				Anions		Ref
			1st coord. shell		2nd coord. shell				
			d/nm	h	d/nm	h	d/nm	h	
Li^+	215	MC[a]	0.210	6					1
	64	MC	0.19						2
Na^+	215	MC	0.235	6					1
	200	MC	0.240		0.520				3
	64	MC	0.23						2
K^+	215	MC	0.270	6					1
	200	MC	0.265		0.505				3
	64	MC	0.28						2
F^-	215	MC					0.260	6.3	1
Cl^-	215	MC					0.325	8.4	1
H_3OCl	3–31	n[b]	0.252	4			0.313	4	4
	4–96	x[c]							
NH_4F	3.6	x					0.287	6	5
NH_4Cl	8.5	x					0.317	6	5
	25	MD[d]	0.305	8			0.322	8	6
NH_4Br	7.6	x					0.336	6	5
NH_4I	8.2	x					0.361	6	5
NH_4NO_3	4.1	x					0.351	9	7
$(NH_4)_2SO_4$	11.7	x					0.379	7.6	8
LiCl	8.2	x	0.195	4			0.310	6	9
	4.0		0.222	4			0.318	6	
	4.0		0.218				0.319		54
	5.6	n	0.195	3.3			0.329	5.3	10
	15.6		0.195	5.5	0.45		0.334	5.9	
	13.9	x	0.208	4			0.308	6	11
	27.8		0.217						
	54.3	n	0.190	4			0.310	6	12
	25	MD	0.210	6			0.27	7	13
	4	MD	0.222				0.315		54
LiBr	25.0	x	0.225	4			0.329	6	14
	10.8		0.214	4					
	8.4		0.216	4					
LiI	25.2	x	0.210	6	0.441	12	0.363	6.9	15
	25	MD	0.206	7			0.30	7	13
	25	MD	0.213	6.1					16
NaCl	10.4	n					0.320	5.5	17
	13.9–27.8	x	0.242	4			0.308–0.316	6	11
	54.3	n	0.250	8			0.310		18
	10.2	x	0.241	6			0.316	6	19
	25	MD	0.23	7			0.27	7	13
	100	MD	0.23	7			0.27	8	20
	25	MD	0.23	6			0.32	6	21
	25	MD	0.230	6			0.330	8	22
NaI	7 M[e]	x	0.24	4			0.360	6	23
$NaNO_3$	6.1, 9.3	x	0.246	6			0.318	3 or 6	24
KCl	13.9, 27.8	x	0.280	6			0.308–0.316	6	11
	53.7	n	0.270	8			0.310		18
RbCl	12.7	n					0.320	5.8	17
CsF	25	MD	0.310	7			0.322	8	13
	25	MD	0.322	7.9			0.264	6.8	55
CsCl	53.1	n	0.295	8			0.310	6	12
	13.9, 27.8	x	0.315	6 to 8					11
	25	MD	0.31	8			0.32	6	13
$MgCl_2$	55.5	x	0.212	6	0.410	12	0.313	6	25
	27.1		0.210	6	0.423	12	0.314	6	
$Mg(NO_3)_2$	24.8	x	0.211	6	0.420	12			26
	10.8		0.210	6	0.430	12			

(continued)

Table 5.12—*continued*

Salt	$R_W = \frac{H_2O}{Salt}$	Method	Cations				Anions		Ref.
			1st coord. shell		2nd coord. shell				
			d/nm	h	d/nm	h	d/nm	h	
$CaCl_2$	55.8	x	0.242	6			0.314	6	27
	26.6		0.241	6			0.314	6	
	12.3		0.242	6			0.315	6	
	12.4	n	0.242	5.5			0.325	5.8	17, 28
$CaBr_2$	44.1	x	0.240	6			0.332	6	29
	26.0		0.244	6			0.334	6	
$BaCl_2$	50.4	x					0.326	6.2	17
$Mn(ClO_4)_2$	21.2	x	0.220	6					31
$Fe(ClO_4)_2$	25.5	x	0.212	6					31
$CoCl_2$	17.9	x	0.210	5.5	(also 0.5 Cl^- at 0.247 nm)		0.311	6	32
$Co(ClO_4)_2$		x	0.209	6.5	0.402				33
	21.0	x	0.208	6					31
$NiCl_2$	27.4–12.6	x	0.205	6			0.313	6	34
	645	n	0.207	6.8					35
	121		0.210	6.8					
	38.0		0.207	5.8					
	12.6		0.207	5.8					
	12.8	n					0.320	5.7	17
	17.9	x	0.207	5.5	(also 0.5 Cl^- at 0.244 nm)		0.306	6	33
$Ni(ClO_4)_2$	14.6	n	0.207	5.8					36
		x	0.205	6	0.406				32
	22.7	x	0.204	6					31
$CuCl_2$	12.9	n	0.205	2.3	(also 4 Cl^- at 0.256 nm)				37
	17.9	x	0.195	2.8	(also 1.2 Cl^- at 0.225 nm)				33
$Cu(ClO_4)_2$	15.7	x	0.194	4	(also 2 H_2O at 0.238 nm)				31
$ZnBr_2$	8.1–15.9	x	0.221	2.4	(also 2.3 Br^- at 0.242 nm)		0.345	4.2	56
$Zn(ClO_4)_2$	19.4	x	0.208	6					31
$Zn(NO_3)_2$	9.0	x	0.217	6.6	0.402	10.8	0.344	17.7	30
$CdCl_2$	63.1–35.1	x	0.237	4	(also 2 Cl^- at 0.257 nm)				38
$Cd(NO_3)_2$	9	x	0.228	5.7	0.431	11.9	0.349	8.8	57
$Cd(ClO_4)_2$	26.6	x	0.231	6					39
$AlCl_3$	54.0	x	0.190	6	0.401	12	0.314	6	40
	23.8		0.188	6	0.402	12	0.311	6	
$Al(NO_3)_2$	14.5	x	0.187	6	0.399	12	0.339		41
$LaCl_3$	31.9–20.8	x	0.248	8.0					42
	27.8–14.6	x	0.258	9.1					43
$LaBr_3$	20.9	x	0.248	8.0					44
$PrCl_3$	14.6	x	0.254	9.2					43
$NdCl_3$	99.1–32.1	x	0.241	8.0					45
	16.5	x	0.251	8.9					43
	19.5	n	0.248	8.5			0.229	3.9	53
$SmCl_3$	38.8	x	0.242	9.9					46
	17.2	x	0.247	8.8					47
$EuCl_3$	17.2	x	0.245	8.3					47
$GdCl_3$	45.1–14.4	x	0.237	8.0					45, 48
	26.8	x	0.240	9.9					46
$TbCl_3$	11.0	x	0.241	8.2					49
$DyCl_3$	11.7	x	0.240	7.9					49
$ErCl_3$	10.8	x	0.237	8.2					49
$TmCl_3$	10.6	x	0.236	8.1					49
$LuCl_3$	10.6	x	0.234	8.0					49
$CrCl_3$	64.8	x	0.199	6	0.405	12	0.308	6	50
$Cr(NO_3)_3$	50.8	x	0.200	6	0.408	12			50, 51
	24.5		0.199	6	0.406	12			
$Fe(NO_3)_3$	68.0	x	0.204	6	0.414	12			52
	32.4		0.205	6	0.412	12			
	18.1		0.204	6	0.412	12			

[a] Monte Carlo method of computer simulation; [b] neutron diffraction method; [c] X-ray diffraction method; [d] Molecular dynamics method of computer simulation.

1. M. Mezei and D. L. Beveridge, *J. Chem. Phys.*, **74**, 6902 (1981).
2. Yu. V. Ergin, O. Ya. Koop and A. M. Khrapko, *Zh. Fiz. Khim.*, **53**, 2109 (1979); *Russ. J. Phys. Chem.*, **53**, 1204 (1979).
3. G. G. Malenkov, L. P. Dyakonova and L. S. Brizhik, *VINITI* 346-80 (1980).
4. R. Triolo and A. H. Narten, *J. Chem. Phys.*, **63**, 3624 (1975).
5. A. H. Narten, *J. Phys. Chem.*, **74**, 765 (1970).
6. G. I. Szasz and K. Heinzinger, *Z. Naturf.*, **A34**, 840 (1979).
7. R. Caminiti, G. Licheri, G. Piccaluga and G. Pinna, *J. Chem. Phys.*, **68**, 1967 (1978).
8. R. Caminiti, G. Paschina and G. Pinna, *Chem. Phys. Lett.*, **64**, 391 (1979).
9. A. H. Narten, F. Vaslow and H. A. Levy, *J. Chem. Phys.*, **58**, 5017 (1973).
10. J. R. Newsome, G. W. Neilson and J. E. Enderby, *J. Phys. C; Solid State Phys.*, **13**, L923 (1980).
11. G. Palinkas, T. Radnai and F. Hajdu, *Z. Naturf.*, **A35**, 107 (1980).
12. N. Ohtomo and K. Arakawa, *Bull. Chem. Soc. Japan*, **52**, 2755 (1979).
13. K. Heinzinger and P. C. Vogel, *Z. Naturf.*, **A31**, 463 (1976).
14. G. Licheri, G. Piccaluga and G. Pinna, *Chem. Phys. Lett.*, **35**, 119 (1975).
15. T. Radnai, K. Heinzinger and I. Szasz, *Z. Naturf.*, **A36**, 1076 (1981).
16. G. I. Szasz, G. Heinzinger and G. Palinkas, *Chem. Phys. Lett.*, **78**, 194 (1981).
17. S. Cummings, J. E. Enderby, G. W. Neilson, J. R. Newsome, R. A. Howe, W. S. Howells and A. K. Soper, *Nature*, **287**, 714 (1980).
18. N. Ohtomo and K. Arakawa, *Bull. Chem. Soc. Japan*, **53**, 1789 (1980).
19. R. Caminiti, G. Licheri, G. Piccaluga and G. Pinna, *Rend. Sem. Fac. Scienze Cagliari*, **XLVV** Suppl. 19 (1977).
20. P. C. Vogel and K. Heinzinger, *Z. Naturf.*, **A31**, 476 (1976).
21. G. Palinkas, W. O. Riede and K. Heinzinger, *Z. Naturf.*, **A32**, 1137 (1977).
22. P. Bopp, W. Dietz and K. Heinzinger, *Z. Naturf.*, **A34**, 1424 (1979).
23. M. Maeda and H. Ohtaki, *Bull. Chem. Soc. Japan*, **48**, 3755 (1975).
24. R. Caminiti, G. Licheri, G. Paschina, G. Piccaluga and G. Pinna, *J. Chem. Phys.*, **72**, 4522 (1980).
25. R. Caminiti, G. Licheri, G. Piccaluga and G. Pinna, *J. Appl. Cryst.*, **12**, 34 (1979).
26. R. Caminiti, G. Licheri, G. Piccaluga and G. Pinna, *Chem. Phys. Lett.*, **61**, 45 (1979).
27. G. Licheri, G. Piccaluga and G. Pinna, *J. Chem. Phys.*, **64**, 2437 (1976).
28. S. Cummings, J. E. Enderby and R. A. Howe, *J. Phys. C; Solid State Phys.*, **13**, 1 (1980).
29. G. Licheri, G. Piccaluga and G. Pinna, *J. Chem. Phys.*, **63**, 4412 (1975).
30. S. P. Dagnall, D. N. Hague and A. D. C. Towl, *J. Chem. Soc. Faraday Trans. 2*, **78**, 2161 (1982).
31. H. Ohtaki, T. Yamaguchi and M. Maeda, *Bull. Chem. Soc. Japan*, **49**, 701 (1976).
32. M. Magini and G. Giubileo, *Gazz. Chim. Ital.*, **111**, 449 (1981).
33. M. Magini, *J. Chem. Phys.*, **74**, 2523 (1981).
34. R. Caminiti, G. Licheri, G. Piccaluga and G. Pinna, *Faraday Disc. Chem. Soc.*, **64**, 62 (1978).
35. G. W. Neilson and J. E. Enderby, *J. Phys. C; Solid State Phys.*, **11**, L625 (1978).
36. J. R. Newsome, G. W. Neilson, J. E. Enderby and M. Sandstrom, *Chem. Phys. Lett.*, **82**, 399 (1981).
37. G. W. Neilson, Priv. Comm., in J. E. Enderby and G. W. Neilson, *Rep. Progr. Phys.*, **44**, 593 (1981).
38. R. Caminiti, G. Licheri, G. Paschina, G. Piccaluga and G. Pinna, *Z. Naturf.*, **A35**, 1361 (1980).
39. H. Ohtaki, M. Maeda and S. Ito, *Bull. Chem. Soc. Japan*, **47**, 2217 (1974).
40. R. Caminiti, G. Licheri, G. Piccaluga, G. Pinna and T. Radnai, J., *Chem. Phys.*, **71**, 2473 (1979).
41. R. Caminiti and T. Radnai, *Z. Naturf.*, **A35**, 1368 (1980).
42. L. S. Smith and D. L. Wertz, *J. Am. Chem. Soc.*, **97**, 2365 (1975); L. S. Smith, D. C. McCain and D. L. Wertz, *ibid.*, **98**, 5125 (1976).
43. A. Habenschuss and F. H. Spedding, *J. Chem. Phys.*, **70**, 3758 (1979).
44. L. S. Smith and D. L. Wertz, *J. Inorg. Nucl. Chem.*, **39**, 95 (1977).
45. M. L. Steele and D. L. Wertz, *Inorg. Chem.*, **16**, 1225 (1977); M. L. Steele, Diss., Univ. Southern Mississippi (1976).
46. A. I. Ryss, M. K. Levositskaya and I. M. Shapovalov, *VINITI* 856-76 (1976); C. A., **89**, 95116 (1978).
47. A. Habenschuss and F. H. Spedding, *J. Chem. Phys.*, **73**, 442 (1980).
48. M. L. Steele and D. L. Wertz, *J. Am. Chem. Soc.*, **98**, 4424 (1976).
49. A. Habenschuss and F. H. Spedding, *J. Chem. Phys.*, **70**, 2797 (1979).
50. R. Caminiti, G. Licheri, G. Piccaluga and G. Pinna, *J. Chem. Phys.*, **69**, 1 (1978).
51. R. Caminiti, G. Licheri, G. Piccaluga and G. Pinna, *Chem. Phys.*, **19**, 371 (1977).
52. M. Magini and R. Caminiti, *J. Inorg. Nucl. Chem.*, **39**, 91 (1977); R. Caminiti and M. Magini, *Chem. Phys. Lett.*, **61**, 40 (1979).
53. A. H. Narten and R. L. Hahn, *J. Phys. Chem.*, **87**, 3193 (1983); S. Biggin, J. E. Enderby, R. L. Hahn and A. H. Narten, *ibid.*, **88**, 3634 (1984).
54. I. Okada, Y. Kitsuno, H.-G. Lee and H. Ohtaki, *Abstr. VIth Int. Symp. Solute-Solute-Solvent Interact., Minoo, 1982*, p. 1A M07.
55. Gy. I. Szasz and K. Heinzinger, *Z. Naturf.*, **38A**, 214 (1983).
56. E. Kalman, I. Serke, G. Palinkas, G. Johansson, G. Kabisch, M. Maeda and H. Ohtaki, *Z. Naturf.*, **38A**, 225 (1983).
57. R. Caminiti, P. Cucca and T. Radnai, *J. Phys. Chem.*, **88**, 2382 (1984).

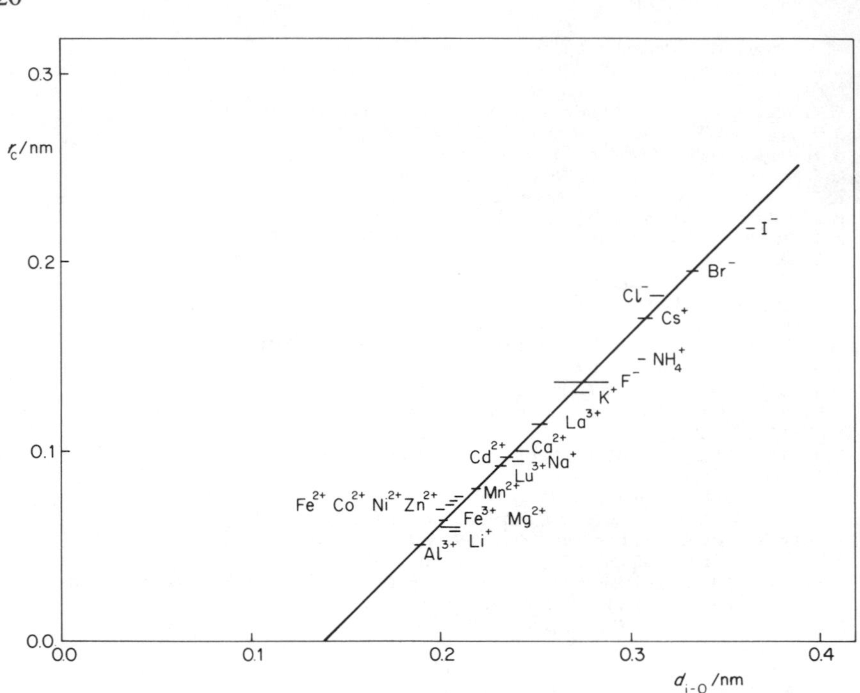

Figure 5.3 A plot of the Pauling crystal ionic radii of ions against the ion–oxygen distances as determined experimentally or by computer simulation (Table 5.12). The line is drawn with a slope of unity and the intercept is 0.139 nm, corresponding to the radius of a water molecule

independent. This expected behaviour is not documented by the extant data, except for one neutron diffraction study of LiCl solutions with 5.6 and 15.6 H_2O/LiCl. This study does, indeed, show the expected increase of h with the dilution.[39]

The geometrical arrangement of the water molecules around the ion has been elucidated in a few cases.[39] It is found that the dipolar axis of the water molecule is at an angle of $50 \pm 10°$ from the line connecting a cation (Li^+, Ca^{2+}, Ni^{2+}, and Cu^{2+} have been tested) with the oxygen atom (Figure 5.4). This value of the angle pertains to fairly concentrated solutions ($\lesssim 10$ H_2O/ion), whereas in more dilute solution the angle tends to decrease. These facts reflect the effects of the neighbouring particles (water molecules or counter ions) on the potential energy of the water in the hydration shell, and on whether the interaction of the cation with the unshared pair of electrons on the oxygen atom or with the dipole as a whole is preferred. The arrangement of water molecules around chloride anions (in solutions of Li^+, Na^+, Rb^+, Ca^{2+}, Ba^{2+}, and Ni^{2+} chlorides) has also been studied. The deviation of the Cl^-—D—O angle from 180° is minimal ($<10°$), the deuterium atom being at a distance of 0.225 ± 0.003 nm from the chloride anion, whereas the second deuterium atom of the same (heavy) water molecule is located at 0.34 to 0.37 nm, see Figure 5.4.

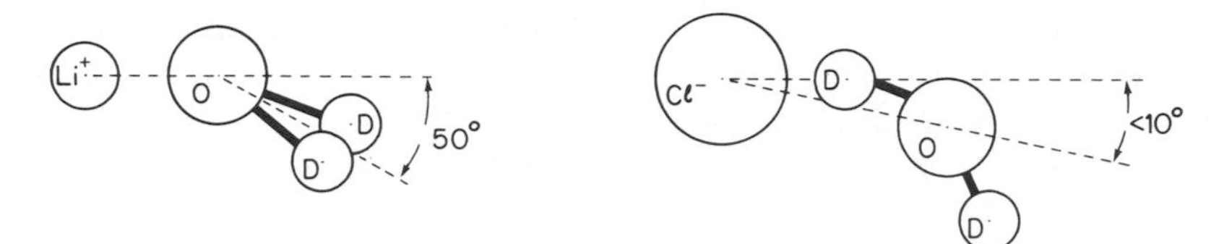

Figure 5.4 The experimentally determined (by differential neutron diffraction of isotopically labelled atoms) geometrical configuration of water molecules around a lithium ion (left-hand side) and a chloride ion (right-hand side)

An attempt at a conceptual definition of the structure of a solvent has been made in Section 3.6, with direct application to the structure of water. The extent of the structure existing in the water is describable in terms of the average number of hydrogen bonds that occur per water molecule. However, no agreement has been reached on the quantitative interpretation of the results of measurements by such methods as infrared, Raman, and NMR spectroscopy or dielectric and ultrasonic relaxation, in terms of the average number of hydrogen bonds. A further concept that has been discussed in Section 3.6 is the change that the structure of the solvent incurs, when solute particles are introduced. This change is superimposed on the orientation of the solvent in the immediate vicinity of the particles into definite solvates (Section 4.1), and can be considered to start just beyond the first coordination sphere.

In the case of water as the solvent, the following effects ought to be considered, when the solute particles are ions:

(i) The remnant of the central electrical field, beyond the shielding caused by the first coordination sphere, if present, has still an orienting influence on the water molecules around the ion. This influence starts at a distance $r_c + 2r_W$ from the centre of the ion, where r_c is the radius of the ion and r_W that of a water molecule, and extends out indefinitely. The field tends to orient the water molecules through ion–dipole interactions, and further ion multipole interactions, but the latter become negligible at the distances involved.

(ii) The water molecules in the first coordination sphere, when one is present, exert their own orienting effect on the outer water molecules. The inner sphere water molecules will have their positive ends pointing outwards from hydrated cations. Their hydrogen atoms will be somewhat more positive than in bulk water, hence more strongly tending towards hydrogen bonding. For hydrated anions the situation should be the opposite one, but anions with a geometrically well defined inner coordination sphere of water molecules are very rare.

(iii) The tetrahedral structure of bulk water will tend to orient the water molecules in conformity with the geometry favourable for the three-dimensional hydrogen-bonded network. This effect will extend as near to the ion as the competition with the other two effects permits. If the ion contains nonpolar ('inert') parts, they may shield the region near the ion from disruption of its structure by the thermal motion of 'free' water molecules, and hence enhance the tetrahedral structure in this region.

These three effects compete with each other in the orienting of the water molecules beyond the first hydration shell. The charge of the ion, its size and shape, its being mono- or polyatomic, and its possible content of 'inert' portions determine the strength of each of these effects in the imaginary concentric shells around the ion. The incompatibility of the structures generated by these effects will cause a certain region around the ion to consist of disoriented water molecules, a 'thawed' region in the 'iceberg' analog of the structured water.

For small, multivalent ions, effect (i) will predominate to a sufficient extent to cause a net structuring of water in a centrally oriented manner in one or two concentric shells round the first hydration shell. This net water structuring is typical for an ion such as Mg^{2+} but less obvious for a small monovalent ion such as Na^+. It gives way to destruction of the structure of water in large monovalent monoatomic cations such as Cs^+ or anions such as Br^-. Certain ions fit into the structure of water without affecting it one way or another: e.g. $NH_4{}^+$ and $N(C_2H_4OH)_4{}^+$. In tetraalkylammonium ions the structure-enhancing effect of the inert alkyl chains may gain the upper hand if they are long enough: $(CH_3)_4N^+$ is considered as a structure breaker, $(C_2H_5)_4N^+$ as indifferent, and $(C_nH_{2n+1})_4N^+$ with $n \geqslant 3$ as structure makers, in the tetrahedral water structure sense of bulk water. Nonsymmetrical alkylammonium ions (e.g., $C_4H_9NH_3{}^+$), carboxylate ions (e.g., $C_4H_9COO^-$) and similar ions have different regions around their different parts. Charge–dipole interactions and ion–water hydrogen bonds near the polar part (the $—NH_3{}^+$ and $—COO^-$ parts in the above examples) form one kind of structure, the structure-enhancing effect near the nonpolar part (the $—C_4H_9$ group in the examples) forms a different kind. A non-structured region results in between. Schematic drawings of these situations are shown in Figure 5.5.

These qualitative statements have arisen from a comprehensive view of experimental facts concerning aqueous solutions of electrolytes involving the various kinds of ions. The question remains whether it is possbile to express the structural effects of the ions, i.e., their effects on the structure of water beyond their first hydration shells, in a quantitative manner. In other words, how can structure-making ions be distinguished quantitatively from structure-breaking ions, and how can the water-structure-affecting properties of ions be measured?

One measure of the structural effect of the ions on water is their entropy of hydration, $\Delta S^0_{\mathrm{i\,hydr}}$. This quantity, derived from the data in Table 5.10 ($\Delta S^0_{\mathrm{i\,hydr}} = (\Delta H^0_{\mathrm{i\,hydr}} - \Delta G^0_{\mathrm{i\,hydr}})/T$ for $T = 298.15$ K) is tabulated in Table 5.13. It does, however, contain contributions from non-structural sources, including electrostatic interactions, changes of volume between standard states, the immobilization of water molecules, and hindered rotations in polyatomic ions. The electrostatic contribution (effect (i) on p. 121) can be expressed by differentiation with respect to the temperature of the corresponding Gibbs free energy change, given by the Born equation 3.34. This equation is taken to apply to $r = r_c + 2r_W$, as discussed above, and the only temperature dependent quantity in it is then ε_W, the dielectric constant of water. Differentiation and

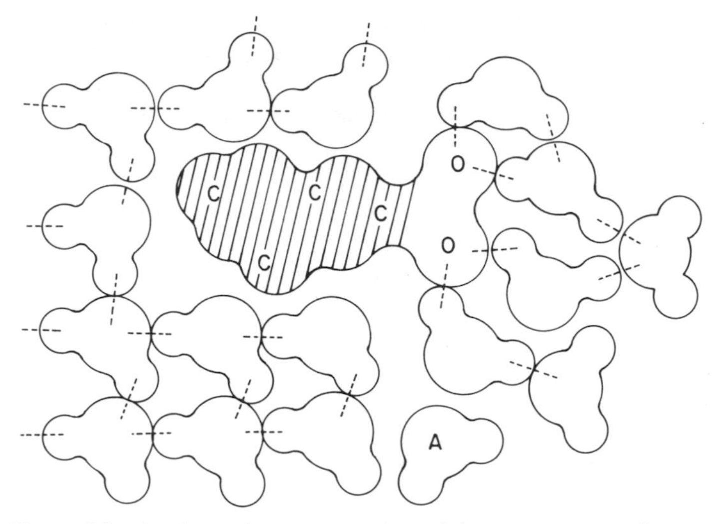

Figure 5.5 A schematic representation of the arrangement of water molecules around a butanoate anion in water. The shaded area is the hydrophobic portion of the anion, around which water molecules have an ice-like tetrahedral arrangement. Near the two oxygen atoms of the carboxylate group the water molecules are oriented by the charge and are hydrogen-bonded to them. A molecule like that marked 'A' belongs to neither category and is in a non-structured region in between the two kinds of structured regions

insertion of the value of $d(\ln \varepsilon_W)/dT$ at 298.15 K from Table 5.5 yields

$$\Delta S^0_{i\,hydr}(\text{Born}) = -4.06\, z_i^2((r_{ci}/\text{nm}) + 0.280)^{-1}\ \text{J K}^{-1}\ \text{mol}^{-1} \qquad (5.26)$$

The entropy change for the transfer of an ion from the gas phase standard state of the hypothetical ideal gas at $P^0 = 1$ atm (101325 Pa) pressure to the aqueous phase standard state of the hypothetical ideal 1 mol dm^{-3} solution is $\Delta S^0_{i\,hydr}(\text{s.s.}) = \mathbf{R} \ln(V_{aq}/V_g)$. The volume V_g accessible to the ion in the gas phase is $V_g = \mathbf{R}T/P^0 = 24.46$ dm^3 at $T = 298.15$ K, but the volume accessible to the ion in the aqueous solution, V_{aq}, is ambiguous, in the sense that the total volume of 1 dm^3 or only a smaller 'free' volume may be meant. The resulting value $|\Delta S^0_{i\,hydr}(\text{st.st.})| \leqslant 26.6$ J K^{-1} mol^{-1} is therefore uncertain to that extent. Contributions to the entropy of hydration arise also from the immobilization of water molecules in the first hydration shell around the ion and from hindered free rotations in polyatomic ions. For the former effect an estimate of -50 J K^{-1} mol^{-1} has been given for monoatomic ions.[40] The latter effect may produce a similar contribution for polyatomic ions, but both estimates are rather uncertain. The non-electrostatic and non-structural contributions may thus be lumped together as -80 J K^{-1} mol^{-1}, with an estimated variability of ± 30 J K^{-1} mol^{-1} among various ions. The entropy change due to the effects of the ions on the structure of water is then the remaining quantity:

$$\Delta S^0_{i\,str} = \Delta S^0_{i\,hydr} - \Delta S^0_{i\,hydr}(\text{Born}) - (-80)\ \text{J K}^{-1}\ \text{mol}^{-1} \qquad (5.27)$$

This quantity is presented in Table 5.13, and it is seen that for some ions it is negative—these are structure-making—whereas for others it is positive—these are structure-breaking. A proviso, that values between -30 and $+30\ \mathrm{J\ K^{-1}\ mol^{-1}}$ may give an unreliable assignment, must be recognized, however.

A definite division into water-structure-making and -breaking ions has been suggested by Samoilov,[41] according to whether the change in the activation energy for water exchange, $\Delta E^{\neq}_{\mathrm{Wi}}$, caused by the ion, is positive or negative. The concepts proposed by Samoilov were, accordingly, 'positive-' and 'negative hydration'. The temperature dependence of the rate of exchange of the water molecules in the vicinity of the ions is obtained from measurements of that of the self diffusion coefficient of pure water, $\mathrm{d}\ln D_{\mathrm{W}}/\mathrm{d}T$, on the one hand, and from the temperature dependence of the equivalent conductivities of the ions, $\mathrm{d}\lambda_{\mathrm{i}}^{\infty}/\mathrm{d}T$, on the other. The operating equation is

$$(\Delta E^{\neq}_{\mathrm{Wi}}/\mathbf{R}T)/(1 + 0.0655\exp(\Delta E^{\neq}_{\mathrm{Wi}}/\mathbf{R}T) = 1 + T(\mathrm{d}\lambda_{\mathrm{i}}^{\infty}/\mathrm{d}T)/\lambda_{\mathrm{i}}^{\infty} - T(\mathrm{d}\ln D_{\mathrm{W}}/\mathrm{d}T) \qquad (5.28)$$

The constant 0.0655 is the ratio of the probabilities for an ion to move together with its hydration shell to that of its moving bare, estimated from the assumption[41] that $\Delta E^{\neq}_{\mathrm{W\,Li}}/\Delta E^{\neq}_{\mathrm{W\,Na}} = \frac{1}{2}\Delta E^{\neq}_{\mathrm{W\,Mg}}/\Delta E^{\neq}_{\mathrm{W\,Ca}}$. The value of $\mathrm{d}\ln D_{\mathrm{W}}/\mathrm{d}T$ used by Samoilov is $0.0265\ \mathrm{K^{-1}}$ for 294.65 K. It becomes $0.0259\ \mathrm{K^{-1}}$ for 298.15 K, at which temperature data have been calculated by the present author from $\mathrm{d}\lambda_{\mathrm{i}}^{\infty}/\mathrm{d}T$ data in the literature. The resulting values of $\Delta E^{\neq}_{\mathrm{Wi}}$, together with those given by Samoilov,[41] are presented in Table 5.13. The Table indeed shows that $\Delta E^{\neq}_{\mathrm{Wi}}$ has positive values for ions known generally to be water-structure-makers and negative values for those which are known to be structure-breakers. For some ions, however, $\Delta E^{\neq}_{\mathrm{Wi}}$ does not have the expected sign: highly negative values (< -4) are obtained for H^{+}, Mn^{2+}, Co^{2+}, Ni^{2+}, and OH^{-}, which are strong structure makers, and also for Pb^{2+} and $(CH_3)_4N^{+}$, which are not expected to be strong structure breakers. A further complication arises from the nature of the left-hand side of equation 5.28, which exhibits a maximum, hence is double-valued in $\Delta E^{\neq}_{\mathrm{Wi}}$ for values of the right-hand side > 0.9 (the lower of the two values has been given in Table 5.13).

A still further approach to this problem is due to Gurney,[42] who based the structure-making and -breaking properties of the ions on their effects on the viscosity of water. Ions that enhance the viscosity in dilute solutions are structure-making, and those that lower it are structure-breaking, since structured water is expected to be more viscous than nonstructed water. The operative quantity is the viscosity B-coefficient of the Jones–Dole equation:

$$\eta = \eta_{\mathrm{W}}(1 + \sum A_{\mathrm{i}}c_{\mathrm{i}}^{1/2} + \sum B_{\mathrm{i}}c_{\mathrm{i}} + \sum D_{\mathrm{i}}c_{\mathrm{i}}^{2} + \ldots) \qquad (5.29)$$

where the summations extend over the cations and anions of an electrolyte. The equation can be rewritten as

$$\sum \nu_i B_i = \lim_{c\to 0}\ [(\eta/\eta_{\mathrm{W}}) - 1 - \sum A_{\mathrm{i}}c_{\mathrm{i}}^{1/2}]/c \qquad (5.30)$$

Table 5.13 Measures of the water-structure-making and -breaking properties of ions (at $T = 298.15$ K)

Ion	$\Delta S^0_{i\,hydr}$ [a] (J K^{-1} mol^{-1})	$\Delta S^0_{i\,str}$ [b] (J K^{-1} mol^{-1})	$\Delta E^{\neq}_{W}$ (kJ mol^{-1})	B_i [g] (dm^3 mol^{-1})	$\Delta\langle g\rangle_i$ [h]
H^+	−137	−32		+0.068	
Li^+	−148	−96	+2.34[e]	+0.146	−0.41
Na^+	−117	−16	+0.59[e]	+0.085	−0.45
K^+	−80	+20	−1.46[e]	−0.009	−0.48
Rb^+	−71	+29	−1.79[c]	−0.033	−0.50
Cs^+	−65	+34	−1.89[d]	−0.047	−0.66
NH_4^+	−95	+5	−1.46[e]	−0.008	−0.53
$(CH_3)_4N^+$	−144[i]	−47	+0.48[f]	+0.123	−0.12
$(C_2H_5)_4N^+$	−222[i]	−125	+1.35[f]	+0.385	
$(C_3H_7)_4N^+$	−327[i]	−231	+2.73[f]	+0.916	
$(C_4H_9)_4N^+$			+3.50[f]	+1.275	+1.02
$(C_6H_5)_4As^+$				+1.09	−0.75
Cu^+	−148	−47			
Ag^+	−123	−23	+0.11[c]	+0.090	+0.27
Tl^+	−78	+22	−1.39[c]		−0.87
Be^{2+}	−310	−158		+0.45	
Mg^{2+}	−331	−82	+3.35[e]	+0.385	
Ca^{2+}	−252	−108	+1.17[e]	+0.298	
Sr^{2+}	−242	−100	+1.42[e]	+0.272	
Ba^{2+}	−205	−65	+0.08[e]	+0.229	−1.29
Ra^{2+}	−167	−28			
Eu^{2+}	−259	−118			
Mn^{2+}	−307	−162		+0.39	
Fe^{2+}	−376	−230	+1.23[d]	+0.42	
Co^{2+}	−340	−194		+0.376	
Ni^{2+}	−365	−219		+0.375	
Cu^{2+}	−316	−180	+3.48[d]	+0.368	
Zn^{2+}	−318	−172	+1.07[d]	+0.361	
Cd^{2+}	−285	−142	+0.82[d]	+0.36	−1.02
Hg^{2+}	−252	−100			
Hg_2^{2+}	−233[i]	−95			
Sn^{2+}	−230	−89			
Pb^{2+}	−209	−68		+0.233	
VO^{2+}	−404[i]				
UO_2^{2+}	−402				
Al^{3+}	−553	−332		+0.67	
Sc^{3+}	−496	−285			
Y^{3+}	−497	−289			
La^{3+}	−467	−262	+8.91[e]	+0.582	
Ce^{3+}	−469	−263		+0.576	
Pr^{3+}	−475	−268		+0.587	
Nd^{3+}	−474	−267		+0.582	
Sm^{3+}	−476	−268		+0.605	
Eu^{3+}	−494	−286		+0.624	
Gd^{3+}	−484	−275		+0.646	
Tb^{3+}	−512	−303		+0.653	
Dy^{3+}	−509	−299		+0.662	

(*continued*)

Table 5.13—*continued*

Ion	$\Delta S^0_{i\,hydr}$ [a] (J K^{-1} mol^{-1})	$\Delta S^0_{i\,str}$ [b] (J K^{-1} mol^{-1})	$\Delta E^{\neq}_W$ (kJ mol^{-1})	B_i [g] (dm^3 mol^{-1})	$\Delta\langle g\rangle_i$ [h]
Ho^{3+}	−502	−292		+0.673	
Er^{3+}	−518	−308		+0.663	
Tm^{3+}	−509	−298		+0.678	
Yb^{3+}	−484	−272		+0.671	
Lu^{3+}	−516	−305		+0.681	
Pu^{3+}	−553	−377			
Fe^{3+}	−578	−362			
Ga^{3+}	−567	−350			
In^{3+}	−391	−180			
Tl^{3+}	−440	−233			
$Co(NH_3)_6{}^{3+}$	−358[i]				
Ce^{4+}	−582	−288			
Th^{4+}	−702	−411			
Zr^{4+}	−780	−475			
Hf^{4+}	−741	−436			
U^{4+}	−708	−416			
F^-	−150	−50	+0.84[e]	+0.127	+0.32
Cl^-	−87	+12	−0.88[e]	−0.005	−0.48
Br^-	−70	+29	−1.63[e]	−0.033	−0.84
I^-	−47	+51	−1.00[e]	−0.073	−1.13
OH^-	−161	−60		+0.120	
SH^-	−101	−2			
SeH^-	−103[i]				
CN^-	−80	+18		+0.024	
$N_3{}^-$	−82	+17		+0.018	
$I_3{}^-$	−73[i]				
SCN^-	−66	+33	−2.91[c]	−0.025	
CNO^-	−90[i]				
$BF_4{}^-$	−66	+32			
$ClO_3{}^-$	−80	+18	−1.34[d]	−0.025	
$BrO_3{}^-$	−95	+2	−3.51[d]	+0.007	
$IO_3{}^-$	−148	−47	+0.66[d]	+0.138	
$ClO_4{}^-$	−59	+38	−1.97[e]	−0.060	−0.40
$NO_2{}^-$	−91	+8			
$NO_3{}^-$	−77	+21	−1.84[e]	−0.045	
$MnO_4{}^-$	−64	+35	−2.86[d]	−0.059	
$ReO_4{}^-$	−71[i]	+12			
$HCO_2{}^-$	−124	−16	−1.62[d]		
$CH_3CO_2{}^-$	−170	−62	+2.08[f]	+0.236	
$HCO_3{}^-$	−145	−36		+0.130	
$HSO_4{}^-$	−129	−44		+0.127	
$H_2PO_4{}^-$	−168	−83		+0.340	
$B(C_6H_5)_4{}^-$				+1.12	−0.75
S^{2-}	−138	−3			
$CO_3{}^{2-}$	−259	−124	+3.23[d]	+0.278	
$SO_3{}^{2-}$	−249[i]				
$SO_4{}^{2-}$	−199	−69	+0.63[e]	+0.206	
$SeO_4{}^{2-}$	−183[i]	−52			

Table 5.13—*continued*

Ion	$\Delta S^0_{i\,hydr}$ [a] (J K^{-1} mol^{-1})	$\Delta S^0_{i\,str}$ [b] (J K^{-1} mol^{-1})	$\Delta E^{\neq}_W$ (kJ mol^{-1})	B_i [g] (dm^3 mol^{-1})	$\Delta\langle g\rangle_i$ [h]
CrO_4^{2-}	−187	−57	−2.71[d]	+0.169	
MoO_4^{2-}	−220[i]				
$S_2O_3^{2-}$	−180[i]				
$C_2O_4^{2-}$	−205	−74		+0.174	
HPO_4^{2-}	−272			+0.382	
PO_4^{3-}	−422[i]				
AsO_4^{3-}	−379[i]				
$Fe(CN)_6^{3-}$	−146			+0.123	
$Co(CN)_6^{3-}$	−166			+0.146	
$Fe(CN)_6^{4-}$	−286			+0.371	

[a] From the data in Table 5.10; [b] from equation 5.27; [c] from equation 5.28 and $d\lambda_i^\infty/dT$ from Falkenhagen (pp. 86, 92 of the book[43]); [d] from equation 5.28 and $\lambda_i^\infty(T)$, data from the same source as [c]; [e] from V. V. Goncharov, I. I. Romanova, O. Ya. Samoilov and V. I. Yashkichev, *Zh. Strukt. Khim.*, **8**, 613 (1967) with 0.33 in equation 5.28 instead of 0.0655; [f] H. Uedaira and H. Uedaira, *Zh. Fiz. Khim.*, **42**, 3024 (1968); [g] from a critical evaluation of data in the literature by the present author (unpublished), with the assumption that $B_{Rb^+} = B_{Br^-}$; [h] from the data in [11] with the assumption that $\Delta\langle g\rangle_{Ph_4As^+} = \Delta\langle g\rangle_{BPh_4^-}$; [i] $S_i^0(g)$ from A. Loewenschuss and Y. Marcus, *Chem. Rev.*, **84**, 89 (1984).

where c is the concentration of the electrolyte and ν_i are the stoichiometric coefficients. The A_i-coefficients are obtained from theory,[43] hence the B_i-coefficients are true infinite dilution properties of the ions and are additive. The use of equation (30) is preferable to the alternative formulation

$$((\eta/\eta_W) - 1)/c^{1/2} = \sum \nu_i A_i + \sum \nu_i B_i c^{1/2} \tag{5.31}$$

where $\sum \nu_i B_i$ is obtained as the slope of a straight line plot of the left-hand side against $c^{1/2}$, since then the possible effect of a D-term is ignored, not necessarily justifiably in the concentration range employed. The quantity obtained experimentally, $\sum \nu_i B_i$, must still be divided into the ionic contributions. Several suggestions have been put forward of how to do this. A common one (used by Gurney)[42] is to assume that $B_{K^+} = B_{Cl^-}$, because $\lambda^\infty_{K^+} \approx \lambda^\infty_{Cl^-}$. The latter approximate equality is off by *ca.* 3% at 25 °C and becomes worse at higher temperatures. The approximation $\lambda^\infty_{Rb^+} = \lambda^\infty_{Br^-}$ is more nearly correct (±1.5%) over a wide temperature range and has been used as the basis for the division into the ionic B_i values recorded in Table 5.13. The equal ion mobility basis for the division of the B-coefficients of electrolytes to their ionic contributions is plausible but not unique. An alternative assumption, that $B_{Ph_4As^+} = B_{BPh_4^-}$, is as plausible and is in line with other applications of the tetraphenylarsonium tetraphenylborate assumption. The consequence of its use is to change the values recorded in Table 5.13 by $0.026 \cdot z_i$ units (mol$^{-1}\cdot$dm^{-3}). The adjustment causes the sign of B_i to change for K^+ and NH_4^+ to positive and of BrO_3^- to negative, the values being near zero in any case. A comparison with the $\Delta E^{\neq}_{Wi}$ values in Table 5.13 shows that on the whole, agreement exists as to the sign (i.e., as to whether ions are structure-makers or -breakers), but not as to the relative magnitudes of the effects.

A final quantity that will be invoked here as a measure for the water-structure-affecting properties of ions is the difference of the solubilities of salts involving these ions in light and heavy water,[44] see Section 3.6. The solubility of the sparingly soluble salt X in heavy water s_X (in D_2O) is related to its standard molar Gibbs free energy of solution by $\Delta G^0_{soln}(\text{X in } D_2O) = v\mathbf{R}T \ln s_x(\text{in } D_2O)$, where v is the number of ions into which the salt dissociates. For the reasons stated in Section 3.1, the molar scale (mol dm^{-3}) should be employed for s_X. A similar relationship exists for $\Delta G^0_{soln}(\text{X in } H_2O)$. The measure for the water-structure-affecting properties of the ions of the salt is[11,44]

$$\langle g \rangle_X = [\Delta G^0_{soln}(\text{X in } D_2O) - \Delta G^0_{soln}(\text{X in } H_2O)]/\Delta u_{hb} \qquad (5.32)$$

where $\Delta u_{hb} = -0.96$ kJ mol^{-1} is the difference in the strength of the hydrogen bond in O—D ... O and in O—H ... O. The problem remains of dividing $\langle g \rangle_X$ of the salt into its constituent $\langle g \rangle_i$ of the ions. The tetraphenylarsonium tetraphenylborate assumption should serve in this case again, i.e., $\langle g \rangle((C_6H_5)_4As^+) = \langle g \rangle(B(C_6H_5)_4^-)$, or alternatively from data for three selected salts:

$$\langle g \rangle((C_6H_5)_4As^+) = \tfrac{1}{2}[\langle g \rangle((C_6H_5)_4AsClO_4) + \langle g \rangle(KB(C_6H_5)_4) - \langle g \rangle(KClO_4)] \qquad (5.33)$$

Values of $\langle g \rangle_i$ of ions calculated on the basis of equations 5.32 and 5.33 are recorded in Table 5.13. A comparison of these values, which represent the change in the average number of the hydrogen bonds in the water molecules affected by these ions, with the other entries in the table shows, again, a general agreement with respect to the sign of the effect but a rather poor quantitative correlation.

References

1. E. Schmidt, *Properties of Water and Steam in SI Units*, Springer, New York (1969).
2. N. E. Dorsey, *Properties of Ordinary Water Substance*, Reinhold, New York (1940).
3. F. Franks, ed., *Water: A Comprehensive Treatise*, Plenum, New York and London, Vol. 1 (1972), Vol. 6(1979).
4. J. A. McMillan and S. G. Los, *Nature*, **206**, 806 (1965).
5. C. A. Angell, *J. Phys. Chem.*, **75**, 3698 (1971).
6. B. J. Mason, *Adv. Phys.*, **7**, 221 (1958).
7. B. V. Zhelevskii, *Zh. Fiz. Khim.*, **42**, 1809 (1968); *ibid.*, **43**, (1969): C. A. Angell, *J. Phys. Chem.*, **77**, 3092 (1973).
8. A. H. Narten and H. A. Levy, *J. Chem. Phys.*, **55**, 2263 (1971); A. H. Narten, *ibid.*, **56**, 5681 (1972); E. Kalman, S. Lengyel, L. Haklik, and E. Eke, *J. Appl. Crystallogr.*, **7**, 442 (1974).
9. G. C. Lie and E. Clementi, *J. Chem. Phys.*, **62**, 2195 (1975); G. C. Lie, E. Clementi, and M. Yoshimine, *ibid.*, **64**, 2314 (1976).
10. A. Rahman and F. H. Stillinger, *J. Chem. Phys.*, **55**, 3336 (1971); F. H. Stillinger and A. Rahman, *ibid.*, **60**, 1545 (1974).
11. Y. Marcus and A. Ben-Naim, *J. Chem. Phys.*, submitted (1985).

12. R. Zana and E. Yeager, *J. Phys. Chem.*, **70**, 954 (1966); *ibid.*, **71**, 521, 4241 (1967).
13. a. B. E. Conway, R. E. Verall, and J. E. Desnoyers, *Z. Phys. Chem.* (*Leipzig*), **230**, 157 (1965); *Trans. Faraday Soc.*, **62**, 2738 (1966); B. E. Conway, J. E. Desnoyers, and R. E. Verall, *J. Phys. Chem.*, **75**, 3031 (1971); C. Jolicoeur, P. R. Philip, G. Perron, P. A. Leduc, and J. E. Desnoyers, *Can. J. Chem.*, **50**, 3167 (1972); b. B. S. Krumgalz, *J. Chem. Soc. Faraday Trans. 1*, **76**, 1887 (1980).
14. F. J. Millero, *J. Phys. Chem.*, **75**, 280 (1971); J. I. Kim, *ibid.*, **82**, 191 (1978); Y. Marcus, *Rev. Anal. Chem.*, **5**, 53 (1980).
15. R. E. Nightingale, *J. Phys. Chem.*, **63**, 1381 (1959); R. A. Robinson and R. H. Stokes, *Electrolyte Solutions*, Butterworth, London (1959).
16. A. Passinsky, *Acta Physicochim. USSR*, **8**, 835 (1938); A. G. Passinsky, *Zh. Fiz. Khim.*, **11**, 606 (1938); *ibid.*, **20**, 982 (1946).
17. J. Padova, *J. Chem. Phys.*, **39**, 2599 (1963); *ibid.*, **40**, 691 (1964).
18. D. S. Allam and W. H. Lee, *J. Chem. Soc.*, **1966A**, 5, 426; K. Tanaka and T. Sasaki, *Bull. Chem. Soc. Japan*, **36**, 975 (1963).
19. Calculated by the Author from data of: J. E. Lind, Jr., H. A. A. Abdel Rehim, and S. W. Rudich, *J. Phys. Chem.*, **70**, 3610 (1966); I. G. Cocker, J. Ambrose, and G. J. Janz, *J. Am. Chem. Soc.*, **92**, 5293 (1970).
20. A. M. Couture and K. J. Laidler, *Can. J. Chem.*, **34**, 1209 (1956); *ibid.*, **35**, 207 (1957).
21. J. W. Akitt, *J. Chem. Soc. Faraday Trans. 1*, **76**, 2259 (1980).
22. a. J. E. Desnoyers, R. E. Verall, and B. E. Conway, *J. Chem. Phys.*, **43**, 243 (1965); b. L. A. Dunn, *J. Soln. Chem.*, **3**, 1 (1974).
23. U.S. National Bureau of Standards, *NBS Tables of Chemical Thermodynamic Properties*, *J. Phys. Chem. Ref. Data*, **11** (1982), Suppl. 2.
24. A. Loewenschuss and Y. Marcus, *Chem. Rev.*, **84**, 89 (1984).
25. D. R. Rosseinsky, *Chem. Rev.*, **65**, 467 (1965).
26. B. E. Conway, *J. Soln. Chem.*, **7**, 721 (1978).
27. M. Eastman, *J. Am. Chem. Soc.*, **50**, 283, 292 (1928); H. D. Cockroft and J. R. Hall, *J. Phys. Chem.*, **54**, 731 (1950); W. G. Breck and J. Lin, *Trans. Faraday Soc.*, **61**, 22 (1965); T. Ikeda, *J. Chem. Phys.*, **40**, 3412 (1965).
28. H. F. Halliwell and S. C. Nyburg, *Trans. Faraday Soc.*, **59**, 1126 (1963); M. W. Lister,, S. C. Nyburg, and R. B. Poyntz, *J. Chem. Soc. Faraday Trans. 1*, **70**, 685 (1974).
29. A. D. Buckingham, *Disc. Faraday Soc.*, **24**, 151 (1957).
30. L. E. Klots, *J. Phys. Chem.*, **85**, 3585 (1981).
31. J. E. B. Randles, *Trans. Faraday Soc.*, **52**, 1573 (1956).
32. C. Shin, I. Worsley, and C. M. Criss, *J. Soln. Chem.*, **5**, 867 (1976); C. Shin and C. M. Criss, *ibid.*, **7**, 205 (1978); Y. Choi and C. M. Criss, *Faraday Disc. Chem. Soc.*, **64**, 27 (1978).
33. C. G. Malmberg and A. A. Maryott, *J. Res. Nat. Bur. Stand.*, **56**, 1 (1956).
34. C. Jolicoeur and J. C. Mercier, *J. Phys. Chem.*, **81**, 1119 (1977).
35. J. L. Fortier, P.-A. Leduc, and J. E. Desnoyers, *J. Soln. Chem.*, **3**, 323 (1974).
36. J. G. Mathieson and B. E. Conway, *J. Soln. Chem.*, **3**, 455 (1974).
37. E. Glueckauf, *Trans. Faraday Soc.*, **64**, 2433 (1968).
38. Y. Marcus, *J. Soln. Chem.*, **12**, 271 (1983).
39. J. E. Enderby and G. W. Nielson, *Rep. Progr. Phys.*, **44**, 593 (1981).
40. H. L. Friedman and C. V. Krishnan, in *Water. A Comprehensive Treatise*, F. Franks, ed., Plenum Press, New York, Vol. 3 (1973), p. 62.
41. O. Ya. Samoilov, *Structure of Electrolyte Solutions and the Hydration of Ions* (English transl.), Consultants Bureau, New York (1965); O. Ya. Samoilon, in *Water and Aqueous Solutions*, R. A. Horne, ed., Wiley, New York (1972), pp. 597–612.
42. R. W. Gurney, *Ionic Processes in Solution*, McGraw-Hill, New York (1953).
43. H. Falkenhagen and M. Dole, *Phys. Z.*, **30**, 611 (1929); H. Falkenhagen, *Theorie der Elektrolyte*, Hirzel, Leipzig (1971).
44. A. Ben-Naim, *J. Phys. Chem.*, **79**, 1268 (1975).

Chapter 6

Ion solvation in nonaqueous solvents

6.1 The relevant properties of nonaqueous solvents

Numerical values of certain properties of nonaqueous solvents are required for the models considered in Chapter 3 and the correlations developed in the present chapter. Many other properties of these solvents, which may be important for their behaviour as liquids in general or as solvents for electrolytes, are known, but have not so far been found to be directly relevant to the solvation of ions. Such are, for example, the critical constants or the self-dissociation equilibrium constant. Of the many solvents in which electrolytes are potentially soluble, only a limited number have attained significant utilization in theoretical studies or practical applications. Most of these are considered in the following, in addition to some representatives of classes of solvents in which only special kinds of electrolytes can be dissolved (e.g., hydrocarbons, dissolving electrolytes with large organic ions).

Table 6.1 presents the names and the elemental formulae of 110 solvents that are of interest with regard to the solvation of ions. It presents also the relative molar mass M, in g mol^{-1}, and the liquid range of the solvents, as delineated by the freezing temperature, t_m, and the normal boiling temperature, t_b, both in degrees Celsius. Most of the latter data are from the compilation of Riddick and Bunger,[1] some are from the *Handbook of Chemistry and Physics*,[2] and a few are from other sources, as shown in the Table.

The other properties considered here can be divided into bulk and molecular properties. The former are pressure and temperature dependent, and the values for 1 atm and 25 °C (101 325 Pa and 298.15 K) are commonly required (except when 25 °C is outside the liquid range at 1 atm). The molecular properties are in principle temperature and pressure independent, but they are generally derived from temperature (and pressure) dependent bulk properties. For the description of ion solvation, the values derived from quantities measured at the standard conditions of 25 °C and 1 atm are generally used.

Of the bulk properties, the following mechanical and thermal ones are of interest in the present context:

- The density d, and the derived molar volume $V = M/d$, given in Table 6.2 in the non-SI units of g cm^{-3} and cm^3 mol^{-1}, respectively.

Table 6.1 Some solvents and their liquid ranges

No.	Solvent	Abbre-viation	Formula	M (g mol^{-1})	t_m (°C)	t_b (°C)
1	n-Hexane		C_6H_{14}	86.18	−95.3	68.7
2	c-Hexane		C_6H_{12}	84.16	6.5	80.7
3	Benzene		C_6H_6	78.12	5.5	80.1
4	Toluene		C_7H_8	92.14	−95.0	110.6
5	p-Xylene		C_8H_{10}	106.18	13.3	138.4
6	Mesitylene		C_9H_{12}	120.20	−44.7	164.7
7	Water	W	H_2O	18.015	0.0	100.0
8	Methanol	MeOH	CH_4O	32.04	−97.7	64.7
9	Ethanol	EtOH	C_2H_6O	46.07	−114.1	78.3
10	1-Propanol	PrOH	C_3H_8O	60.10	−126.2	97.2
11	2-Propanol	iPrOH	C_3H_8O	60.10	−88.0	82.3
12	1-Butanol	BuOH	$C_4H_{10}O$	74.12	−88.6	117.7
13	2-Methyl-1-propanol	iBuOH	$C_4H_{10}O$	74.12	−108.0	107.7
14	2-Butanol	2-BuOH	$C_4H_{10}O$	74.12	−114.7	99.6
15	t-Butyl alcohol	tBuOH	$C_4H_{10}O$	74.12	25.8	82.4
16	1-Pentanol	PnOH	$C_5H_{12}O$	88.15	−78.2	137.8
17	3-Methyl-1-butanol	iPnOH	$C_5H_{12}O$	88.15	−117.2	130.5
18	1-Hexanol	HxOH	$C_6H_{14}O$	102.18	−44.6	157.0
19	1-Octanol	OcOH	$C_8H_{18}O$	130.23	−15.0	195.2
20	Benzyl alcohol	BzOH	C_7H_8O	108.14	−15.3	205.5
21	Phenol	PhOH	C_6H_6O	94.11	40.9	181.8
22	2-Methylphenol		C_7H_8O	108.12	11.5[b]	202.2[b]
23	2-Chlorophenol		C_6H_5ClO	128.56	9.0	174.9
24	2-Chloroethanol		C_2H_5ClO	80.52	−67.5	128.6
25	2,2,2-Trifluoroethanol	TFE	$C_2H_3F_3O$	100.04	−43.5[b]	73.8[a]
26	Ethylene glycol	$En(OH)_2$	$C_2H_6O_2$	62.07	−13.0	197.3
27	Glycerol		$C_3H_8O_3$	92.10	18.2	290.0
28	2-Methoxyethanol		$C_3H_8O_2$	76.10	−85.1	124.6
29	2-Ethoxyethanol		$C_4H_{10}O_2$	90.12	< -90.0	135.6
30	2-Butoxyethanol		$C_6H_{14}O_2$	118.18		170.2
31	Diethylene glycol		$C_4H_{10}O_3$	106.12	−6.5	244.8
32	Diethyl ether	Et_2O	$C_4H_{10}O$	74.12	−116.3	34.6
33	Di-1-propyl ether	Pr_2O	$C_6H_{14}O$	102.18	−123.2	89.6
34	Di-2-propyl ether	iPr_2O	$C_6H_{14}O$	102.18	−85.5	68.3
35	1,2-Dimethoxyethane	DME	$C_4H_{10}O_2$	90.12	−58	93.0
36	Tetrahydrofuran	THF	C_4H_8O	72.11	−108.5	66.0
37	Dioxane		$C_4H_8O_2$	88.11	11.8	101.3
38	Anisole		C_7H_8O	108.14	−37.5	153.8
39	Bis(2-chloroethyl)ether		$C_4H_8Cl_2O$	143.01	−46.8	178.8
40	Benzaldehyde		C_7H_6O	106.13	−55.6	178.9
41	Acetone	Me_2CO	C_3H_6O	58.08	−94.7	56.3
42	2-Butanone		C_4H_8O	72.11	−86.7	79.6
43	c-Hexanone		$C_6H_{10}O$	98.15	−32.1	155.7
44	Methyl isobutyl ketone		$C_6H_{12}O$	100.16	−84	116.5
45	Acetophenone		C_8H_8O	120.15	19.6	202.0
46	Formic acid		CH_2O_2	46.03	8.3	100.6
47	Acetic acid	AcOH	$C_2H_4O_2$	60.05	16.7	117.9
48	Propanoic acid		$C_3H_6O_2$	74.08	−20.7	140.8
49	Butanoic acid		$C_4H_8O_2$	88.11	−5.2	163.3
50	Pentanoic acid		$C_5H_{10}O_2$	102.13	−33.7	185.5
51	Hexanoic acid		$C_6H_{12}O_2$	116.16	−4.0	205.7
52	Trifluoroacetic acid		$C_2HF_3O_2$	114.02	−15.3	71.8
53	Acetic anhydride	Ac_2O	$C_4H_6O_3$	102.09	−73.1	140.0
54	Methyl acetate	MeOAc	$C_3H_6O_2$	74.08	−98.1	56.3

(continued)

Table 6.1—*continued*

No.	Solvent	Abbreviation	Formula	M (g mol^{-1})	t_m (°C)	t_b (°C)
55	Ethyl acetate	EtOAc	$C_4H_8O_2$	88.11	−84.0	77.1
56	Butyl acetate	BuOAc	$C_6H_{12}O_2$	116.16	−73.5	126.1
57	Ethyl butanoate		$C_6H_{12}O_2$	116.16	−98.0	121.6
58	Diethyl carbonate		$C_5H_{10}O_3$	118.13	−43.0	126.8
59	γ-Butyrolactone		$C_4H_6O_2$	86.09	−43.5	204.0
60	Ethylene carbonate	EC	$C_3H_4O_3$	88.06	36.4	238.0
61	Propylene carbonate	PC	$C_4H_6O_3$	102.08	−54.5[c]	242[c]
62	Methyl salicylate		$C_8H_8O_3$	152.15	−8.6	233.3
63	Diethyl sulphite		$C_4H_{10}O_3S$	138.19		157[b]
64	Trimethyl phosphate	TMP	$C_3H_9O_4P$	140.00	−46.0[b]	197.2[b]
65	Triethyl phosphate	TEP	$C_6H_{15}O_4P$	182.16	−56.4[b]	215[b]
66	Tributyl phosphate	TBP	$C_{12}H_{27}O_4P$	266.32	< −80	289
67	Methylene chloride		CH_2Cl_2	89.93	−95.1	39.8
68	Chloroform		$CHCl_3$	119.38	−63.6	61.2
69	Carbon tetrachloride		CCl_4	153.82	−23.0	76.8
70	1,1-Dichloroethane	1,1DClE	$C_2H_4Cl_2$	98.96	−97.0	57.3
71	1,2-Dichloroethane	1,2DClE	$C_2H_4Cl_2$	98.96	−35.7	83.5
72	Chlorobenzene	PhCl	C_6H_5Cl	112.56	−45.6	131.7
73	o-Dichlorobenzene		$C_6H_4Cl_2$	147.01	−17.0	180.5
74	Carbon disulphide		CS_2	76.14	−111.6	46.2
75	Dimethyl sulphoxide	DMSO	C_2H_6OS	78.13	18.5	189.0
76	Tetramethylene sulphone	TMS	$C_4H_8O_2S$	120.17	28.5	287.3
77	Ammonia		NH_3	17.03	−77.7[b]	−33.4[b]
78	Hydrazine		N_2H_4	32.05	1.4[b]	113.5[b]
79	Cyclohexylamine		$C_6H_{13}N$	99.18	−17.7	134.8
80	Dipropylamine		$C_6H_{15}N$	101.19	−63.0	109.2
81	Triethylamine	TEA	$C_6H_{15}N$	101.19	−114.7	89.5
82	Aniline	$PhNH_2$	C_6H_7N	93.13	−6.0	184.4
83	Pyridine	py	C_5H_5N	79.10	−41.6	115.3
84	Quinoline		C_9H_7N	129.16	−14.9	237.1
85	2-Chloroaniline		C_6H_6ClN	127.57	−1.9	208.8
86	Ethylene diamine	en	$C_2H_8N_2$	60.10	11.3	117.3
87	2-Aminoethanol		C_2H_7ON	61.08	10.5	171.0
88	Diethanolamine		$C_4H_{11}O_2N$	105.14	28.0	268.4
89	Triethanolamine		$C_6H_{15}O_3N$	149.19	21.6	335.4
90	Morpholine		C_4H_9ON	87.12	−3.1	128.9
91	Nitromethane	$MeNO_2$	CH_3O_2N	61.04	−28.6	101.2
92	Nitrobenzene	$PhNO_2$	$C_6H_5O_2N$	123.11	5.8	210.8
93	Acetonitrile	MeCN	C_2H_3N	41.05	−43.8	81.6
94	Propionitrile		C_3H_5N	55.08	−92.8	97.4
95	Butyronitrile		C_4H_7N	69.11	−111.9	117.9
96	Benzonitrile	PhCN	C_7H_5N	103.13	−12.8	191.1
97	Benzyl cyanide	BzCN	C_8H_7N	117.15	−23.8	233.5
98	Succinonitrile		$C_4H_4N_2$	80.09	57.9	267.0
99	Formamide	FA	CH_3ON	45.04	2.6	210.5
100	N-Methylformamide	NMF	C_2H_5ON	59.07	−3.8	~180
101	N,N-Dimethylformamide	DMF	C_3H_7ON	73.10	−60.4	153.0
102	N,N-Diethylformamide	DEF	$C_5H_{11}ON$	101.13		177[b]
103	Acetamide		C_2H_5ON	59.07	80.0	222.2
104	N-Methylacetamide	NMA	C_3H_7ON	73.10	30.6	206.0
105	N,N-Dimethylacetamide	DMA	C_4H_9ON	87.12	−20.0	166.1
106	N,N-Diethylacetamide	DEA	$C_6H_{13}ON$	115.18		184[c]
107	N-Methylpropionamide	NMP	C_4H_9ON	87.12	−30.9	~193
108	N-Methylpyrrolidinone	NMPy	C_5H_9ON	99.13	−24.4	202

Table 6.1—*continued*

No.	Solvent	Abbre-viation	Formula	M (g mol^{-1})	t_m (°C)	t_b (°C)
109	Tetramethylurea	TMU	$C_5H_{12}ON_2$	116.16	−1.2	175.2
110	Hexamethylphosphoric triamide	HMPT	$C_6H_{18}ON_3P$	179.20	7.2	233

[a] L. M. Mukherjee and E. Grunwald, *J. Phys. Chem.*, **62**, 1311 (1958); [b] Ref. 2; [c] J. Barthel, H. J. Gores, G. Schmeer, and R. Wachter, *Topics Curr. Chem.*, **111**, 33 (1983).

— The derived thermal expansibility $\alpha_P = (\partial \ln V/\partial T)_P = -(\partial \ln d/\partial T)_P$ (calculated as $-\Delta d/d_{25}\Delta T$, where d_{25} is the density at 25 °C), and the isothermal compressibility $\kappa_T = -(\partial \ln V/\partial P)_T$.
— The molar heat of vaporization, ΔH^V, in kJ mol^{-1}.
— The molar heat capacity at constant pressure C_P, in J K^{-1} mol^{-1}.

These quantities, obtained mostly from Ref. 1, are presented in Table 6.2 for the same set of 110 solvents, and are valid for 1 atm and 25 °C unless otherwise noted.

Table 6.2 Thermodynamic properties of solvents as 25 °C[a]

No.	Solvent	d (g cm^{-3})	V (cm^3 mol^{-1})	α_P (10^{-3} K^{-1})	κ_T (10^{-10} Pa^{-1})	ΔH^V (kJ mol^{-1})	C_P (J K^{-1} mol^{-1})
1	n-Hexane	0.6548	131.6	1.361	16.27[b]	31.55	195.4
2	c-Hexane	0.7739	108.7	1.573	12.3[b]	33.04	155.6
3	Benzene	0.8737	89.9	1.202	9.38[b]	33.85	136.1
4	Toluene	0.8623	106.9	1.074	9.40[b]	37.99	151.0
5	p-Xylene	0.8567	123.9	1.015	7.26[b]	42.38	176.5[c]
6	Mesitylene	0.8611	139.6	0.943	6.8[b]	47.48	212.9
7	Water	0.9970	18.07	0.257	4.52	43.99	75.3
8	Methanol	0.7866	40.7	1.182	11.95[d]	37.43	80.2
9	Ethanol	0.7850	58.7	1.103	11.87[d]	42.30	109.9[e]
10	1-Propanol	0.7998	75.1	1.000	9.70[d]	47.32	143.8
11	2-Propanol	0.7813	76.9	1.073	11.29[d]	45.52	154.6
12	1-Butanol	0.8060	92.0	0.918	8.97[d]	52.47	177.0
13	2-Methyl-1-propanol	0.7978	92.9	0.953	9.97[d]	51.17	181.0
14	2-Butanol	0.8026	92.3	1.072	9.76[d]	49.66	213.8
15	t-Butyl alcohol	0.7812[f]	94.9[f]	1.408[f]	10.82[df]	46.82[f]	220.0
16	1-Pentanol	0.8115	108.5	0.887	8.63[d]	56.94	208.3
17	3-Methyl-1-butanol	0.8071	109.2	0.818	8.95[d]	55.6	211.3
18	1-Hexanol	0.8159	125.2	0.907	8.26[d]	61.9	232.5[g]
19	1-Octanol	0.8221	158.4	0.842	7.83[d]	72.0	284.5[g]
20	Benzyl alcohol	1.0413	103.9	0.695	4.74[d]	61.55[h]	219.0
21	Phenol	1.0722[f]	87.8[f]	0.802[i]		57.8[f]	201.4[i]
22	2-Methylphenol	1.0302	105.0	0.759		61.71	224.9
23	2-Chlorophenol	1.2634[ej]	101.8[e]	0.922[q]		67.31[q]	
24	2-Chloroethanol	1.1965	67.3	0.896		29.6[h]	
25	2,2,2-Trifluoro-ethanol	1.3826	72.4	1.187	10.0	44.0	

(*continued*)

Table 6.2—*continued*

No.	Solvent	d (g cm^{-3})	V (cm^3 mol^{-1})	α_P (10^{-3} K^{-1})	κ_T (10^{-10} Pa^{-1})	ΔH^V (kJ mol^{-1})	C_P (J K^{-1} mol^{-1})
26	Ethylene glycol	1.1100	55.9	0.618	3.82[d]	61.1	151.0
27	Glycerol	1.2582	73.2	0.494	2.46[d]	85.8	223.0
28	2-Methoxyethanol	0.9602	79.3	0.906		54.0	167.0
29	2-Ethoxyethanol	0.9252	97.4	0.925		47.2	209.0
30	2-Butoxyethanol	0.8964	131.8	0.980		49.0	276.0
31	Diethylene glycol	1.1127	95.4	0.674	4.08[d]	52.2	245.0
32	Diethyl ether	0.7076	104.7	1.603	14.4[b]	27.2	167.0[g]
33	Di-1-propyl ether	0.7419	137.7	1.267		35.7	221.6[s]
34	Di-2-propyl ether	0.7182	142.3	1.476		32.6	216.0[e]
35	1,2-Dimethoxyethane	0.8621	104.5	1.021		32.1[j]	193.3[t]
36	Tetrahydrofuran	0.8842	81.6	1.142	8.0[r]	32.0	141.0[e]
37	Dioxane	1.0280	85.7	1.097		35.7[j]	153.0[e]
38	Anisole	0.9893	109.3	0.950		45.0	209.0[c]
39	Bis(2-chloroethyl) ether	1.2130	117.9	0.965		45.2[h]	221.0[c]
40	Benzaldehyde	1.0434	101.7	0.249		39.6	171.3
41	Acetone	0.7844	74.0	1.423	12.55[b]	30.8	129.0[c]
42	2-Butanone	0.7997	90.2	1.301		34.1[j]	158.9
43	c-Hexanone	0.9421	104.2	0.945	6.94[i]	42.0[j]	200.0[c]
44	Methyl isobutyl ketone	0.7961	125.8	1.181		41.7	192.0[e]
45	Acetophenone	1.0238	117.4	0.836		53.4	228.0[c]
46	Formic acid	1.2141	37.9	1.021		19.9	99.1
47	Acetic acid	1.0437	57.5	1.073	8.64[b]	23.0	124.8[e]
48	Propanoic acid	0.9880	75.0	1.073		54.9	154.3[e]
49	Butanoic acid	0.9532	92.4	1.049		60.5	174.0[e]
50	Pentanoic acid	0.9345	109.3	0.963		69.3	210.4
51	Hexanoic acid	0.9230	125.9	1.014		78.8	248.1
52	Trifluoroacetic acid	1.4785	77.1	1.587		36.3	221.3
53	Acetic anhydride	1.0752	95.0	1.125		48.9	192.0[c]
54	Methyl acetate	0.9279	79.8	1.358	10.1[b]	32.3	155.6
55	Ethyl acetate	0.8946	98.5	1.342	10.5[b]	35.1	169.0[e]
56	Butyl acetate	0.8764	132.5	1.164		43.6	242.0[e]
57	Ethyl butanoate	0.8739	132.9	1.197	10.1	42.0	220.2
58	Diethyl carbonate	0.9693	121.9	1.100		41.1[c]	210.8
59	γ-Butyrolactone	1.1254	76.5	0.861		52.2[h]	143.9
60	Ethylene carbonate	1.3214[i]	66.6[i]	0.737[i]		50.1[i]	138.7[i]
61	Propylene carbonate	1.198	85.2	0.958	5.9	42.8	167.6
62	Methyl salicylate	1.1782	129.1	0.849		53.0[j]	248.8
63	Diethyl sulphite	1.0829[ej]	127.6[e]			45.1[j]	
64	Trimethyl phosphate	1.2144[ej]	115.3[e]			53.8[j]	
65	Triethyl phosphate	1.0695[ej]	170.3[e]			48.3[j]	
66	Tributyl phosphate	0.9727	273.8	0.807		61.5[h]	
67	Methylene chloride	1.3168	64.5	1.367	6.87	28.6	167.2
68	Chloroform	1.4799	80.7	1.255	10.2[b]	32.2[e]	117.1[e]
69	Carbon tetrachloride	1.5844	97.1	1.211	10.91	32.4	133.0
70	1,1-Dichloroethane	1.1680	84.7	1.336	5.9	31.0	126.3
71	1,2-Dichloroethane	1.2458	79.4	1.156	7.8	34.3	129.1
72	Chlorobenzene	1.1009	102.2	0.981	6.14	42.6	150.2
73	o-Dichlorobenzene	1.3003	113.1	0.855		50.2	221.7
74	Carbon disulphide	1.2555	60.6	1.162	10.73	27.5	76.0
75	Dimethyl sulphoxide	1.0958	71.3	0.982	5.2	52.9	152.3
76	Tetramethylene sulphone	1.2614[c]	95.3[c]	0.729[c]		73.2[c]	159.0[c]
77	Ammonia	0.7601[h]	22.4[h]	1.88[h]	0.13[k]	23.3[h]	76.5[l]
78	Hydrazine	1.0036[q]	31.9[q]	0.867[q]		46.31[q]	98.8[q]
79	Cyclohexylamine	0.8622	115.0	1.045		45.7	223

Table 6.2—*continued*

No.	Solvent	d (g cm^{-3})	V (cm^3 mol^{-1})	α_P (10^{-3} K^{-1})	κ_T (10^{-10} Pa^{-1})	ΔH^V (kJ mol^{-1})	C_P (J K^{-1} mol^{-1})
80	Dipropylamine	0.7329	138.1	1.255	10.5[r]	40.8	253
81	Triethylamine	0.7230	140.0	1.273	11.5[r]	35.0[e]	224.5
82	Aniline	1.0175	91.5	0.849	4.49	55.8	192.0
83	Pyridine	0.9782	80.9	1.012	5.50[r]	40.4	136.0[g]
84	Quinoline	1.0898	118.5	0.729		64.1[h]	200.0
85	2-Chloroaniline	1.2078	105.6	0.788		56.8	
86	Ethylene diamine	0.8895	67.6	0.787		46.6[e]	177
87	2-Aminoethanol	1.0116	60.4	0.781	3.45[r]	66.1	127[c]
88	Diethanolamine	1.0899[c]	96.5[c]	0.587[c]		48.3[c]	233.5[c]
89	Triethanolamine	1.1196	133.3	0.482		112.5	310.0[c]
90	Morpholine	0.9955	87.5	0.952	5.07[r]	44.0	165.0[u]
91	Nitromethane	1.1313	54.0	1.217		38.3	105.8
92	Nitrobenzene	1.1984	102.7	0.825	4.79	52.5	177.0[c]
93	Acetonitrile	0.7766	52.9	1.388	10.7	33.2	91.6
94	Propionitrile	0.7768	70.9	1.326	8.72[r]	36.1	119.7
95	Butyronitrile	0.7865	87.9	1.154		37.0	
96	Benzonitrile	1.0006	103.1	0.881	5.23[r]	55.5	190.3
97	Benzyl cyanide	1.0125	115.7	0.593		51.9[h]	
98	Succinonitrile	0.9867[l]	81.2[l]	0.822[l]		62.4[l]	161.0[l]
99	Formamide	1.1292	39.9	0.746	4.0	65.0	107.6
100	N-Methylformamide	0.9988	59.1	0.868		59.6[v]	123.8[t]
101	N,N-Dimethylformamide	0.9440	77.4	1.009	6.22	47.5	156.7
102	N,N-Diethylformamide	0.9017	111.4	0.860		49.0	
103	Acetamide	0.9901[m]	59.7[m]	0.881[m]		59.0[h]	166.0[m]
104	N-Methylacetamide	0.9498[fc]	77.0[fc]	0.884[fc]		69.0	151.4[t]
105	N,N-Dimethylacetamide	0.9366	93.0	0.918		53.1	176[e]
106	N,N-Diethylacetamide	0.9080	127.3	0.866		45.3[v]	
107	N-Methylpropionamide	0.9305	93.6	0.850		54.4	179[t]
108	N-Methylpyrrolidone	1.0279	96.4	0.904		56.3	
109	Tetramethylurea	0.9619	120.3	1.41		45.5[l]	
110	Hexamethyl phosphoric triamide	1.0202	175.7	0.819	7.9	56.6	

[a] Ref. 1; [b] Landoldt-Börnstein, Vol. II/1 (1971); [c] at 30 ± 2 °C; [d] G. W. Marks, *J. Acoust. Soc. Am.*, **41**, 103 (1967); [e] at 20 ± 2 °C; [f] supercooled liquid; [g] at 15 ± 2 °C; [h] at normal boiling point; [i] at 40 ± 2 °C; [j] Ref. 2; [k] at 0 °C; [l] at 60 °C; [m] at 90 °C; [n] at 115 °C; [p] E. Wilhelm and R. Battino, *Chem. Rev.*, **73**, 1 (1973), p. 7; [q] *Adv. Chem. Ser.*, Vol. 15 (1955) and Vol. 29 (1961), *Am. Chem. Soc.*, Washington; [r] K. J. Patil, *Indian J. Pure Appl. Phys.*, **16**, 608 (1978); [s] R. J. L. Andon, J. F. Counsell, D. A. Lee, and J. F. Martin, *J. Chem. Thermod.*, **7**, 587 (1975); [t] J. Konicek and I. Wadsö, *Acta Chem. Scand.*, **25**, 1541 (1971), K. Kusano, J. Suurkuusk, and I. Wadsö, *J. Chem. Thermod.*, **5**, 757 (1973), R. Skold, J. Suurkuusk, and I. Wadsö, *ibid.*, **8**, 1075 (1976); [u] V. I. Lyashkov, *Izv. Vyshch. Ucheb. Zaved., Khim Khim Teknnol.*, **23**, 1085 (1980); [v] B. M. Rode and A. Pontani, *Monatsh. Chem.*, **108**, 1153 (1977) and private comm., 1983.

Other bulk properties that are of importance in the present context include the following flow, surface, optical and electrical ones:

— The coefficient of viscosity η, in mPa · s (numerically equal to centipoise).
— The coefficient of surface tension γ, in 10^{-2} N m^{-1}.
— The refractive index (for the sodium D line) n_D.
— The relative permittivity (dielectric constant) ε and $-(\partial\varepsilon/\partial T)_P$.

The values for 25 °C and 1 atm are presented in Table 6.3 for the same set of 110 solvents, and are taken mostly from Ref. 1. The values for η and γ are generally interpolated to 25 °C from data given at 15 and 30 °C. For some applications the temperature and pressure derivatives of the relative permittivity, $(\partial \ln \varepsilon/\partial T)_P$ and $(\partial \ln \varepsilon/\partial P)_T$, may be required, as for the description of interionic interactions in electrolyte solutions of finite concentrations, but the latter is available only for a minority of the solvents discussed.

Table 6.3 Some flow, surface, optical and electrical properties at 25 °C[a]

No.	Solvent	η (10^{-3} Pa·s)	γ (10^{-2} N m^{-1})	n_D	ε	$-d\varepsilon/dT$[x] (K^{-1})
1	n-Hexane	0.299	1.791	1.3723	2.023[b]	0.00155
2	c-Hexane	0.898	2.438	1.4235	1.880	0.00155[y]
3	Benzene	0.603	2.818	1.4979	2.275	0.00199[y]
4	Toluene	0.552	2.792	1.4941	2.379	0.00243
5	p-Xylene	0.605	2.776	1.4933	2.270[b]	0.00160
6	Mesitylene	1.039	2.831	1.4968	2.279	
7	Water	0.890	7.181	1.3325	78.39	0.3595[y]
8	Methanol	0.545	2.212	1.3265	32.70	0.197
9	Ethanol	1.078	2.190	1.3594	24.55	0.147[y]
10	1-Propanol	1.956	2.330	1.3837	20.33	0.142[y]
11	2-Propanol	2.073	2.124	1.3752	19.92	0.131
12	1-Butanol	2.593	2.416	1.3973	17.51	0.132
13	2-Methyl-1-propanol	3.91	2.255	1.3939	17.93	0.154
14	2-Butanol	3.66	2.305	1.3950	16.56	0.154
15	t-Butyl alcohol	5.12[c]	2.00[c]	1.3851[c]	12.47[c]	
16	1-Pentanol	3.35	2.516	1.4079	13.9	0.114
17	3-Methyl-1-butanol	3.48	2.388	1.4052	14.7	0.122
18	1-Hexanol	4.59	2.405	1.4161	13.3	0.107
19	1-Octanol	7.36	2.546	1.4275	10.34[b]	0.098
20	Benzyl alcohol	5.52	3.945	1.5384	13.1[b]	
21	Phenol	4.60[cd]	3.87[cd]	1.5427[cd]	11.43[cdg]	0.072
22	2-Methylphenol	15	3.703	1.5376	11.8	0.11
23	2-Chlorophenol	4.11	4.23[e]	1.5524[b]	6.16[f]	0.027
24	2-Chloroethanol	3.064	3.89[b]	1.4399	25.8	
25	2,2,2-Trifluoro-ethanol	1.651			26.67	
26	Ethylene glycol	16.9	4.80	1.4306	37.7	0.194
27	Glycerol	945.0	6.3	1.4730	42.5	0.204
28	2-Methoxyethanol	1.60	3.084	1.4002	16.93	
29	2-Ethoxyethanol	1.85	2.82	1.4057	29.6	
30	2-Butoxyethanol	3.15	2.74	1.4177	9.30	
31	Diethylene glycol	30.0	4.48	1.4461	31.7[b]	
32	Diethyl ether	0.238	1.650	1.3495	4.34	0.020
33	Di-1-propyl ether	0.399	1.994	1.3780	3.39	
34	Di-2-propyl ether	0.379	1.734	1.3655	3.88	0.018
35	1,2-Dimethoxyethane	0.455		1.3781	7.20	0.0410[y]
36	Tetrahydrofuran	0.460	2.64	1.4050	7.58	0.0299[y]
37	Dioxane	1.194	3.295	1.4203	2.209	0.0170
38	Anisole	0.895	3.483	1.5143	4.33	0.011
39	Bis(2-chloroethyl) ether	2.14	3.70	1.4553	21.2[b]	
40	Benzaldehyde	1.321	3.9[e]	1.5428	17.8[b]	
41	Acetone	0.304	2.267	1.3560	20.70	0.0977
42	2-Butanone	0.383	2.397	1.3764	15.45	0.00207
43	Cyclohexanone	1.998	3.45[b]	1.4499	18.2[b]	

Table 6.3—*continued*

No.	Solvent	η (10^{-3} Pa·s)	γ (10^{-2} N m^{-1})	n_D	ε	$-\mathrm{d}\varepsilon/\mathrm{d}T$[x] ($K^{-1}$)
44	Methyl isobutyl ketone	0.542	2.329	1.3933	13.11[b]	0.0633
45	Acetophenone	1.642	3.884	1.5322	17.39	0.04
46	Formic acid	1.966	3.703	1.3694	58.5[e]	
47	Acetic acid	1.124	2.688	1.3698	6.15[b]	−0.0011
48	Propanoic acid	1.025	2.621	1.3843	3.37[g]	
49	Butanoic acid	1.515	2.616	1.3958	2.90[g]	−0.0023
50	Pentanoic acid	1.951	2.684	1.4060	2.66[b]	
51	Hexanoic acid	2.81	2.755	1.4148	2.63[h]	
52	Trifluoroacetic acid	0.855	13.6	1.2850[b]	8.55[b]	−0.50
53	Acetic anhydride	0.841	3.190	1.3884	20.7[b]	
54	Methyl acetate	0.364	2.41	1.3614[b]	6.68	0.022
55	Ethyl acetate	0.426	2.315	1.3698	6.02	0.015
56	Butyl acetate	0.672	2.454	1.3873	5.1	0.014
57	Ethyl butyrate	0.613	2.392	1.3905	5.10[b]	0.010
58	Diethyl carbonate	0.748	2.595	1.3829	2.820[b]	
59	γ-Butyrolactone	1.7		1.4348	39[b]	
60	Ethylene carbonate	1.925[rd]		1.4199[d]	89.6[d]	0.408[z]
61	Propylene carbonate	2.513[q]	4.14[t]	1.4209	66.1	0.240[z]
62	Methyl salicylate		3.86[g]	1.5333	9.53[g]	0.031
63	Diethyl sulphite	0.839		1.4144[b]	15.6[b]	
64	Trimethyl phosphate	2.03		1.3939	22.3	
65	Triethyl phosphate	1.55	3.06[e]	1.4053[b]	13.3	
66	Tributyl phosphate	3.39	2.715	1.4071	8.05	0.0274
67	Methylene chloride	0.411	2.733	1.4212	8.93	0.0387
68	Chloroform	0.540	2.653	1.4429	4.90[b]	0.0177
69	Carbon tetrachloride	0.905	2.615	1.4574	2.238	0.00200
70	1,1-Dichloroethane	0.505	2.419	1.4138	10.0	0.0480[y]
71	1,2-Dichloroethane	0.779	3.154	1.4421	10.36	0.0560
72	Chlorobenzene	0.756	3.270	1.5221	5.62	0.0168
73	o-Dichlorobenzene	1.324	2.620	1.5491	9.93	0.0444
74	Carbon disulphide	0.347	3.152	1.6241	2.63	0.00268
75	Dimethyl sulphoxide	1.996	4.286	1.4773	46.68	0.106[y]
76	Tetramethylene sulphone	10.286[f]	3.55[f]	1.4820[f]	43.3[f]	
77	Ammonia	0.166[i]	2.34[i]	1.325[e]	23.9[j]	0.0780[y]
78	Hydrazine	0.905[n]	7.531[n]	1.5568[n]	52.9[bg]	0.256
79	Cyclohexylamine	1.097	3.090	1.4565	4.73[b]	
80	Dipropylamine	0.500	2.228	1.4018	2.8[g]	
81	Triethylamine	0.363	2.014	1.4010[bg]	2.42[g]	
82	Aniline	3.770	4.279	1.5836	6.98[g]	0.00235
83	Pyridine	0.884	3.633	1.5075	12.3	
84	Quinoline	3.37	4.524	1.6248	9.00	
85	2-Chloroaniline	2.60	4.310	1.5859	13.4	
86	Ethylene diamine	1.54	4.013	1.4543	12.9	0.010
87	2-Aminoethanol	19.346	4.832	1.4521	37.72	
88	Diethanolamine	380[f]	4.53[u]	1.4747[f]		
89	Triethanolamine	613.6		1.4835	29.36	
90	Morpholine	2.011	3.694	1.4511	7.42	
91	Nitromethane	0.610	3.648	1.3796	35.8[f]	0.161
92	Nitrobenzene	1.795	4.276	1.5500	34.82	0.180
93	Acetonitrile	0.341	2.760	1.3416	37.5[b]	0.160
94	Propionitrile	0.410	2.582	1.3636	27.2[b]	
95	Butyronitrile	0.549	2.679	1.3820	20.3[b]	
96	Benzonitrile	1.213	3.846	1.5259	25.20	0.091
97	Benzyl cyanide	1.961	4.082	1.5209	18.7	

(*continued*)

Table 6.3—*continued*

No.	Solvent	η (10^{-3} Pa·s)	γ (10^{-2} N m^{-1})	n_D	ε	$-d\varepsilon/dT^x$ (K^{-1})
98	Succinonitrile	2.591[k]	4.678[k]	1.4173[k]	56.5[k]	0.20
99	Formamide	3.30	5.791	1.4468	111.0	0.72
100	N-Methylformamide	1.65	3.87	1.4300	182.4	1.62[y]
101	N,N-Dimethylformamide	0.802	3.52	1.4282	36.71	0.178[y]
102	N,N-Diethylformamide	1.254[g]		1.4318		
103	Acetamide	2.203[l]	3.846[l]	1.4233[l]	59[l]	
104	N-Methylacetamide	3.64[fc]	3.367[cf]	1.4277[cf]	191.3[cf]	
105	N,N-Dimethylacetamide	0.88	3.315	1.4356	37.78	
106	N,N-Diethylacetamide	1.226[p]		1.4401[b]		
107	N-Methylpropionamide	5.22	3.175	1.4345	172.2	
108	N-Methylpyrrolidinone	1.666	4.05[s]	1.4680	32.0	
109	Tetramethylurea	1.395		1.4493	23.06	
110	Hexamethyl phosphoric triamide	3.245	3.38[b]	1.4570	30.0	

[a] Ref. 1; [b] at 20 °C; [c] as a supercooled liquid; [d] at 40 °C; [e] at 15 °C; [f] at 30 °C; [g] from ref. 2 or Landoldt-Börnstein, Vol. X/X (1900); [h] at 70 °C; [i] at 0 °C; [j] at the normal boiling temperature; [k] at 60 °C; [l] at 90 °C; [m] E. Wilhelm and R. Battino, *Chem. Rev.*, **73**, 1 (1973); [n] *Adv. Chem. Ser.*, **15** (1955) and **29** (1961), *Am. Chem. Soc.*, Washington; [p] J. Barthel, H. J. Gores, G. Schmeer, and R. Wachter, *Topics Curr. Chem.*, **111**, 33 (1983); [q] M. L. Jansen and H. L. Yeager, *J. Phys. Chem.*, **77**, 3089 (1973); [r] G. Petrella and A. Sacco, *J. Chem. Soc. Faraday Trans. 1*, **74**, 2070 (1978); [s] E. Sacher, *J. Coll. Interf. Sci.*, **83**, 649 (1981); [t] R. Aveyard and Y. Thompson, *Can. J. Chem.*, **57**, 856 (1979); [u] V. G. Lundina, V. I. Komenenko, I. I. Kurnikova, and M. A. Bulatov, *Koll. Zh.*, **41**, 1102 (1980); [x] A. A. Maryott and E. R. Smith, *Tables of Dielectric Constants of Pure Liquids*, NBS Circular 514 (1951); [y] M. H. Abraham, *J. Chem. Soc. Faraday Trans. 1*, **74**, 2858 (1978) and references therein; [z] W. H. Lee, in *Chemistry of Nonaqueous Solvents*, J. J. Lagowski, ed., Vol. IV, Academic Press, New York (1976).

The following molecular properties for the set of 110 solvents are presented in Table 6.4:

— The dipole moment μ, in Debye units ($D = 3.33564 \cdot 10^{-30}$ C·m), obtained in favourable cases from measurements on gaseous samples, but more commonly from measurements on dilute solutions in nonpolar solvents, using the Debye theory.
— The polarizability α, in 10^{-30} m^3, obtained from the refractive index and the molar volume by means of the Lorentz–Lorenz relation.
— The (volume) diamagnetic susceptibility $-\chi$, in 10^{-35} m^3, obtained from the mass susceptibility by means of the molar volume.
— The solubility parameter $\delta = [(\Delta H^V - \mathbf{R}T)/V]^{1/2}$, in J$^{1/2}$ cm$^{-3/2}$.
— The depth of the intermolecular interaction potential well, u, divided by Boltzmann's constant, $\mathbf{k}$ (so that $u/\mathbf{k}$ is in the units K^{-1}) and the molecular collision diameter σ, in nm.

The value of μ is generally from Ref. 1 and α and δ are calculated from the values of n_D, ΔH^V and V given in Tables 6.2 and 6.3, traceable generally to the same source. The values of χ and $u/\mathbf{k}$ and σ are from various sources, as noted in Table 6.4.

Table 6.4 Some molecular properties of solvents

No.	Solvent	μ^a (D)	α^b (10^{-30} m^3)	$-\chi^c$ (10^{-35} m^3)	δ^d ($J^{1/2}$ cm$^{-3/2}$)	$u/\mathbf{k}^e$ (K)	σ^e (nm)
1	n-Hexane	0.09	11.78	12.29	15.0	423	0.592
2	c-Hexane	0.00	10.87	11.31	16.8	313	0.614
3	Benzene	0.00	10.32	9.10	18.8	335	0.563
4	Toluene	0.31	12.26	10.97	18.3	575	0.565
5	p-Xylene	0.02	14.92	12.78	18.1		
6	Mesitylene	0.00	16.19	15.33	18.1		
7	Water	1.834	1.456	2.145	47.9	79	0.274
8	Methanol	2.87	3.26	3.556	29.3	234	0.371
9	Ethanol	1.66	5.13	5.57	26.0	324	0.436
10	1-Propanol	3.09	6.96	7.50	24.4		
11	2-Propanol	1.66	6.98	7.58	23.7		
12	1-Butanol	1.75	8.79	9.32	23.3		
13	2-Methyl-1-propanol	1.79	8.81	9.50	22.9	408	0.529
14	2-Butanol	1.55	8.77	9.51	22.6		
15	t-Butyl alcohol	1.66	8.82	9.53	21.6		
16	1-Pentanol	1.7	10.62	11.19	22.4		
17	2-Methyl-1-butanol	1.82	10.61	11.45	22.1		
18	1-Hexanol	1.55	12.46	13.15	21.8		
19	1-Octanol	1.76	16.14	16.97	20.9	662[k]	0.662[k]
20	Benzyl alcohol	1.66	12.89	11.93	23.8		
21	Phenol	1.45	11.12	10.00	25.1		
22	2-Methylphenol	1.54	13.08	11.96	24.2		
23	2-Chlorophenol	1.33	12.73	12.85	20.0		
24	2-Chloroethanol	1.88	7.03		20.1		
25	2,2,2-Trifluoroethanol	2.52			23.9		
26	Ethylene glycol	2.28	5.73	6.46	32.4		0.437[k]
27	Glycerol	2.67	8.14	9.47	33.7		
28	2-Methoxyethanol	2.04	7.62		25.5		
29	2-Ethoxyethanol	2.08	9.48		21.4		
30	2-Butoxyethanol	2.08	13.16		18.8		
31	Diethylene glycol	2.31	10.09		22.8		
32	Diethyl ether	1.15	8.92	9.15	15.4	351	0.525
33	Di-1-propyl ether	1.32	12.58		15.5		
34	Di-2-propyl ether	1.22	12.62	13.18	14.6		
35	1,2-Dimethoxyethane	1.71	9.55	9.16[l]	16.8		
36	Tetrahydrofuran	1.75	7.93		19.0		
37	Dioxane	0.45	8.60	8.48	19.7	500[k]	0.522[k]
38	Anisole	1.25	13.05	11.98	19.7		
39	Bis(2-chloroethyl) ether	2.58	12.69		19.0		
40	Benzaldehyde	2.77	12.70	10.09	19.1		
41	Acetone	2.69	6.41	5.64	22.1	362	0.476
42	2-Butanone	2.76	8.21	7.57	18.7		
43	Cyclohexanone	3.01	11.10	10.30	19.7		
44	Methyl isobutyl ketone	2.79	11.91	11.63	17.7		
45	Acetophenone	2.96	14.43	12.05	20.8		
46	Formic acid	1.82	3.39	3.31	21.4		
47	Acetic acid	1.68	5.15	5.27	18.9		
48	Propanoic acid	1.68	6.96	7.21	26.4		
49	Butanoic acid	1.65	8.80	9.15	25.1		
50	Pentanoic acid	1.61	10.64	11.10	24.7		
51	Hexanoic acid	1.13	12.49	13.04	24.6		
52	Trifluoroacetic acid	2.28	5.45	7.21[f]	20.9		
53	Acetic anhydride	2.82	8.89	8.77	22.1		0.572

(continued)

Table 6.4—*continued*

No.	Solvent	μ^a (D)	α^b (10^{-30} m^3)	$-\chi^c$ (10^{-35} m^3)	δ^d (J$^{1/2}$ cm$^{-3/2}$)	$u/\mathbf{k}^e$ (K)	σ^e (nm)
54	Methyl acetate	1.61	7.0	7.01	19.3		0.491
55	Ethyl acetate	1.88	8.83	8.98	18.2	531	0.532
56	Butyl acetate	1.84	12.37	12.86	17.6		
57	Ethyl butyrate	1.74	12.50	12.86	17.2		
58	Diethyl carbonate	0.90	11.27	12.52	17.8		
59	γ-Butyrolactone	4.12	7.91		25.5		
60	Ethylene carbonate	4.87	6.69		26.9		0.504
61	Propylene carbonate	4.98	8.56	9.05[f]	21.8	400[k]	0.536
62	Methyl salicylate	2.47	15.89	14.33	19.8		
63	Diethyl sulphite	2.96	12.48	12.54	18.3		
64	Trimethyl phosphate	3.02	11.00		21.1		
65	Triethyl phosphate	3.07	16.33	18.53[j]	16.4		
66	Tributyl phosphate	3.07	27.59	30.08[j]	15.3		0.816
67	Methylene chloride	1.14	6.49	7.74	20.1		
68	Chloroform	1.15	8.48	9.85	19.5	327	0.496
69	Carbon tetrachloride	0.00	10.49	11.09	17.6	536	0.538
70	1,1-Dichloroethane	1.98	8.39	9.53	18.3		
71	1,2-Dichloroethane	1.86	8.33	9.90	20.0		0.501
72	Chlorobenzene	1.54	12.36	11.56	19.8	610	0.561
73	o-Dichlorobenzene	2.27	14.26	14.16	20.5		0.574
74	Carbon disulphide	0.06	8.48	7.02	20.3	466	0.453
75	Dimethyl sulphoxide	3.9	7.99	7.34[f]	26.6	333	0.491
76	Tetramethylene sulphone	4.81	10.77		27.2		0.581
77	Ammonia	1.82	2.26	2.71[i]	30.5	528	0.302
78	Hydrazine	1.85[g]	4.01		37.1	138	0.362
79	Cyclohexylamine	1.26	12.40	13.04[h]	19.4		
80	Dipropylamine	1.03	13.32	14.23[h]	16.7		
81	Triethylamine	0.66	13.30	13.66	15.2		
82	Aniline	1.51	12.13	10.36	24.1		
83	Pyridine	2.37	9.55	8.06	21.7		0.513
84	Quinoline	2.18	16.60	14.26	22.8		
85	2-Chloroaniline	1.77	14.05	13.19	22.7		
86	Ethylene diamine	1.90	7.26	7.56	25.5		
87	2-Aminoethanol	2.27	6.46	7.06	32.5		
88	Diethanolamine	2.81	10.76		31.7		
89	Triethanolamine	3.57	15.10		28.7		
90	Morpholine	1.50	9.34	9.13	21.8		
91	Nitromethane	3.56	4.95	3.47	25.7	298	0.431
92	Nitrobenzene	4.03	12.97	10.26	22.1	609	0.574
93	Acetonitrile	3.44	4.41	4.58	24.1	275[k]	0.427
94	Propionitrile	3.57	6.26	6.44	21.8		0.477
95	Butyronitrile	3.57	8.11	8.37	19.8		0.532
96	Benzonitrile	4.05	12.54	10.82	22.7	520[k]	0.574
97	Benzyl cyanide	3.47	13.96	12.76	20.7		
98	Succinonitrile	3.68	8.10		27.2		
99	Formamide	3.37	4.23	3.83	39.6		
100	N-Methylformamide	3.86	6.05	9.81[f]		320[k]	
101	N,N-Dimethylformamide	3.86	7.90	6.44[f]	24.1	380[k]	0.517
102	N,N-Diethylformimde		11.45		20.4		
103	Acetamide	3.44	6.03	5.66	30.8		
104	N-Methylacetamide	4.39	7.85		29.4	392	0.496
105	N,N-Dimethylacetamide	3.72	9.63	9.31	23.3	450[k]	0.548
106	N,N-Diethylacetamide	3.69	13.30				

Table 6.4—*continued*

No.	Solvent	μ[a] (D)	α[b] (10^{-30} m^3)	$-\chi$[c] (10^{-35} m^3)	δ[d] ($J^{1/2}$ $cm^{-3/2}$)	$u/\mathbf{k}$[e] (K)	σ[e] (nm)
107	N-Methylpropionamide	3.59	9.69		24.1		
108	N-Methylpyrrolidinone	4.09	10.62	10.63[f]	23.6		0.569
109	Tetramethylurea	3.47	12.76	12.57	18.9		0.544
110	Hexamethyl phosphoric triamide	5.54	16.03	19.56[f]	19.1	670[k]	0.698

[a] Ref. 1; [b] from n_D of Table 6.3 and V of Table 6.2, according to: $\alpha = (n_D^2 - 1)(n_D^2 + 2)^{-1}(3/4\pi)(V/\mathbf{N})$; [c] from Landoldt-Börnstein, Vol. II/10 (1967), pp. 66ff, from $\chi = \chi_m \cdot \rho/\mathbf{N}$ where χ_M is the tabulated mass susceptibility and ρ is from Table 6.2, cf. also ref. 2, 47th ed., p. E-108ff (1966/7); [d] from ΔH^V and V of Table 6.2 according to: $\delta = [(\Delta H^V - \mathbf{R}T)/V]^{1/2}$; [e] from gas solubilities, according to E. Wilhelm and R. Battino, *J. Chem. Phys.*, **55**, 4012 (1971); [f] from W. Gerger, U. Mayer, and V. Gutmann, *Monatsh. Chem.*, **108**, 417 (1977), in the same manner as in note c above; [g] A. L. McClellan, *Tables of Experimental Dipole Moments*, Freeman, San Francisco (1963); [h] V. Shanmugasundaram, R. Sabesan, and K. Krishnan, *Z. Phys. Chem. (Leipzig)*, **251**, 407 (1972); [i] W. S. Glaunsinger, S. Zolotov, and M. J. Sienko, *J. Chem. Phys.*, **56**, 4757 (1972); [j] N. Baetman and J. Baudet, *Compt. Rend.*, **265c**, 288 (1967); [k] N. Bruckl and J. I. Kim, *Z. Phys. Chem. (NF)*, **126**, 133 (1981); [l] A. K. Burnham, J. Lee, T. G. Schmalz, P. Beak, and W. H. Flyggare, *J. Am. Chem. Soc.*, **99**, 1836 (1977).

The polarity and polarizability of a solvent is describable not only in terms of the dipole moment μ and the polarizability α, but also by suitable combinations of them. One such expression is

$$[(\varepsilon - 1)\cdot(2\varepsilon + 1)^{-1}]\cdot[(n_D^2 - 1)\cdot(2n_D^2 + 1)^{-1}]$$

that turns out to be well correlated[3b] ($r_{corr} = 0.989$) with the π^* polarity index of Abboud, Kamlet and Taft.[3] (This is true for 'select' solvents; the correlation is considerably worse for the entire body of solvents for which π^* data are available, taken as a whole). This polarity index is a solvatochromic parameter, and measures the ability of a solvent to stabilize a charge or a dipole in solution by virtue of its dielectric effect. The π^* polarity index has been normalized to the values 0.00 for cyclohexane and 1.00 for dimethylsulphoxide, and its values are shown in Table 6.5.

Certain quantities that relate to the electron pair donation and acceptance abilities of solvents are also of great importance for ion solvation. These include, explicitly or implicitly, also measures of the Lewis basicity and acidity of their solvents and, where relevant, their abilities to accept and donate a hydrogen atom towards the formation of a hydrogen bond. Many of the quantities, often called polarity indices, that have been proposed for the description of these abilities are quite well correlated with each other. Hence only four will be discussed here: DN and β for donicity and E_T and α for acceptance.

Gutmann's donor number DN is the negative of the molar enthalpy of reaction of the solvent in question in a dilute solution in 1,2-dichloroethane with antimony pentachloride. The units, according to the original definition, are in

Table 6.5 Polarity indices of solvents

No.	Solvent	DN[a] (kcal mol^{-1})	DN^{N}[b]	E_T[c] (kcal mol^{-1})	E_T^N[d]	π^*[e]
1	n-Hexane	(0)	(0.00)	30.9	0.00	−0.08
2	c-Hexane	(0)	(0.00)	31.2	0.01	0.00
3	Benzene	0.1	0.00	34.5	0.11	0.59
4	Toluene	0.1	0.00	33.9	0.09	0.54
5	p-Xylene	(5)	(0.12)	33.2	0.07	0.43
6	Mesitylene	(10)	(0.26)	33.1	0.07	0.41
7	Water	(33, 18)	(0.85)	63.1	1.00	1.09
8	Methanol	(30, 19)	(0.77)	55.4	0.76	0.60
9	Ethanol	(32)	(0.82)	51.9	0.65	0.54
10	1-Propanol			50.7	0.62	0.52
11	2-Propanol	(36)	(0.93)	48.6	0.55	0.48
12	1-Butanol	(29)	(0.75)	50.2	0.60	0.47
13	2-Methyl-1-propanol			49.0	0.56	
14	2-Butanol			47.1	0.50	
15	t-Butyl alcohol	(38)	(0.98)	43.9	0.41	0.41
16	1-Pentanol	(25)	(0.64)	50.0	0.59	
17	2-Methyl-1-butanol	(32)	(0.82)	47.0	0.50	
18	1-Hexanol			49.4	0.58	
19	1-Octanol	(32)	(0.82)	48.5	0.55	
20	Benzyl alcohol	(23)	(0.59)	50.8	0.62	0.98
21	Phenol	(11)	(0.28)	61.4	0.95	
22	2-Methylphenol			53.5	0.70	
23	2-Chlorophenol			55.4	0.76	
24	2-Chloroethanol			55.5	0.77	
25	2,2,2-Trifluoroethanol			59.5	0.89	0.73
26	Ethylene glycol	(20)	(0.52)	56.3	0.79	0.92
27	Glycerol	(19	(0.49)	57.0	0.81	
28	2-Methoxyethanol			52.3	0.67	(0.71)
29	2-Ethoxyethanol			50.8	0.62	
30	2-Butoxyethanol			50.2	0.60	
31	Diethylene glycol			53.8	0.71	
32	Diethyl ether	19.2	0.49	34.7	0.12	0.27
33	Di-1-propyl ether	(18)	(0.46)	34.0	0.10	(0.27)
34	Di-2-propyl ether	(19)	(0.49)	34.1	0.10	0.27
35	1,2-Dimethoxyethane	23.9	0.62	38.2	0.23	0.53
36	Tetrahydrofuran	20.0	0.52	37.4	0.20	0.58
37	Dioxane	14.8	0.38	36.0	0.16	0.55
38	Anisole	7.5	0.19	37.2	0.20	0.73
39	Bis(2-chloroethyl) ether					
40	Benzaldehyde	(16)	(0.41)	53.1	0.69	(0.92)
41	Acetone	17.0	0.44	42.2	0.35	0.71
42	2-Butanone	17.4	0.45	41.3	0.33	0.67
43	Cyclohexanone	19.4	0.50	40.8	0.31	0.76
44	Methyl isobutyl ketone	(16)	(0.41)	39.4	0.27	
45	Acetophenone	(15)	(0.39)	41.3	0.33	0.90
46	Formic acid	(19)	(0.49)	57.7	0.83	
47	Acetic acid	(20)	(0.52)	57.2	0.82	0.64

Table 6.5—*continued*

No.	Solvent	DN[a] (kcal mol^{-1})	DN^{N}[b]	E_T[c] (kcal mol^{-1})	E_T^N[d]	π^*[e]
48	Propanoic acid			55.9	0.78	
49	Butanoic acid			55.4	0.76	
50	Pentanoic acid			55.3	0.76	
51	Hexanoic acid			55.7	0.77	
52	Trifluoroacetic acid					0.50
53	Acetic anhydride	10.5	0.27	43.9	0.41	0.76
54	Methyl acetate	16.5	0.43	40.0	0.28	0.60
55	Ethyl acetate	17.1	0.44	38.1	0.23	0.55
56	Butyl acetate	11.0	0.28			0.46
57	Ethyl butyrate	16.8	0.43	38.5	0.24	
58	Diethyl carbonate	16.0	0.41	36.2	0.17	0.45
59	γ-Butyrolactone	(18)	(0.46)	44.3	0.42	0.87
60	Ethylene carbonate	16.4	0.42	48.6	0.55	
61	Propylene carbonate	15.1	0.39	46.6	0.49	0.83
62	Methyl salicylate			45.4	0.45	
63	Diethyl sulphite					
64	Trimethyl phosphate	23.0	0.59	43.6	0.40	
65	Triethyl phosphate	(26)	(0.67)	41.7	0.34	0.72
66	Tributyl phosphate	23.7	0.61	39.6	0.27	0.65
67	Methylene chloride	(1)	(0.03)	41.1	0.32	0.82
68	Chloroform	(4)	(0.10)	39.1	0.26	0.58
69	Carbon tetrachloride	(0)	(0.00)	32.5	0.05	0.28
70	1,1-Dichloroethane			39.6	0.27	
71	1,2-Dichloroethane	0.0	0.00	41.9	0.34	0.81
72	Chlorobenzene	3.3	0.09	37.5	0.21	0.71
73	o-Dichlorobenzene	(3)	(0.08)	38.1	0.23	0.80
74	Carbon disulphide	(2)	(0.05)	32.6	0.06	
75	Dimethylsulphoxide	29.8	0.77	45.1	0.44	1.00
76	Tetramethylene sulphone	14.8	0.38	44.0	0.41	0.98
77	Ammonia	(59, 42)	(1.52)			
78	Hydrazine	(44)	(1.13)			
79	Cyclohexylamine					
80	Dipropylamine	(40)	(1.03)			
81	Triethylamine	(25.8, 61)	(1.57)	33.3	0.08	0.14
82	Aniline	(35)	(0.90)	44.3	0.42	
83	Pyridine	33.1	0.85	40.2	0.29	0.87
84	Quinoline	(32)	(0.82)	39.4	0.27	(0.92)
85	2-Chloroaniline	(31)	(0.80)	45.5	0.46	
86	Ethylene diamine	(55)	(1.42)	42.0	0.35	
87	2-Aminoethanol			51.7	0.65	
88	Diethanolamine					
89	Triethanolamine					
90	Morpholine			41.0	0.32	
91	Nitromethane	2.7	0.07	46.2	0.48	0.85
92	Nitrobenzene	4.4	0.11	41.9	0.34	1.01
93	Acetonitrile	14.1	0.36	46.0	0.47	0.75
94	Propionitrile	16.1	0.41	43.7	0.40	0.71

(*continued*)

Table 6.5—*continued*

No.	Solvent	DN[a] (kcal mol^{-1})	DN^{N}[b]	E_T[c] (kcal mol^{-1})	E_T^N[d]	π^*[e]
95	Butyronitrile	16.6	0.43	43.0	0.39	0.71
96	Benzonitrile	12.0	0.31	41.9	0.34	0.90
97	Benzyl cyanide	15.1	0.39	42.9	0.37	
98	Succinonitrile					
99	Formamide	(36)	(0.93)	56.6	0.80	0.97
100	N-Methylformamide	(49)	(1.26)	54.1	0.72	
101	N,N-Dimethylformamide	26.6	0.69	43.9	0.41	0.88
102	N,N-Diethylformamide	31.0	0.80			
103	Acetamide					
104	N-Methylacetamide			52.0	0.66	
105	N,N-Dimethylacetamide	27.8	0.72	43.7	0.40	0.88
106	N,N-Diethylacetamide	32.1	0.83			
107	N-Methylpropionamide					
108	N-Methylpyrrolidinone	27.2	0.70	42.2	0.35	0.92
109	Tetramethylurea	29.6	0.76	41.0	0.32	0.83
110	Hexamethylphosphoric triamide	38.8	1.00	40.9	0.32	0.87

[a] Y. Marcus, *J. Soln. Chem.*, **13**, 599 (1984), values in parentheses have been obtained indirectly or are uncertain; [b] $DN^N = DN/38.8$; [c] Ref. 53, Ch. Reichardt, *Solvent Effects in Organic Chemistry*, Verlag Chemie, Weinheim (1979); [d] $E_T^N = (E_T - 30.8)/32.3$; [e] M. J. Kamlet, J. L. M. Abboud, M. H. Abraham, and R. W. Taft, *J. Org. Chem.*, **48**, 2877 (1983), values in parentheses are uncertain.

kcal mol^{-1}, 1 cal = 4.184 J.[4] The β-scale is established by the deviation of hydrogen-bond-accepting solvents from linear relationships established for non-hydrogen-bond-accepting solvents, for the absorption peaks of similar indicator substances, one of which is a hydrogen bond donor and the other is not. One such pair is 4-nitroaniline and N,N-diethyl-4-nitroaniline, but several other pairs have been used. The scales are averaged and normalized to give a value of 1.000 for hexamethylphosphoric triamide.[5a] The relationships of the DN and β-scales with other donicity scales have been discussed in detail, as have some uncertain DN values for protic solvents (e.g., water) and amines (e.g., triethylamine).[5b] The β-scale for hydrogen-bond-acceptance ability is linear with DN: $\beta = 0.026(DN/\text{kcal mol}^{-1}) + 0.005$, with a correlation coefficient $r_{corr} = 0.976$ for the 13 solvents for which both kinds of quantities are available. A normalized donicity scale, $DN^N = (DN/\text{kcal mol}^{-1})/38.8$ is obtained on the basis of assigning the value $DN^N = 1.000$ to hexamethylphosphoric triamide and $DN^N = 0.000$ to 1,2-dichloroethane. The values of DN^N and β are quite similar, the former being listed, along with DN, in Table 6.5.

The electron pair acceptance polarity index E_T is the lowest energy transition of the indicator solute 2,6-diphenyl-4-(2′,4′,6′-triphenyl-1-pyridino) phenoxide dissolved in the given solvent, expressed in kcal mol^{-1}. Many other common polarity indices, such as Kosower's Z or Gutmann, Mayer, and Gerger's AN, are linearly correlated with Dimroth and Reichardt's E_T.[6] Although not specifically designed to measure it, E_T is also a measure of the hydrogen-bond-donating ability of the solvent. An alternative scale for this parameter, designed to measure the hydrogen-bond-donating ability, is Taft and Kamlet's α,[7] based on the solvatochromic shift of hydrogen-bond-accepting indicators. Although originally the scale was normalized to a value of 1.00 for methanol, the latest implementation [3c] assigns to methanol the value 0.93. Non-zero values are assigned to only 18 solvents altogether, including some (such as acetone, nitromethane, and acetonitrile) that cannot be considered as hydrogen bond donors. The α-scale is fairly well linearly correlated with E_T: $\alpha = 0.054(E_T/\text{kcal mol}^{-1}) - 2.08$, with a correlation coefficient $r_{corr} = 0.860$. This improves to $r_{corr} = 0.985$ if only the eight aliphatic monohydric solvents for which there are data are included. A normalized E_T scale,

$$E_T^N = [E_T - E_T(\text{TMSi})]/[E_T(\text{water}) - E_T(\text{TMSi})]$$

where TMSi = tetramethylsilane, $E_T(\text{TMSi}) = 30.8\ \text{kcal mol}^{-1}$, and $E_T(\text{water}) = 63.1\ \text{kcal mol}^{-1}$, is obtained by reference to the solvents with the least and (almost) the highest E_T. Values of E_T and E_T^N are presented in Table 6.5.

Ideally, the quantities π^*, DN (or β), and E_T (or α) should be orthogonal to each other, since they should measure different properties of the solvents. In practice this is not so: for example, $\pi^* = -1.83 + 0.00364(E_T/\text{kcal mol}^{-1})$ with $r_{corr} = 0.907$. This arises from the fact that polar solvents of appreciable polarizability are often also amphoteric, i.e., may act under appropriate circumstances as Lewis bases or as Lewis acids. They have then high values of all three of these polarity indices.

In addition to the acid–base properties according to the Lewis concept, that can be expressed by means of the electron pair donicity and acceptance discussed above, solvents can be characterized by their tendency to self ionization. This is expressed by the auto-ionization constant, pK_{ai}, that is the ionic concentration product, generally on the molar scale. A further manner of characterizing the acid–base properties is by the equilibrium constant for the donation of a proton or the acceptance of a proton from a standard base, respectively acid, in a reference solvent. The solvent water is the base, H_3O^+ in water is the acid, for which most information is available. Table 9.3, p. 252, presents the data, in connection with a discussion of the transfer of hydrogen ions from water to solvents or their mixtures with water.

A rather ephermeral property of solvents that is relevant to ion solvation is their structure. It is difficult to quantify, and even to define, this property. In the case of the solvent water, structural studies by means of X-ray and neutron diffraction or computer simulation have provided information that is very useful; see Section 5.1. Such information concerning nonaqueous solvents is almost totally lacking. Notable exceptions are the structures obtained for formamide,[8] liquid ammonia,[9] and 2-methyl-2-propanol (tertiary butyl alcohol)[9] among the hydrogen-bonded solvents, acetonitrile as a polar non-hydrogen-bonded solvent,[10] and carbon disulphide, carbon tetrachloride, benzene, and 2,2-dimethylpropane (neopentane) among the nonpolar ones.[11] In the case of formamide, each molecule is hydrogen-bonded to four neighbours, on the average, and for liquid ammonia and tertiary butyl alcohol the corresponding numbers are <6 and 2, respectively. However, complete statements of the structures are not yet possible for any of these solvents on the basis of the experimental material so far available.

There are some other indications that solvent molecules associate to form transient structures in the liquid state. Two related properties are the proton relaxation times T_1 measured by NMR and the viscosity of the solvents, and the effects of ions on them, Compared with the T_1 value of 3.60 s at 25 °C for water, the values for some other solvents are 0.0335 s for glycerol, 0.395 s for ethylene glycol, 2.54 s for formamide, 3.10 s for dimethylsulphoxide, 4.57 s for N-methylformamide, 4.84 s for ethanol, 8.43 s for methanol, and 16.50 s for acetone.[12] These values are related to the viscosities of the solvents by the empirical equation $T_1 \cdot \eta^{0.776} = 5.3$, and hence are not independent quantities. It is interesting to note that among these solvents, only water, ethylene glycol, and glycerol have enough structure for it to be broken by electrolytes composed of structure-breaking ions (see Section 5.4).[12,13] This structure-breaking manifests itself in the lengthening of the proton relaxation times and the lowering of the viscosities in dilute solutions of such electrolytes relative to those of the pure solvents.

These observations do not negate the expectation that polar solvents will associate to some extent by dipole–dipole interactions, and those polar solvents that consist of molecules capable of hydrogen bonding will associate also through the formation of such bonds. Such associations affect the structural properties of solvents, categorized as 'stiffness', 'openness', and 'ordering'.[14] The first of these is measurable by the energy that must be expended for hole-formation, roughly measured by the solubility parameter δ (see Table 6.4). The second is related both to this and to the free volume of the solvent, and is roughly measured by the viscosity η (see Table 6.3).

The last property, 'ordering', is related to the deficit in entropy a solvent may have relative to a similar one, composed of molecules of the same size and shape, but devoid of any structure, i.e., one that is completely disordered (non-associated).[14] A commonly used expression of (the negative of) this deficit in entropy is the Trouton constant, i.e., the entropy of vaporization at constant pressure, chosen arbitrarily as 1 atm (101 325 Pa), corresponding to the normal boiling point. Enthalpies of vaporization $\Delta H^{\mathrm{V}}(t_{\mathrm{b}})$ from Ref. 1 yield the dimensionless quantity $\Delta S^{\mathrm{V}}/\mathbf{R} = \Delta H^{\mathrm{V}}(t_{\mathrm{b}})/\mathbf{R} \cdot (273.15 + t_{\mathrm{b}})$, presented in Table 6.6 for the solvents dealt with in the previous Tables. Table 6.6 shows that 45 nonassociated solvents have values of $\Delta S^{\mathrm{V}}/\mathbf{R} = 10.7 \pm 0.7$. Very few solvents have lower values than this: these are the carboxylic acids that are associated both in the vapour and in the liquid phases—formic, acetic, and propanoic.

Table 6.6 Some measures of the structuredness of solvents

No.	Solvent	$S^V/\mathbf{R}$	g
1	n-Hexane	10.1	—
2	c-Hexane	10.2	—
3	Benzene	10.5	—
4	Toluene	10.4	—
5	p-Xylene	10.5	—
6	Mesitylene	10.7	—
7	Water	13.1	2.61
8	Methanol	12.6	0.99
9	Ethanol	13.3	3.01
10	1-Propanol	13.6	0.88
11	2-Propanol	13.5	3.08
12	1-Butanol	13.3	2.79
13	2-Methyl-1-propanol	13.3	2.78
14	2-Butanol	13.2	3.37
15	2-Methyl-2-propanol	13.3	2.24
16	1-Pentanol	13.2	2.66
17	3-Methyl-1-butanol	13.2	2.50
18	1-Hexanol	13.8	3.47
19	1-Octanol	13.0	2.50
20	Benzyl alcohol	12.7	2.01
21	Phenol (40 °C)	12.1	1.98
22	2-Methylphenol	11.7	2.10
23	2-Chlorophenol	10.8	1.11
24	2-Chloroethanol	12.4	2.50
25	2,2,2-Trifluoroethanol		
26	Ethylene glycol	13.0	2.13
27	Glycerol	13.0	2.16
28	2-Methoxyethanol	11.9	1.70
29	2-ethoxyethanol	11.9	3.60
30	2-Butoxyethanol	13.3	1.34
31	Diethylene glycol	12.1	2.89
32	Diethyl ether	10.4	1.39
33	Di-1-propyl ether	11.0	0.79
34	Di-2-propyl ether	10.3	1.32
35	1,2-Dimethoxyethane		1.23
36	Tetrahydrofuran	10.5	0.93
37	Dioxane	11.4	−0.03
38	Anisole	10.4	0.78
39	Bis(2-chloroethyl) ether	12.0	1.84
40	Benzaldehyde	11.4	0.99
41	Acetone	10.6	1.21
42	2-Butanone	10.8	0.99
43	Cyclohexanone	11.3	1.02
44	Methyl isobutyl ketone	11.0	1.10
45	Acetophenone	9.8	0.99
46	Formic acid	7.5	3.90
47	Acetic acid	7.5	0.58
48	Propanoic acid	9.4	0.26
49	Butanoic acid	11.6	0.20
50	Pentanoic acid	11.6	0.16
51	Hexanoic acid	13.8	0.33
52	Trifluoroacetic acid	11.8	0.74
53	Acetic anhydride	11.1	1.35
54	Methyl acetate	11.1	0.99
55	Ethyl acetate	11.1	0.77
56	Butyl acetate	10.8	0.81
57	Ethyl butyrate	11.1	0.91
58	Diethyl carbonate	10.9	0.86
59	γ-Butyrolactone	13.2	0.92
60	Ethylene carbonate (40 °C)	11.8	1.44
61	Propylene carbonate	11.2	1.23
62	Methyl salicylate	11.1	0.78
63	Diethyl sulphite		1.16
64	Trimethyl phosphate		1.53
65	Triethyl phosphate		1.22
66	Tributyl phosphate	13.1	1.09
67	Methylene chloride	10.8	2.06
68	Chloroform	10.6	1.05
69	Carbon tetrachloride	10.3	—
70	1,1-Dichloroethane	10.4	1.04
71	1,2-Dichloroethane	10.8	1.09
72	Chlorobenzene	10.4	0.77
73	o-Dichlorobenzene	10.5	0.83
74	Carbon disulphide	10.1	—
75	Dimethylsulphoxide	11.2	1.08
76	Tetramethylene sulphone (30 °C)	12.9	0.88
77	Ammonia (−33 °C)	11.7	0.78
78	Hydrazine	12.5	2.16
79	Cyclohexylamine	10.7	1.13
80	Dipropylamine	11.0	0.65
81	Triethylamine	10.6	0.69
82	Aniline	11.7	0.88
83	Pyridine	11.3	0.75
84	Quinoline	11.7	0.72
85	2-Chloroaniline	11.1	1.71
86	Ethylene diamine	12.9	1.13
87	2-Aminoethanol	13.5	2.25
88	Diethanolamine	14.5	
89	Triethanolamine	13.3	
90	Morpholine	11.1	1.21
91	Nitromethane	11.1	0.86
92	Nitrobenzene	10.1	0.96
93	Acetonitrile	10.1	1.01
94	Propanenitrile	10.1	0.87
95	Butanenitrile	10.6	0.77
96	Benzonitrile	11.9	0.70
97	Benzel cyanide	12.5	0.79
98	Succinonitrile (60 °C)	10.8	2.05
99	Formamide		2.04
100	N-Methylformamide		3.90
101	N,N-Dimethylformamide	10.8	1.00
102	N,N-Diethylformamide		
103	Acetamide (90 °C)	13.6	1.95
104	N-Methylacetamide (30 °C)	14.9	4.21
105	N,N-Dimethylacetamide	11.9	1.32
106	N,N-Diethylacetamide		
107	N-Methylpropanamide		6.70
108	N-Methylpyrrolidinone		0.91
109	Tetramethylurea	12.2	1.15
110	Hexamethylphosporic triamide	13.5	0.86

R R R
H—C—H H—C—H H—C—H
(a) ···O—H ···O—H ···O—H ···O—H ···O—H ···O—H
H—C—H H—C—H H—C—H
R R R

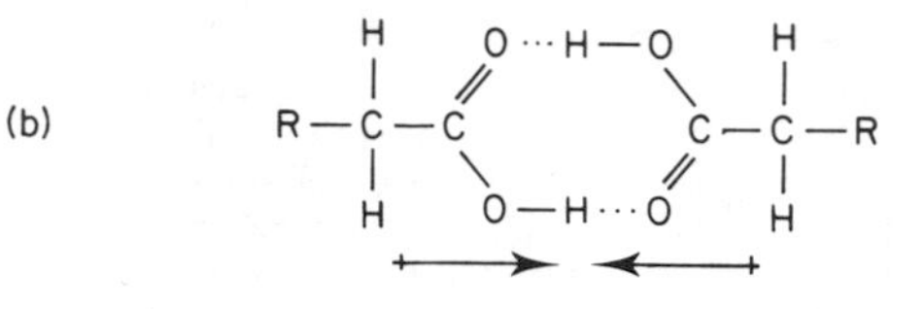

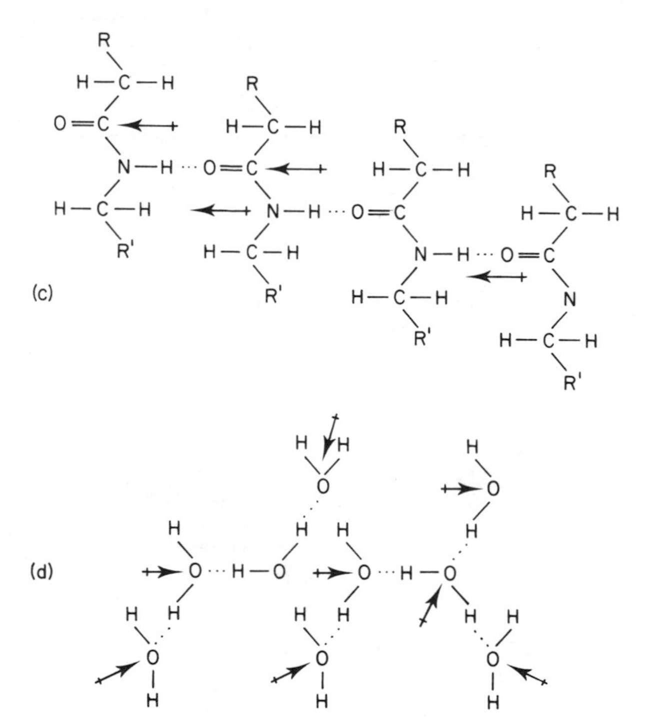

Figure 6.1 Two-dimensional representations of some associated solvents: (a) alcohols, (b) carboxylic acids, (c) N-alkylamides, and (d) water. The arrows denote the directions of the molecular dipole moments, that of the associated oligomer being their vectorial resultant. (After Y. Marcus, *Introduction to Liquid State Chemistry*, Wiley, Chichester (1977), by permission.)

Associated liquids have values of $\Delta S^V/\mathbf{R}$ higher than the upper limit of this average value, and those that are hydrogen-bonded have generally values above 13. This is the case for water and the alkanols (except methanol: 12.6), alkanedi- and -triols, γ-butyrolactone, tributyl phosphate, the aminoethanols, amides and N-methylamides, and hexamethyl phosphoric triamide. (The unusually high values 13.3 for 2-butoxyethanol and 13.8 for hexanoic acid are suspect.)

A measure for the structure of the solvent that is distinct from the above is the dipole angular correlation parameter g of polar liquids. It is a measure of the deviation of the dielectric constant from the value it would have, with the given polarizability and dipole moment of the solvent, if the solvent were not associated. It represents also the hindrance to free rotation of the molecules with respect to one another due to their dipole–dipole interactions. For nonassociated molecules $g = 1$, but if the dipoles are aligned in a parallel manner $g > 1$, whereas if in an antiparallel manner $g < 1$ (see Figure 6.1). The values of g can be calculated from the modified Kirkwood equation [15]

$$g = (9\mathbf{k}\varepsilon_0/4\pi\mathbf{N})VT\mu^{-2}(\varepsilon - 1.1n_D^2)(2\varepsilon + 1.1n_D^2)/\varepsilon(1.1n_D^2 + 2)^2 \qquad (6.1)$$

and are shown in Table 6.6 for the solvents dealt with in this Chapter. It is seen that nonassociated solvents have, indeed, g values between 0.7 and 1.4, whereas associated ones have $g > 2$, except for the carboxylic acids from acetic on, with $g < 0.7$. The dimers that these acids form in the liquid have their dipoles in an effectively antiparallel orientation, therefore the low g values. The alcohols have generally $g > 2.5$, due to their chain-like association that leads to a cooperative action of the dipole moments. Notable exceptions are methanol and 1-propanol, due to their outstandingly high dipole moments. Amides and N-alkylamides have large values of g, again because of a cooperative action of the dipoles. A correlation is expected between the excess $\Delta S^V/\mathbf{R}$ over 10.7 and g, but actually the correlation is rather poor, since association may yield average dipole orientations which are neither parallel nor antiparallel, hence near unity. Examples are 1-propanol ($\Delta S^V/\mathbf{R} = 13.6$, $g = 0.88$), hexamethyl phosphoric triamide ($\Delta S^V/\mathbf{R} = 13.5$, $g = 0.86$), formic acid ($\Delta S^V/\mathbf{R} = 7.5$, $g = 3.90$), and 2-ethoxyethanol ($\Delta S^V/\mathbf{R} = 11.9$, $g = 3.60$). A value of $g < 2$ is therefore an insufficient criterion for nonassociation, whereas $\Delta S^V/\mathbf{R} \geqslant 13$ and/or $g \geqslant 2$ are sufficient criteria for the existence of some structure in the solvent, due to association.

6.2 Thermodynamics of ion solvation in nonaqueous solvents

The solvation of an ion in any solvent, as in the case of its solvation in water, its hydration, is represented by the process of transfer of the ion from its ideal gas standard state into its standard state in solution (see equation 5.15). The problematics of dealing with single ions in this context (in distinction from dealing with complete electrolytes) have been discussed in Section 5.3. Conventional standard thermodynamic quantities of solvation by nonaqueous solvents can be defined relative to some arbitrarily chosen reference ion, but, as will be shown below, this is not necessarily the most useful approach.

The statistical thermodynamic aspects of the transfer of a particle of solute into a solvent have been discussed in Section 3.1. For a solution of the solute i in some solvent S, the chemical potential of the solute can be written as

$$\mu_{i,S} = \mu_{i,S}^{\infty} + \mathbf{R}T \ln (c_i/c_i^{\theta}) + \mathbf{R}T \ln y_{i,S} \tag{6.2}$$

where

$$\mu_{i,S}^{\infty} = \lim_{c_i \to 0} (\mu_{i,S} - \mathbf{R}T \ln (c_i/c_i^{\theta})) \tag{6.3}$$

is its standard chemical potential, $y_{i,S}$ is its activity coefficient, and c_i^{θ} is the unit concentration ($c^{\theta} = 1 \text{ mol dm}^{-3}$). The definition in equation 6.3 has as a corollary that $y_{i,S}$ tends to unity as c_i tends to zero. The standard chemical potential, in turn, is related to the work $W(\mathrm{i} \mid \mathrm{S})$ of 'coupling' the particle of solute (the ion i) to the solvent S. This coupling means its solvation by the solvent surrounding it as well as the reorganization of the affected solvent molecules, when the ion is transferred from a fixed position in the (ideal) gaseous phase to a fixed position in the solvent. Since the momentum partition function will be the same in these two phases, the standard Gibbs free energy of solvation is (see equations 3.19 and 3.20):

$$\Delta G^0_{\mathrm{solv},i,S} = \mu_{i,S}^{\infty} - \mu_{i,\mathrm{gas}}^{\infty} = \mathbf{N}W(\mathrm{i} \mid \mathrm{S}) - \mathbf{R}T \ln (q_{i,S}/q_{i,\mathrm{gas}}) \tag{6.4}$$

where q is the internal partition function (which equals unity for monoatomic ions). If it is assumed that the internal degrees of freedom of the ion are identical in the gaseous and the solution phases, the last term on the right-hand side of equation 6.4 vanishes. This is the case for monoatomic ions, but may be a rather crude approximation for polyatomic ones.

The changes that a solute particle undergoes internally when it is transferred from the ideal gas phase, where it partakes of no interactions, to the solution standard state, where it interacts fully with the solvent (but not with other solute particles) are quite drastic. These profound changes are reflected by the changes in the internal degrees of freedom of polyatomic ions. If, however, the solute particle is transferred from some solvent W to a solvent S, it is expected that the changes in the internal degrees of freedom are only moderate. The thermodynamic quantities of transfer from one solvent to the next thus reflect more nearly the changes in the solute–solvent interactions. If these interactions with the solvent W are well known, those with the solvent S can then be understood better, than if transfer occurred from the gas phase directly.

The *standard molar Gibbs free energy of transfer* of the ion i from the solvent W to the solvent S is

$$\Delta G_t^0(\mathrm{i}, \mathrm{W} \to \mathrm{S}) = \mu_{i,S}^{\infty} - \mu_{i,W}^{\infty} = \mathbf{N}[W(\mathrm{i} \mid \mathrm{S}) - W(\mathrm{i} \mid \mathrm{W})]$$
$$- \mathbf{R}T \ln(q_{i,S}/q_{i,W}) \approx \mathbf{N}[W(\mathrm{i} \mid \mathrm{S}) - W(\mathrm{i} \mid \mathrm{W})] \tag{6.5}$$

the last approximation holding if the changes in the internal partition functions are, indeed, negligible.

Since experimentally the transfer is carried out of electrolytes at finite concentrations, rather than that of single ions at infinite dilution, it is of interest to refer the activity coefficient $y_{i,S}$ of equation 6.2 not to infinite dilution in the solvent S via equation 6.3, but rather to the other solvent, W, chosen as reference. This can be done formally by the addition and subtraction on the right-hand side of equation 6.2 of the quantity $\mu_{i,W}^{\infty}$:

$$\begin{aligned}\mu_{i,S} &= \mu_{i,W}^{\infty} + \mathbf{R}T\ln(c_i/c^{\theta}) + \mathbf{R}T\ln y_{i,S} + (\mu_{i,S}^{\infty} - \mu_{i,W}^{\infty}) \\ &= \mu_{i,W}^{\infty} + \mathbf{R}T\ln(c_i/c^{\theta}) + \mathbf{R}T\ln y_{i,S} + \mathbf{R}T\ln y_t^{\infty}(\text{i, W} \rightarrow \text{S}) \\ &= \mu_{i,W}^{\infty} + \mathbf{R}T\ln(c_i/c^{\theta}) + \mathbf{R}T\ln y_{i,S(W)} \end{aligned} \tag{6.6}$$

Here

$$y_{i,S(W)} = y_{i,S}\cdot\exp(\mu_{i,S}^{\infty} - \mu_{i,W}^{\infty})/\mathbf{R}T = y_{i,S}\cdot y_t^{\infty}(\text{i, W} \rightarrow \text{S})$$

is the activity coefficient referred to the solvent W, and is the product of two factors: the so-called *salt effect* $y_{i,S}$ and the so-called *medium effect* or *transfer activity coefficient* y_t^{∞}(i, W → S). As the concentration is allowed to decrease and $y_{i,S}$ tends to unity, $y_{i,S(W)}$ will tend to y_t^{∞}(i, W → S), which is a constant characteristic of the solute (ion) i and the two solvents W and S.

A possible experimental method to obtain the necessary data is to make EMF measurements on suitable electrochemical cells. Such a cell may comprise two otherwise identical cells provided with electrodes reversible to the ions ij of the electrolyte studied and a given concentration of the electrolyte, but with either solvent W or solvent S. When the electrodes reversible to one kind of ions are interconnected, then the EMF E measured between the other pair is

$$-E = (\mu_{ij,S} - \mu_{ij,W})/n\mathbf{F} = \Delta G_t^{\infty}/n\mathbf{F} + \mathbf{R}T\ln(y_{ij,S}/y_{ij,W}) \tag{6.7}$$

where n is the number of electrons involved in the cell reaction and $\mathbf{F}$ is the Faraday constant. The EMF E approaches $\Delta G_t^{\infty}/n\mathbf{F}$ as the (equal) concentrations in the two cells decrease, since the ratio of the activity coefficients approaches unity, even more rapidly then does each individually. (Note that in equation 6.7 ij denotes the complete electrolyte and not any one of its ions.)

Another experimental method employs the measurement of the distribution of an electrolyte ij between two virtually immiscible solvents W and S. Distribution equilibrium yields

$$\begin{aligned} 0 = \Delta G_t(\text{ij, W} \rightarrow \text{S})_{\text{equil}} &= \Delta G_t^{\infty} + \mathbf{R}T\ln(c_{ij,S}/c_{ij,W})_{\text{equil}} \\ &+ \mathbf{R}T\ln(y_{ij,S}/y_{ij,W})_{\text{equil}} \end{aligned} \tag{6.8}$$

or if the equilibrium distribution ratio of the electrolyte is D_{ij}

$$\lim_{c_{ij}\rightarrow 0} D_{ij} = \exp(-\Delta G_t^{\infty}(\text{ij, W} \rightarrow \text{S})/\mathbf{R}T) \tag{6.9}$$

If the two solvents W and S are not virtually immiscible, but the electrolyte ij is sparingly soluble in both of them, then the ratio of the solubility products (corrected, if necessary, for solute–solute interaction in each solvent) can serve instead of the distribution ratio in equation 6.9. It must then be ascertained that no crystal solvates are formed between ij and the solvents, i.e., that the solid unsolvated electrolyte is in equilibrium with the saturated solutions in solvent W as in solvent S.

There are other experimental methods that can be used for obtaining $\Delta G_t^\infty(\text{ij}, \text{W} \rightarrow \text{S})$, and in all cases care must be taken to correct appropriately for solute–solute interactions, either by extrapolation to infinite dilution or by the application of theoretical expressions for the salt effect activity coefficients.

In addition to the standard molar Gibbs free energy of transfer, the corresponding enthalpy and entropy of transfer are also of interest, as well as the heat capacity and the volume of transfer. For the latter two quantities the data are scarce, but there exists a considerable body of data on the enthalpy and entropy of transfer. The enthalpy of transfer of a solute between immiscible solvents is measurable directly, but in general ΔH_t^∞ is obtained as the difference between the standard molar heats of solution of the solute in the solvents W and S. It must be ascertained, as in the case of ΔG_t^∞, that the same solid solute is at equilibrium with the saturated solutions in solvents W and S, i.e., that no crystal solvates are formed. The entropy of transfer is generally obtained from $\Delta S_t^\infty = (\Delta H_t^\infty - \Delta G_t^\infty)/T$, but it has also been obtained as $\Delta S_t^\infty = \mathbf{F}(\partial E^\infty/\partial T)_P$ from EMF measurements or as $-(\partial \Delta G_t^\infty/\partial T)_P$, although less reliably, as has also $\Delta H_t^\infty = T^2(\partial(\Delta G_t^\infty/T)/\partial T)_P$. The volumes and heat capacities of transfer have generally been obtained as the difference between the corresponding standard partial molar quantities in S and in W.

6.3 Single ion standard molar Gibbs free energies of transfer

The significance of $\Delta G_t^\infty(\text{ij}, \text{W} \rightarrow \text{S})$ of an electrolyte for the understanding of the interactions of its constituent ions with the solvent S relative to those with W have been established above.

There remain two problems concerning the use of $\Delta G_t^\infty(\text{i}, \text{W} \rightarrow \text{S})$ for the elucidation of the interactions of the ion i with the solvent S. One is the choice of the reference solvent W and the other is the splitting of the experimental data obtained for complete electrolytes ij into the ionic contributions.

There are several aspects to the choice of the reference solvent: the feasibility of the use of W in the experimental set-up envisaged, its inertness towards the electrolyte beyond solvation (avoidance of extensive solvolysis), experimental convenience, the ready availability of W in pure form, and the extensiveness of the knowledge of the interactions it undergoes with ions. From some of these points of view some polar aprotic solvent such as acetonitrile may be preferred. These include the better reproducibility of liquid junctions and the better stability of the measured EMF's in certain cells than if water is employed. In the case of solubility measurements of a reference electrolyte such as tetraphenylarsonium tetraphenylborate (see below, p. 162), one argument in favour of the use of a polar aprotic solvent for the reference solvent W has been the relative instability of the anion in water, and another the extremely low solubility of this salt in the solvent water, that is difficult to measure. However, some of these objections to water can be overcome with more advanced experimental methodology, and it is certainly true that water is more readily purified from other solvating impurities than are most solvents from water. Furthermore, there has been no universal agreement as to which polar aprotic solvent to choose on the grounds of experimental convenience, acetonitrile being a favourite but by no means generally endorsed choice.

The better criterion for the choice is, therefore, the availability of extensive knowledge of the solvation characteristics of ions in the solvent W, together with theoretical insight into the interactions involved. From this standpoint water is unique as the reference solvent W, in view of the thorough knowledge of hydration that already exists (see Chapter 5) and that is constantly being augmented. This holds true, notwithstanding the imperfect understanding at present of some features of ion hydration, such as the role of the structure of the water.

Henceforth, therefore, the symbol W will represent the reference solvent water.

Whereas the decision concerning the choice of the reference solvent is, after all, arbitrary, not so is the decision regarding the division of ΔG_t^∞ of an electrolyte ij into the ionic contributions, if these are to throw some light on the interactions involved. It is clear that for the *standard* molar Gibbs free energy of transfer the solute–solute interactions make no contribution at all, so that $\Delta G_t^\infty(\mathrm{ij}, \mathrm{W} \rightarrow \mathrm{S})$ is perforce additive with respect to the ionic contributions, weighted according to the stoichiometric coefficients. This is an important criterion by which any *absolute* scale for $\Delta G_t^\infty(\mathrm{i}, \mathrm{W} \rightarrow \mathrm{S})$, where i is a single ion, must be judged. Scales of $\Delta G_t^\infty(\mathrm{i}, \mathrm{W} \rightarrow \mathrm{S})$ have been proposed that provide values for ions of one kind of charge only. For example, polarographic half-wave potentials versus a certain reference electrode (see p. 161) have been converted to $\Delta G_t^\infty(\mathrm{i}, \mathrm{W} \rightarrow \mathrm{S})$ values of metallic cations, e.g., ref. 16. Such scales are less useful than those based on data for complete electrolytes, since they cannot be checked with this criterion.

A large number of schemes have been proposed for effecting the division of standard molar Gibbs free energies of transfer of electrolytes into their ionic contributions. They may be classified as follows.

(i) The determination of 'real' potentials, in cells with a gas gap between streaming electrolyte solutions in the two solvents.
(ii) The assumption that a specified liquid junction between the two halves of an electrochemical cell involving the two solvents makes a negligible contibution to the EMF of the cell.
(iii) Electrostatic models, based essentially on the Born equation modified with respect to the radii of the ions and/or the dielectric constants of the media, and by the inclusion of a 'neutral term'.
(iv) Extrapolation methods, where in series of electrolytes having a common ion, values of ΔG_t^∞ are extrapolated to infinite size of the other ions.

(v) The assumption that a poorly solvated reference ion has the same molar Gibbs free energy of solvation in all solvents.
(vi) The assumption that a reference couple consisting of a large ion and its uncharged structural analogue have the same difference in ΔG_t^∞ to all solvents, which is negligible.
(vii) The assumption that a reference pair of ions of opposite sign and large size and similar structure have the same ΔG_t^∞ to all solvents.

The following is a discussion of the merits and drawbacks of the methods and extrathermodynamic assumptions listed above, and a more detailed reasoning behind the choice of the method from category (vii) finally selected.

(i) *Real potentials*[17,18] are obtained from the measurement of the compensating voltage in a cell such as

$$Ag/AgNO_3 \text{ in } W/gas/AgNO_3 \text{ in } S/Ag \qquad (A)$$

for Ag^+ ions or

$$Ag/AgCl/NaCl \text{ in } W/gas/NaCl \text{ in } S/AgCl/Ag \qquad (B)$$

for Cl^- ions. In a cell like these, one of the solutions flows down along the walls of a vertical glass tube, the other flows as a jet down the axis of the tube, while a gas (air or nitrogen) flows between the solutions, carrying along with it vapours that might otherwise condense into the wrong solvent. The EMF E_A measured in cell A, for instance, is related to the 'real' standard Gibbs free energy $\alpha_{Ag^+}^\infty$ of the silver ion by

$$\mathbf{F}E_A = [\alpha_{Ag^+}^\infty(S) + \mathbf{R}T \ln a_{Ag^+}(S)] - [\alpha_{Ag^+}^\infty(W) + \mathbf{R}T \ln a_{Ag^+}(W)] \quad (6.10)$$

where $\mathbf{F}$ is the Faraday constant and a the thermodynamic activity. This quantity is the product of the concentration of the silver ion and its activity coefficient, so that a suitable expression for the latter quantity in each of the solvents W and S is required in order for $\Delta\alpha_t^\infty(Ag^+, W \rightarrow S)$ to be evaluated. Since

$$\alpha_i^\infty(S) = \mu_i^\infty + z_i\mathbf{F}\chi(S) \qquad (6.11)$$

and similarly for i in W, where z_i is the charge on the ion i and $\chi(S)$ is the surface potential of the solvent S (against the gas used in the gap between the solution streams), it follows that ($z_{Ag^+} = 1$)

$$\Delta G_t^\infty(Ag^+, W \rightarrow S) = \Delta\alpha_t^\infty(Ag^+, W \rightarrow S) + \mathbf{F}\Delta\chi \qquad (6.12)$$

The difference in surface potentials must still be estimated in order to obtain from the measurements of the compensating voltage the standard molar Gibbs free energy of transfer. Estimates of χ have been given for S being methanol or ethanol, based essentially on other methods for the estimation of ΔG_t^∞. The limits of uncertainty estimated for χ(W) are ± 0.1V, those for χ(methanol) are ± 0.13V (corresponding to ± 10 and ± 13 kJ mol^{-1} in ΔG_t^∞, respectively). The most direct method for the estimation of χ is that applied where S is acetonitrile.[17] For several multinuclear aromatic hydrocarbons (such as anthracene or pyrene), the charge distribution and the size are the same for the positive and negative ions produced on oxidation or reduction. If, for such ions, the assumption ΔG^{∞}_{solv} (+ve ion) = ΔG^{∞}_{solv}(−ve ion) is made, the value of χ(acetonitrile) = -0.10 ± 0.06V is obtained for half a dozen of such ionized large aromatic hydrocarbons. The relative constancy of χ(acetonitrile) speaks for the validity of the assumption.[17]

In conclusion, the use of the 'real' potentials for the estimation of $\Delta G_t^{\infty}(\mathrm{i, W} \rightarrow \mathrm{S})$ depends on the proper selection of expressions for the activity coefficients of ions in the two solvents W and S and on an estimate of the surface potential between them. For the latter quantity, an assumption of category (vii) has been proposed, so that altogether the class (i) method does not seem to be the most useful one.

(ii) A *negligible liquid junction* was assumed a long time ago by Bjerrum and Larsson[1] to exist when a 3.5 M aqueous KCl–calomel reference electrode is used in aqueous, ethanolic, or mixed aqueous–ethanolic solutions. Although the partial suppression of the liquid junction potential between two rather dissimilar aqueous solutions by means of the 3.5 M aqueous KCl salt bridge, by virtue of the high concentration and nearly equal mobilities of K^+ and Cl^- ions in this bridge solution, is well documented and supported, the extension to nonaqueous and mixed media is a mere assumption. Its validity has been tested only by a comparison of the consequences of this assumption with results obtained by a method[19] from category (iii), which are not necessarily trustworthy. A revival of this kind of assumption was later made by Parker[20] in the suggestion that 0.1 M tetraethylammonium picrate could make a suitable salt bridge in any solvent, that practically eliminates the liquid junction potential. It does so by virtue of the 'inertness' and similar mobilities of the constituent ions. This assumption is supported by the fact that in cells of the type

$$\mathrm{Ag/0.01\ M\ AgClO_4\ in\ S_1/0.1\ M\ (C_2H_5)_4NC_6H_2(NO_2)_3O\ in\ S_2/0.01\ M\ AgClO_4\ in\ S_3/Ag} \qquad (C)$$

where the bridge solvent S_2 is either the same as one of the solvents S_1 or S_3 or a different one altogether, the EMF is virtually independent of the nature of S_2. For S_1 = acetonitrile and S_2 = almost any one of ten solvents tested, the EMF's are -0.155 ± 0.005V for S_3 = dimethylsulphoxide, 0.066 ± 0.008V for S_3 = formamide, 0.182 ± 0.023V for S_3 = water, and 0.262 ± 0.009V for S_3 = methanol. Thus, both protic and polar aprotic solvents can be separated with such a salt bridge and a practically constant EMF is produced. It is unlikely that the same appreciable liquid junction potential would occur with bridge solvents as diverse as acetonitrile, dimethylsulphoxide, nitromethane, methanol, acetone, dimethylformamide, and formamide. Hence it is concluded that within the errors quoted above the liquid junction potential is negligible. Apart from the case of transfer between acetonitrile and water, the above error limits correspond to $+0.7$ kJ mol^{-1} only; in the case of transfer to water they are ± 2.2 kJ mol^{-1}.

It should be noted, however, that in every case there is one or more nonconforming bridge solvent (e.g., S_2 = formamide for S_3 = dimethylsulphoxide, S_2 = dimethylsulphoxide, N-methylpyrrolidinone, and dimethylformamide for S_3 = formamide), that cause larger errors, especially in the case of S_3 = water.

In conclusion, the negligible liquid junction potential assumed with an 0.1 M tetraethylammonium picrate salt bridge is convenient to apply to cell EMF measurements of good accuracy. Its main drawbacks are its completely empirical nature, the necessity to examine each new solvent S_3 in cell (C) separately with a group of bridge solvents S_2 to establish its applicability, and its partial failure with water as the solvent.

(iii) *Electrostatic models* have been used for the calculation of the work done on the transfer of an ion from one solvent to the other. The Born equation gives the electrostatic work[19]:

$$\Delta G_t^{\infty}(\mathrm{i}, \mathrm{W} \rightarrow \mathrm{S})_{\mathrm{el}} = \tfrac{1}{2}\mathbf{N} z_i^2 \mathbf{e}^2 \varepsilon_0^{-1}[(r_{\mathrm{i,S}}\varepsilon_{\mathrm{S}})^{-1} - (r_{\mathrm{i,W}}\varepsilon_{\mathrm{W}})^{-1}] \qquad (6.13)$$

where $\mathbf{e}$ is the unit charge and, in general, the radii of the ion in the two solvents are taken to be equal: $r_{\mathrm{i,S}} = r_{\mathrm{i,W}}$. If the crystal ionic radius is used for r_i in both solvents and the bulk dielectric constants ε_S and ε_W of the solvents are employed, then the sum of the calculated quantities $\Delta G_{\mathrm{t\,el}}^{\infty}$ for a cation and an anion does not yield the measured ΔG_t^{∞} of the corresponding electrolyte. Equation 6.13 requires, therefore, modification in order to provide acceptable ΔG_t^{∞} values for ions.

The first modification that has been proposed[19] involves the addition of a 'neutral term' to the electrostatic term given by equation 6.13, to express the difference in the work required to produce a cavity in the solvents W and S for the accommodation of the transferred ion, disregarding its charge, and the unequal compensation of this work by the nonelectrostatic interactions of the ion with these solvents. The experimental distribution ratio D(neut, S/W) of a neutral analogue of the ion (e.g., Ar for K^+ or Cl^-, benzoic acid for the benzoate ion) is used[19] to provide this term:

$$\Delta G_t^{\infty}(\mathrm{i}, \mathrm{W} \rightarrow \mathrm{S})_{\mathrm{neut}} = \mathbf{R}T \ln D(\mathrm{neut}, \mathrm{S/W}) \qquad (6.14)$$

The sum of $\Delta G_{\mathrm{t\,el}}^{\infty}$ and $\Delta G_{\mathrm{t\,neut}}^{\infty}$ should then be a better approximation for the individual ionic ΔG_t^{∞}. Alternatively, the neutral solute is sought not as the isoelectronic analogue of the ion, but as a particle of the same radius, obtained by interpolation in the (almost) linear plot of ln D(neut, S/W) against r_{neut}^2 for a series of noble gas or other nonpolar solutes.[21]

A major difficulty with this treatment arises from the fact that for ions as small as the crystal ionic radius implies, dielectric saturation sets in near the ion, and the bulk dielectric constant is not the appropriate quantity to be used in the electrostatic expression 6.13. Either the radii $r_{\mathrm{i,S}}$ and $r_{\mathrm{i,W}}$ or the dielectric constants ε_S and ε_W or both pairs of quantities must be modified to take this effect into account. Various schemes have been proposed for this purpose (see Ref. 22, for instance), but because of the lack of information concerning dielectric saturation in nonaqueous solvents (i.e., the effect of the ionic electrical field on the dielectric constant), most of the schemes have tackled the ionic size instead. In the most useful scheme proposed by Strehlow and co-workers[23] a quantity is empirically added to r_i, individually for the two solvents W and S. In equation 6.13 each $(r_i\varepsilon)^{-1}$ is replaced by $-(r_i + \Delta r)^{-1}(1 - 1/\varepsilon)$, where Δr is the same for all cations (only the alkali metal cations have been considered) and has another value for all anions (only the halide anions have been considered) in a given solvent. If the Pauling ionic radii are used for r_i then for water $\Delta r_+ = 0.085$ nm and $\Delta r_- = 0.025$ nm, for methanol $\Delta r_+ = 0.079$ nm and $\Delta r_- = 0.039$ nm, for acetonitrile $\Delta r_+ = 0.072$ nm and $\Delta r_- = 0.061$ nm, and so on.[23] These values have been selected empirically so that the deviation of the calculated $\Delta G^{\infty}_{\mathrm{tel}}$ from the experimental $\Delta G^{\infty}_{\mathrm{t}}$ for an electrolyte ($\Delta G^{\infty}_{\mathrm{tneut}}$ being disregarded) is divided equally between the largest ions considered, Cs^+ and I^-. Different choices of the addends are, however, equally plausible, as for instance $\Delta r_+ = 0.072$ nm and $\Delta r_- = 0.045$ nm for water,[24] which then yield different values for solvents S, e.g., $\Delta r_+ = 0.082$ nm for acetonitrile and $\Delta r_+ = 0.074$ nm for acetone. This indefiniteness may possibly be due to the disregard of $\Delta G^{\infty}_{\mathrm{tneut}}$ in the procedure.

In conclusion, it is seen that this modelling attempt involves a number of arbitrary choices, such as the addends to the ionic radii in the electrostatic term and the neutral analogue of the ion for the neutral term, if used at all. Even for the reference solvent, water, these quantities (pertaining to $\Delta G^{\infty}_{\mathrm{hydr}}$) are not definitely known (see also Chapter 3), and it is unlikely that a compelling theoretical breakthrough will be made in this regard. A comparable success with nonaqueous solvents is even less likely.

(iv) *Extrapolation methods* also employ models, in which $\Delta G^{\infty}_{\mathrm{t}}$ of a series of electrolytes with a common ion are plotted against a smoothly varying quantity characterizing the counter ion and is extrapolated to an infinitely large size of the latter. In its more primitive form, this method has been applied by Izmailov[25] to the alkali halides and hydrohalic acids. The quantities characterizing the counter ions are either some reciprocal power of the crystal ionic radius, r_i^{-p}($p = 1, 2, 3, 4$, or 6) or the square of the reciprocal of the main quantum number, n^{-2}. These two kinds of extrapolation give discordant values to the extent of 4 to 6 kJ mol^{-1}, and that against n^{-2} has been considered as the more reliable.[25] This approach runs into the difficulty that if a neutral term is considered as essential in the electrostatic modelling dealt with in category (iii), then this term should also be added in the extrapolations that implicitly are based on such an electrostatic model. However, the neutral term depends on a positive power of the ionic radius, so that the extrapolation to $(1/r_i)$ necessarily diverges to infinity.

In a more sophisticated approach DeLigny and co-workers[26] take this into account by first substracting a calculated neutral term, before making the extrapolation. Also, not a single power p is used, but a series of values:

$$\Delta G^{\infty}_{\mathrm{t}} - \Delta G^{\infty}_{\mathrm{tneut}} = Ar_i^{-1} + Br_i^{-2} + Cr_i^{-3} + Dr_i^{-4} + \cdots \qquad (6.15)$$

The coefficients $A, B, \ldots$ are estimated from electrostatic theory: A is the Born coefficient, B is zero, and C and D are ion–quadrupole interaction coefficients. The latter depend on the quadrupole moment of the solvent (not generally known) and the coordination number of the ion (neither generally known). The extrapolation is carried out either with the crystal ionic radius r_i or with its sum with the diameter of the solvent molecules in different version of this method.[26]

In conclusion, the extrapolation methods do not provide a sound theoretical basis for the selection of the property of the counter ion against which the ΔG_t^∞ of electrolytes (or sums or differences of them) should be plotted. The more sophisticated approach involves a host of unknown coefficients, including an estimate of the neutral term, which is wrongly omitted in the simpler approach, but is not known definitely either.

(v) The *reference ion* assumption considers $\Delta G_t^\infty(i_R, W \rightarrow S)$ of a judiciously selected reference ion i_R to be the same for all solvents S. In order to fulfill this requirement i_R should be poorly solvated but not so large that differences in the work of the creation of a cavity for its accomodation in various solvents ensues. These two requirements are contradictory, and it is, therefore, impossible in principle to find such a reference ion. On the practical side, Pleskov[27] considered that Rb^+ might fill this role of i_R, but it soon transpired that it was not sufficiently large ($r_i = 0.148$ nm) for differences in its solvation not to be manifested (see Figure 6.2).

A more sophisticated idea is behind the use of a complexing ligand that 'shields' the reference ion from interactions with its environment. The cryptand tricyclo-N,N'-bis(3,6-dioxa-1,8-octadiyl)(3-oxa-1,5-pentadiyl) envelops an alkali metal cation such as Na^+ completely. It has been assumed by Villermaux and Delpuech[28a] that the cryptated cation does not contribute to ΔG_t^∞ of the electrolyte, so that $Na(cryptand)^+$ is a reference ion with $\Delta G_t^\infty(i_R, W \rightarrow S) = 0$. Similarly, the cryptated Ag^+ cation has been assumed[28b] to have the same ΔG_t^∞ value as the cryptand ligand itself.

This idea has been tested on very few systems, and although promising, the results are not conclusive.

Crown ethers, admittedly, envlop alkali metal ions less completely than do cryptands, and may permit interactions with solvent molecules located along the axis perpendicular to the plane of the crown ether. Still, it has been shown empirically by Marcus and Asher[29] and rationalized on the basis of an electrostatic model, that the distribution of KCl between a dilute aqueous solution and a suitable crown ether (e.g., eicosahydro- or octahydrodibenzo-2,5,8,15,18,21-hexaoxacyclo-octadecin, i.e., dicyclohexo- or dibenzo-18-crown-6, respectively) in a variety of water-immiscible solvents can be used for the evaluation of $\Delta G_t^\infty(Cl^-, W \rightarrow S)$. The distribution ratio yields a distribution equilibrium constant, hence the corresponding standard Gibbs free energy change, $\Delta G_{distr}^\infty(KCl, S/W)$. A semi-empirical relation[29] then relates this to the transfer of the chloride ion:

$$\Delta G_t^\infty(Cl^-, W \rightarrow S) = \Delta G_{distr}^\infty(KCl, S/W) + A + B/\varepsilon_S \tag{6.16}$$

where $A = 29 \text{ kJ mol}^{-1}$ and $B = 116 \text{ kJ mol}^{-1}$. The last two terms on the right-

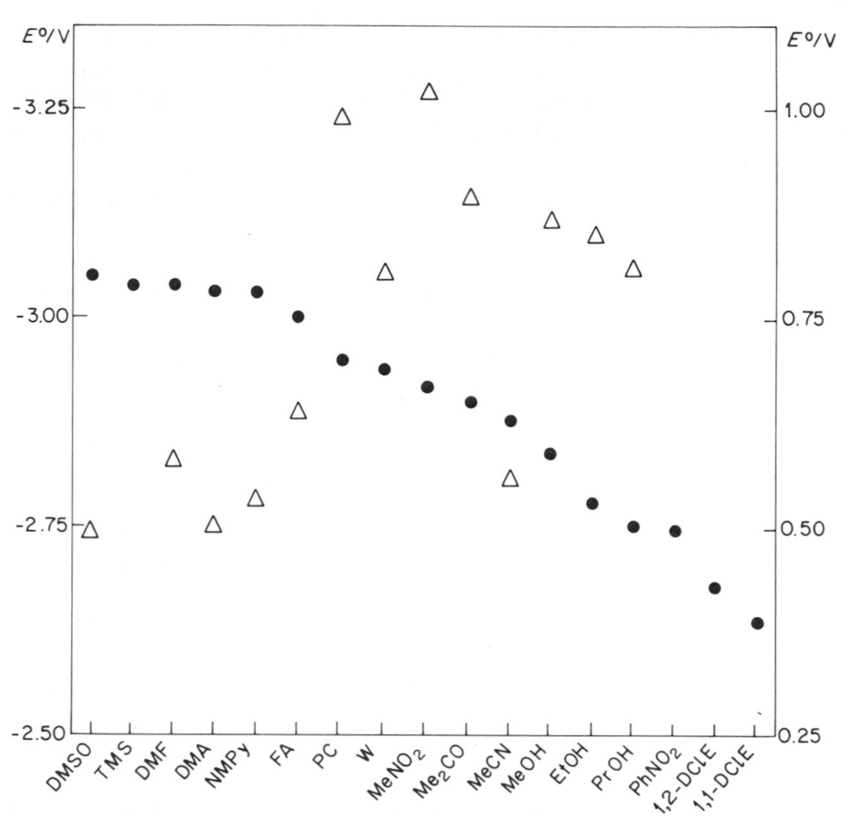

Figure 6.2 The standard electrode potential of the Rb^+/Rb (●, left-hand ordinate) and the Ag^+/Ag (△, right-hand ordinate) electrodes against the aqueous standard hydrogen electrode, $E^0(\text{SHE}) = 0$. The TATB assumption (see p. 162) is used for obtaining the required single ion Gibbs free energies of transfer. The solvents are ordered so as to yield a monotonous curve for $E^0(Rb^+/Rb)$, otherwise there is no significance to the spacing of the abcissa. According to the Pleskov assumption $E^0(Rb^+/Rb)$ should be solvent-independent

hand side represent $-\Delta G_t^\infty(K^+, W \rightarrow S)$, and if the potassium cation were completely shielded by the ligand, then B would presumably be zero (and A would have a larger value, the same for all solvents S). As it is, the finite value of B takes into account the imperfect shielding of the potassium cation by the crown ligand.

In conclusion, it seems that the reference ion method can be accepted, provided effective shielding of this ion from all solvents can be achieved. This has so far not been done completely satisfactorily, but promising approaches have been suggested. The experimental applications are limited, however, by the laborious vapour pressure measurements that have been used with the cryptand[28] and the limitation to water-immiscible solvents for the distribution measurements used for the crown ethers.[29]

(vi) The *reference ion/molecule* assumption involves an ion that can be oxidized (or reduced) or deprotonated (or protonated) to a neutral species of essentially the same size and structure. If the ion is large and symmetrical, then the interactions of the ion and the neutral molecular species with any solvent would be the same, except for a presumably negligibly small electrostatic interaction term for the ion, absent for the molecule $\Delta G_t^{\infty}(I^{\pm}, W \to S) \approx \Delta G_t^{\infty}(I^0, W \to S)$.

An early reduction of this idea to practice involved the Hammett acidicity function H_0 measurable with the Hammett indicators, i.e. nitro- and/or halo-substituted anilines. It is known that in aqueous acid solutions H_0 is independent of the particular indicator employed, due to the cancellation of activity coefficient effects. The transfer activity coefficient of the hydrogen ion becomes

$$\begin{aligned}\log y_t^{\infty}(H^+, W \to S) = & -H_0 - \log c_{H^+} - \log y_{H^+,S} \\ & + \log(y_{BH^+,S}/y_{B,S}) \\ & + \log[y_t^{\infty}(BH^+, W \to S)/y_t^{\infty}(B, W \to S)] \quad (6.17)\end{aligned}$$

where $H_0 = \log([B]/[BH^+]) + pK_{BH^+}$ is obtained spectrophotometrically with the indicator B employed. Log $y_{H^+,S}$ is generally unknown, hence neglected, $(y_{BH^+,S}/y_{B,S})$ is set equal to unity because of the similarity in size of the protonated and unprotonated forms of the indicator B, and so is the ratio of their transfer activity coefficients, on the same grounds. Therefore, equation 6.17 gives an approximate value of $y_t^{\infty}(H^+, W \to S)$ from the measured H_0 and c_{H^+}. This approach has been applied mainly to mixed aqueous–alcoholic media, but the unsatisfactoriness of the required assumptions is apparent.[22] Not the least reason for this is the fact that while substituted anilines and anilinium ions are undoubtedly large, the charge is effectively localized on the $-NH_3^+$ group, which is small and interacts strongly with its surroundings.[23]

A much more elaborate approach due to Wells (see also Section 7.3) that has again been applied to mixed aqueous–organic media, but wisely limited to the water-rich part of the composition range, involves the following reasoning.[30] The hydrogen ion in aqueous solutions is assumed to be the tetrahedral species $H_3O(H_2O)_4^+$, which can be approximated by a sphere with a radius $r_{H^+,aq} = 3r_{H_2O} = 0.41$ nm. This is assumed to be sufficiently large for the bulk dielectric constant to apply in the Born equation (6.13) for both the aqueous and the mixed aqueous–organic media. To the electrostatic term

$$\Delta G_{t,el}^{\infty} = \tfrac{1}{2}\mathbf{N}\mathbf{e}^2\varepsilon_0^{-1}(3r_{H_2O})^{-1}(\varepsilon_S^{-1} - \varepsilon_W^{-1})$$

is added a proton exchange term, $\Delta G_{t,exch}^{\infty}$, and together they make up $\Delta G_t^{\infty}(H^+, W \to W + S)$. The proton exchange term is calculated from the equilibrium constant

$$K_{exch} = [SH^+](55.5 - [S]_T)/([H^+]_T - [SH^+])([S]_T - [SH^+]) \quad (6.18)$$

where 55.5 mol dm^{-3} is the concentration of water in the absence of the cosolvent S, the amounts with subscript T being the total concentrations of the cosolvent S and acid. The amount of proton exchange that has taken place per unit concentration of H^+ is $[SH^+]$, hence $\Delta G^{\infty}_{t,\text{exch}} = [SH^+](-\mathbf{R}T \ln K_{\text{exch}})$. It remains only to find the concentration $[SH^+]$, and this is made by using a Hammett acidity function type experiment with 4-nitroanline as the indicator. This procedure involves the assumptions that the ratio of the concentrations of free (i.e., not protonated) water in the presence and absence of the cosolvent S is essentially unity and also that the activity coefficient ratio $y_{BH^+}y_{H_2O}/y_B y_{H^+,aq}$ is not affected by the presence of S. These assumptions are tenable only if the concentration of S is very small. This limits the applicability of this approach to very water-rich mixtures, provided that the other assumptions, concerning the electrostatic term, are valid too.

The reference ion/molecule method that has been most widely used involves complexed metal ions in two oxidation states. The experimentally measured quantity is the EMF of a cell such as (D), in which the redox couple constitutes the reference electrode, and another electrode is reversible to the ion i, of which $\Delta G^{\infty}_t(\text{i, W} \rightarrow \text{S})$ is thus estimated:

$$\text{Ag/AgClO}_4 \text{ in W or in S, red, ox}^+\text{/Pt} \tag{D}$$

Here 'red' and 'ox^+' are examples of the molecule (the reduced state of the metal complex, which is uncharged) and the ion (the corresponding charged oxidized form). When the cell solvent is W in one experiment and S in another, then

$$\Delta G^{\infty}_t(\text{i, W} \rightarrow \text{S}) = \mathbf{F}(E^{\infty}_S - E^{\infty}_W) \approx \mathbf{F}(E_{\frac{1}{2}S} - E_{\frac{1}{2}W}) \tag{6.19}$$

where $E_{\frac{1}{2}}$ is the polarographically determined half-wave potential, which is a sufficiently good approximation for the standard EMF's E^{∞}. It turns out that for the most frequently used reference couples, W as water is a poor choice for the reference solvent, from the standpoints of solubility and electrochemical stability. The reference solvent usually preferred in this case is acetonitrile.

In principle, a variant of the method involving two ions of different charge, e.g. tris(4,7-dimethyl-1,10-phenantrolino)iron(II) and -iron(III), can be used.[31] Since, however, the electrostatic contribution to ΔG^{∞}_t is proportional to the square of the charge on the ion, the discrepancy between the contributions of $[Fe(DiMephen)_3]^{2+}$ and $[Fe(DiMephen)_3]^{3+}$ is considerably larger than would be obtained from a singly charged/uncharged couple. The criteria for the choice of the ion and the molecule that constitute the redox couple are[23] a large size and nearly spherical shape, and an invariant structure on oxidation and reduction. The redox equilibrium must be established reversibly and rapidly, and the limits of the redox potential are such that the solvent is neither oxidized nor reduced by the constituents of the redox couple. Among the many redox couples that have been examined, only the so-called 'fic^+/foc', i.e., dicyclopentadienyl-iron(III) (ferricinium)/ -iron(II) (ferrocene), one advocated by Strehlow has gained wide acceptance, although the corresponding cobaltocene couple[23] and the bis(biphenyl)chromium(0)/ -chromium(I) couple[32] have also been used. The suitability of the 'fic^+/foc' couple stems from its nearest approach to the criteria listed above, although some apprehensions have been voiced regarding the insufficiently large size of these species. The dimensions of ferrocene are 0.51 nm across the rings and 0.40 nm thickness, corresponding to a mean radius of only 0.23 nm. Furthermore, some specific solvation of the iron atom between the two 'covers' of the 'sandwich' complex seems to be possible.

In conclusion, it must be admitted that whereas the 'fic$^+$/foc' couple answers fairly well the criteria for the redox reference ion/molecule couple, the contributions of the reduced and oxidized forms to ΔG_t^∞ do differ by the electrostatic term for 'fic$^+$', that is absent for 'foc'. With the mean radius given above, the Born expression gives for transfer from water to a solvent with $\varepsilon_S = 10$ an electrostatic contribution of 26 kJ mol^{-1}, and even for transfer to a solvent with $\varepsilon_S = 40$ this is still 4 kJ mol^{-1}. These quantities are not entirely negligible, and may even not be the full manifestation of the difference.

(vii) The *reference electrolyte* assumption states that if the cation and the anion of an electrolyte are large, spherical, and equal in size, then ΔG_t^∞ of this reference electrolyte should be divided equally between the cation and the anion. On close examinationn it is realized that this should be true for the contributions to ΔG_t^∞ from cavity formation work and dispersion and similar interactions, i.e., $\Delta G_{\text{t neut}}^\infty$, as well as for the major components of $\Delta G_{\text{t el}}^\infty$: the Born changing and the ion–dipole interactions, since these are independent of the sign of the charge on the ion. However, there is a minor contribution to $\Delta G_{\text{t el}}^\infty$ arising from ion–quadrupole interactions, that changes sign when the charge of the ion is flipped from positive to negative, hence is not the same for the cation and anion.[33,34]

If the ion–quadrupole interaction is disregarded for the moment, the plausibility of the reference electrolyte assumption is evident, since it is difficult to imagine reasons for differing interactions of large, spherical cations and anions of equal size. Note that it is not claimed that the interactions that each kind of ions undergoes are small (although $\Delta G_{\text{t el}}^\infty < \Delta G_{\text{t neut}}^\infty$, the latter is appreciable), only that they are equal. The problem is thus reduced to the finding of a suitable reference electrolyte, since large ions are polyatomic, and these are not strictly spherical, with a central charge. Close approximations to such ions are provided by the constituents of three electrolytes that have been used for the purpose: the tetraphenylborates of tetraphenylphosphonium (Ph_4PBPh_4), arsonium (Ph_4AsBPh_4), and tris(3-methylbutyl)butylammonium ($iPn_3BuNBPh_4$). Of these, mainly the second, Ph_4AsBPh_4 (the so-called TATB), has gained wide acceptance. The idea of the reference electrolyte was first suggested by Grunwald and co-workers with the use of Ph_4PBPh_4 in a solution in 50 mass% dioxane in water, compared with a solution of tetraphenyl methane in the same solvent, in vapour pressure measurements.[35] Recently, a more convenient solubility method has been used, where, for instance

$$\Delta G_t^\infty(Ph_4As^+, W \to S) = \Delta G_t^\infty(BPh_4^-, W \to S)$$

$$= -\mathbf{R}T \ln(s_{Ph_4AsBPh_4,S}/s_{Ph_4AsBPh_4,W}) \qquad (6.20)$$

where s is the solubility, provided that as in water the solubility in the solvent S is so small that no activity coefficient corrections are required, i.e., $y_{Ph_4AsBPh_4}$ is very near unity. If all the three reference electrolytes listed above were equally suitable, then necessarily

$$\Delta G_t^\infty(Ph_4P^+) = \Delta G_t^\infty(Ph_4As^+) = \Delta G_t^\infty(iPn_3BuN^+)$$

for transfer from water to any given solvent S. For the former two, the differences found are within the experimental errors (± 0.4 kJ mol^{-1}) for transfer into methanol, ethanol, or acetonitrile, whereas for the less symmetrical substituted ammonium ion the difference is somewhat larger.[36] The validity of the Ph_4AsBPh_4 reference electrolyte assumption has been thoroughly studied and affirmed by Kim,[34] with two minor reservations, one conerning the ion–quadrupole interactions mentioned above, the other the actual nonequality of the sizes of the cation and the anion. The significance of the nonequality of the sizes is seen in the differences in ΔG_t^∞ of the neutral tetraphenyl compounds Ph_4C, Ph_4Si, and Ph_4Ge, that can be considered as the neutral analogues of Ph_4As^+ or BPh_4^- for transfer into acetonitrile or acetone.[34,36] In order to decide which of the three neutral molecules is the analogue of which ion, it is necessary to define the concept of 'size' to be applied in this problem, which is not as straightforward as it seems.[37] The van der Waals volumes V_{vdW} (corresponding to the van der Waals radii $r_{vdW} = (3V_{vdW}/4\pi N)^{1/3}$) are calculated as

$$V_{vdW}(MPh_4) = 4V_{vdW}(C_6H_5-) + (r_{cov}(M)/r_{cov}(C))^3 V_{vdW} \begin{array}{c} | \\ -C- \\ | \end{array}$$

$$= 183.36 + 7294.1(r_{cov}(M)/nm)^3 \text{ cm}^3 \text{ mol}^{-1} \qquad (6.21)$$

where r_{cov} is the tetrahedral covalent Pauling radius.[38] The molar volumes and radii of the tetraphenyl ions and molecules discussed here are shown in Table 6.7. It is seen that Ph_4C is, indeed, the correct analogue of BPh_4^-, but that Ph_4Si is a better analogue to Ph_4As^+ than Ph_4Ge, advocated by Kim.[34] Since Ph_4Si is intermediate in size between the other two neutral molecules, the discrepancy in

Table 6.7 The sizes of tetraphenyl species, MPh_4

	Ph_4C	Ph_4Si	Ph_4Ge	Ph_4B^-	Ph_4P^+	Ph_4As^+
$r_{cov}(M)$/nm	0.077	0.117	0.122	0.088	0.110	0.118
$V_{vdW}(M)$/cm^3 mol^{-1}	3.3	11.7	13.2	5.0	9.7	12.0
$V_{vdW}(MPh_4)$/ cm^3 mol^{-1}	186.7	195.0	196.6	188.3	193.1	195.3
$r_{vdW}(MPh_4)$/nm	0.420	0.426	0.427	0.421	0.425	0.426

In the first two rows, M refers to the tetrahedrally bound central atom; the size of a Ph-group (C_6H_5-) is taken as 45.8 cm^3 mol^{-1}.

ΔG_t^∞ between Ph_4As^+ and BPh_4^- on account of the size differences

$$V_{vdW}(Ph_4As^+)/V_{vdW}(BPh_4^-) = 1.0372,$$

$$r_{vdW}(Ph_4As^+)/r_{vdW}(BPh_4^-) = 1.0122$$

may be even smaller than the 1.7% in the Born term and 3.3% in the neutral term according to Kim.[34] (These are proportional to r_{vdW}^{-1} and r_{vdW}^2, respectively). Whereas the problem of the size difference can be handled satisfactorily by means of suitable neutral analogs, not so that of the ion–quadrupole interactions. An attempt to calculate the difference in ΔG_t^∞ of Ph_4As^+ and BPh_4^- arising from this effect in terms of the charge on the surface of the tetraphenyl moiety, the sizes of the solvent molecules, and their quadrupole moments involves too many unverifiable assumptions to be useful. In all the cases examined, the deviation of the ionic ΔG_t^∞ on account of both the size and the ion–quadrupole interaction effects from the equipartition value is less than the percent experimental error (except for acetone and perhaps dimethylformamide). Thus to a good first approximation the equipartition of $\Delta G_t^\infty(Ph_4AsBPh_4, W \rightarrow S)$ between the constituting ions should be accepted.

The criteria for the acceptance or the rejection of ΔG_t^∞ values for single ions that have been recorded in the literature depend, however, not only on the validity of the extrathermodynamic assumptions made, but also on other factors. Not least among them, of course, is the reliability of the experimental work that has gone into their establishment, including the purity of the non-aqueous solvent employed, in particular its freedom from more strongly solvating substances, such as water. Another criterion that must be met is the proper conversion of the experimental data obtained at finite solute concentrations to the standard state of infinite dilution, by extrapolation or the application of acceptable theoretical expressions for the activity coefficients. A most important criterion, that has already been discussed above, but is sometimes overlooked, is the additivity of the cation and anion values (weighted with the stoichiometric coefficients in the case of multivalent ions). If values for one kind of charge only are available (cations, in some of the polarographic data), then the alternative criterion which should be used is the independence of the differences between the ΔG_t^∞ of pairs of such ions from the nature of the counter ion. A final criterion that may be applied (cautiously) for the rejection of reported values is the requirement that ΔG_t^∞ should depend 'smoothly' on monotonically varying properties of series of related solvents or ions, provided that nonrelated properties are kept essentially constant.

With the application of these criteria it has been possible to evaluate on a common concentration scale (mol dm^{-3}) and at a common temperature (25 °C) the ΔG_t^∞ values recorded in the literature for single ions, and to obtain 'selected values' as weighted means.[37,39] For this weighting, reliable data belonging to category (vii) have been given full weight, as have those belonging to category (ii), provided independence of the EMF from the nature of the bridge solvent has been demonstrated. Data from categories (i), (iii), and (iv) have generally been given zero weight, since the criteria for validity have not been fulfilled. Data from categories (v) and (vi), where the additivity criterion has been kept, have been adjusted by the algebraic addition of a constant amount to the data for the cations and its subtraction from those of the anions for a given solvent. This adjustment has been done in a manner minimizing the error of the weighted mean, the weight given to these data ranging between 0.2 and 0.8, depending on the other criteria. The selected data[39] are shown in Table 6.8. In those cases where the data from reliable independent sources agree within 1 to 2 kJ mol^{-1}, the selected value is given to one place beyond the decimal point. In cases where no confirmation from independent sources is available for otherwise reliable data, or where the data from independent sources agree only within 2 to 5 kJ mol^{-1} but are otherwise given full weight, the selected value is reported as an integer. Values which cannot be judged as reliable, though given nonzero weight, but where no better alternatives are available, are reported within parentheses. Cases where data have been reported in the literature, but that are judged unreliable and given zero weight are not reported by a figure in the Table, but are marked with an asterisk. The compiled data in ref. 39 should be consulted if access to the literature concerning these cases is desired.

A quantity that is closely related to $\Delta G_t^\infty(\mathrm{i, W \rightarrow S})$ is the standard electrode potential of an electrode reversible to the ion i in a nonaqueous solvent, with respect to the aqueous standard hydrogen electrode (SHE) as reference. This topic is developed further in Section 9.2.

6.4 Single ion standard molar enthalpies, entropies, and volumes of transfer

As in the case of the standard molar Gibbs free energies of transfer ΔG_t^∞, it is necessary also to split ΔH_t^∞ and ΔS_t^∞ values obtained for electrolytes into the contribution from individual ions, in order to relate these quantities to the interactions the ions undergo in a solvent S relative to those in W. The arguments presented for the choice of W as water in the case of ΔG_t^∞ are valid also for ΔH_t^∞ and ΔS_t^∞, and this identification will be used in the following.

The methods for splitting ΔH_t^∞ of electrolytes into the ionic contributions are as varied as those used for splitting ΔG_t^∞ (see p. 153). A comparison of the results of some of them has been made,[40] from which it has been concluded that the assumption of a negligible liquid junction potential with tetraethylammonium picrate as the bridge electrolyte does not yield a sufficiently precise temperature coefficient of the cell EMF, and is, therefore, not recommended. The reference electrolyte assumption, and specifically that $\Delta H_t^\infty(\mathrm{Ph_4As^+, W \rightarrow S}) = \Delta H_t^\infty(\mathrm{BPh_4^-, W \rightarrow S})$ for all solvents S, is recommended, and is shown to yield values not deviating much from those obtained from the reference ion/molecule assumption. The latter has been tested with the 'fic$^+$/foc' couple as well as with tetraphenylmethane versus either tetraphenylarsonium or tetraphenylborate.[40]

Attempts have been made to calculate $\Delta H_t^\infty(\mathrm{i, W \rightarrow S})$ using the electrostatic model. The differentiation of the Born equation 6.13 with respect to the

Table 6.8 $\Delta G_t^0(X, H_2O \rightarrow S)$/kJ mol^{-1}, mol dm^{-3} scale, 25 °C. Solvents with selected values for cations[a]

No.*	8	9	10	25	26	41	61	99	101
Solvent*	MeOH	EtOH	PrOH	TFE	$En(OH)_2$	Me_2CO	PC	FA	DMF
H^+	10.4	11.1	9		5		50		−18
Li^+	4.4	11	11		0		23.8	−10	−10
Na^+	8.2	14	17		−2		14.6	−8	−9.6
K^+	9.6	16.4	18	39	−2	4	5.3	−4.3	−10.3
Rb^+	9.6	16	19			4	−1.0	−5	−9.7
Cs^+	8.9	15	17			4	−7.0	−6.0	−10.8
Ag^+	6.6	4.9	1	50	1	9	18.8	−15.4	−20.8
Tl^+	4.1	7					11.0		−11.5
MH_4^+	(5)	7							
Me_4N^+	6	10.9	11			3	−11		−5.3
Et_4N^+	1	6					−13		−8.0
Pr_4N^+		(−6)					−22		−17
Bu_4N^+	−21	(−8)					−31		−29
Ph_4As^+	−24.1	−21.2	−25		−21	−32	−36.0	−23.9	−38.5
Cu^{2+}	(26)	(46)	(43)				(75)		(−18)
Zn^{2+}									
Cd^{2+}									

[a] From Y. Marcus, *Pure Appl. Chem.*, **53**, 955 (1983); * See list of solvents, Table 6.1; ** N,N-Dimethylthioformamide; † 30 °C.

temperature, and the use of an addend to the crystal ionic radius (which is temperature independent) to account for the dielectric saturation near the ion, yields

$$\Delta H_{\text{tel}}^{\infty} = -\tfrac{1}{2}\mathbf{N}z_i^2\mathbf{e}^2\varepsilon_0^{-1}[(r_i + \Delta r_S)^{-1}\varepsilon_S^{-1}(1 - \mathrm{d}\ln\varepsilon_S/\mathrm{d}\ln T) - (r_i + \Delta r_W)^{-1}\varepsilon_W^{-1}(1 - \mathrm{d}\ln\varepsilon_W/\mathrm{d}\ln T)] \tag{6.22}$$

Whether the addends used for $\Delta G_{\text{tel}}^{\infty}$ are valid also for $\Delta H_{\text{tel}}^{\infty}$ is not self-evident, since the Δr's are mere empirical fitting parameters. The not completely satisfactory results obtained[41] with the application of equation 6.22 to transfers

5	**	108	93	91	92	75	76†	110	70	71
IA	DMThF	NMPy	MeCN	$MeNO_2$	$PhNO_2$	DMSO	TMS	HMPT	1.1DClE	1,2DClE
		−25	46.4		33	−19.4				
	55	−35	25	48	38	−15				
2.1	39	−15	15.1	(26)	34	−13.4	−3		29	25
1.7	27	−11	8.1	19	23	−13.0	−4	−16	30	26
8		−8	6.3	(5)	19	−10.4	−9		29	25
7)	14	−10	6.0	(1)	15	−13.0	−10		28	24
9.0	−102	−26	−23.2	21		−34.8	−4	−44		
	−16	−15	8.0		(15)	−21.4				
		−24			27					
		−3	3		4	−2			18	16
			−7		−5	(−9)			11	5
			−13		−16					
			−31		−24					
8.7		−40	−32.8		−36	−37.4	−36	−39	−27	−33
			(95)			(−49)	(71)			
						(−45)				
						(−58)				

(*continued*)

into propylene carbonate may be due to this reason, or to the non-use of a corresponding neutral term, $\Delta H^{\infty}_{\mathrm{t\,neut}}$.

An extrapolation method, that is completely analogous to that used for aqueous solutions by Halliwell and Nyburg,[42] based on the electrostatic approach of Buckingham,[33] has been used for obtaining the standard molar enthalpies of solvation of single ions in nonaqueous solvents. However, for obtaining $\Delta H^{\infty}_{\mathrm{t}}(\mathrm{i}, \mathrm{W} \rightarrow \mathrm{S})$ values, a small difference between large numbers must be calculated, hence the values resulting are not accurate. If, instead of extrapolation, the differences between $\Delta H^{\infty}_{\mathrm{t}}$ of ions of equal sizes and opposite

Table 6.8—*continued*

Solvents with selected values for anions

No.*	8	9	10	25	26	41	61	99	1
Solvent*	MeOH	EtOH	PrOH	TFE	En(OH)$_2$	Me$_2$CO	PC	FA	D
F^-	16						56	25	
Cl^-	13.2	20.2	26	−10	9	57	39.8	13.7	
Br^-	11.1	18.2	22	−8	7	42	30.0	10.7	
I^-	7.3	12.9	19	−8	3	25	13.7	7.3	
I_3^-	−12.6							−7	−
N_3^-	9.1	17.0			7	43	27	11	
CN^-	8.6	7				48	36	13.3	
SCN^-	5.6				5		7.0	7	
NO_3^-		14							
ClO_4^-	6.1	10	17				−3	−12	
$CH_3CO_2^-$	16.0							20	
Pic$^-$**	−6	0.5			−7		−6	−7	−
BPh_4^-	−24.1	−21.2	−25		−21	−32	−36.0	−23.9	−

* See list of solvents, Table 6.1; ** Pic$^-$ = picrate; † 30 °C.

charges are considered, only the ion-quadrupole interactions remain, so that[26,43]

$$\begin{aligned}\Delta H_t^\infty(\mathrm{Na}^+, \mathrm{W} \to \mathrm{S}) &+ \tfrac{1}{2}\Delta H_t^\infty[\mathrm{A}^-(r_\mathrm{A} = r), \mathrm{W} \to \mathrm{S}] \\ &- \tfrac{1}{2}\Delta H_t^\infty[\mathrm{M}^+(r_\mathrm{M} = r), \mathrm{W} \to \mathrm{S}] \\ = \Delta H_t^\infty(\mathrm{Na}^+, \mathrm{W} \to \mathrm{S}) &- \tfrac{1}{2}\mathrm{N}C_3\mathbf{e}[\theta_\mathrm{S}(r + \Delta r_\mathrm{S})^{-3} - \theta_\mathrm{W}(r + \Delta r_\mathrm{W})^{-3}] \\ &- \tfrac{1}{2}\mathrm{N}C_4[\mu_\mathrm{S}\theta_\mathrm{S}(r + \Delta r_\mathrm{S})^{-4} - \mu_\mathrm{W}\theta_\mathrm{W}(r + \Delta r_\mathrm{W})^{-4}]\end{aligned} \quad (6.23)$$

where θ is the quadrupole moment. The left-hand side is experimentally accessible (equalling $\frac{1}{2}[2\Delta H_t^\infty(\mathrm{NaA}) - \Delta H_t^\infty(\mathrm{MA})]$), provided values for M^+ and A^- of equal radius can be obtained by interpolation, but the parameters C_3, C_4, θ_S and Δr_S are generally not known (it being assumed that θ_W and Δr_W are). The constants C are specific for a solvent in the sense that they depend on the coordination number of the solvent around the ion. If these parameters are estimated, then $\Delta H_t^\infty(\mathrm{Na}^+, \mathrm{W} \to \mathrm{S})$ (or that of another ion, depending on the

05	108	93	91	92	75	76†	110	70	71
MA	NMPy	MeCN	$MeNO_2$	$PhNO_2$	DMSO	TMS	HMPT	1,1DClE	1,2DClE
		71							
54.9	51	42.1	37	35	40.3	47	58	58	52
44.0	37	31.3	30	29	27.4	35	46	43	38
21	19	16.8	17	18	10.4	21	30	31	25
30		−15		−23	(−41)				
40	46	37	28		25.8	41	49		
		35			35				
21	18	14.4	15		9.7	22	20		
		21							7
	−12	2	(−5)	7			−7	22	16
70		61			(50)				
		−4		−5					
38.7	−40	−32.8		−36	−37.4	−36	−39	−27	−33

experimental electrolyte data used) can be estimated, hence by the additivity rule those of other ions. No satisfactory method for estimating the necessary parameters has been provided, however.

The methods for splitting ΔS_t^∞ of electrolytes into the ionic contributions depend generally on those applied to ΔG_t^∞ and ΔH_t^∞, since the entropy of transfer of electrolytes is commonly derived from these two quantities. Care must, of course, be taken to use the same method for both these quantities. Two independent methods have also been proposed. In one, standard partial molar entropies of electrolytes in a nonaqueous solvent S are split, and are compared with the corresponding ionic quantities in water. The split is effected in such a manner that when plotted against the aqueous quantities, the values for the cations and those for the anions lie on the same straight line. This is equivalent to writing according to Criss and co-workers[44]

$$\Delta S_t^\infty(\text{i, W} \rightarrow \text{S}) = a_\text{S} + (b_\text{S} - 1)\bar{S}_\text{i}^\infty(\text{W}) \tag{6.24}$$

where the constants (a_S and b_S) are specific for each solvent: (0.8, 1.04) for heavy water, (−1.6, 0.64) for formamide, (−5.7, 0.72) for N-methylformamide,

(-15.9, 0.79 for N,N-dimethylformamide, (-22.4, 0.82) for ammonia, (-10.9, 0.82) for methanol, and (-16.0, 0.79) for ethanol.[44] The values of $\bar{S}_i^\infty(W)$ can be taken from Table 5.10. The linearity shown by equation 6.24 is an empirical observation for the solvents listed, and holds to within ± 1.6 to ± 9 J K^{-1} mol^{-1}. There exists no valid theoretical basis for its extension to solvents generally, and it must be validated for each solvent individually. A similar method essentially assumes b_S of equation 6.24 to be solvent-independent, but then $(b_S - 1)\cdot\bar{S}_i^\infty(W)$ is replaced by an ion-specific constant $-I_W$, and the identification $E_S = a_S$ is made in equation 6.24 to give according to Abraham[45]

$$\Delta S_t^\infty(i, W \to S) = -E_S - I_W \tag{6.25}$$

The solvent-specific constant E_S has the following values (in J K^{-1} mol^{-1}): ethanol 113, liquid ammonia 146, formamide 71, N-methylformamide 84, N,N-dimethylformamide 134, dimethylsulphoxide 105, acetonitrile 126, and acetone 146. The values for I_W, in J K^{-1} mol^{-1} are: Li^+ 63, Na^+ 50, K^+ 38, Rb^+ 42, Cs^+ 46, $(CH_3)_4N^+$ 113, $(C_2H_5)_4N^+$ 177, $(C_3H_7)_4N^+$ 243, $(C_4H_9)_4N^+$ 297, $(C_6H_5)_4P^+$ 222, $(C_6H_5)_4As^+$ 218, F^- 75, Cl^- 50, Br^- 42, I^- 38, ClO_4^- 38, and $(C_6H_5)_4B^-$ 209. The fact that these values of I_W are not valid for the transfer of the ions into methanol[45] poses a *caveat* against the extension of the method to solvents not examined in the original paper without specific validation.

Conflicting results naturally arise from the application of various methods for obtaining single ion standard molar enthalpies and entropies of transfer. However, since the reference electrolyte method, with the specific assumption that $\Delta G_t^\infty(Ph_4As^+, W \to S) = \Delta G_t^\infty(BPh_4^-, W \to S)$ for all S has been adopted at a given temperature, 25 °C, it does not make sense to restrict its validity to just this one temperature. Accepting it for all temperatures has as a corollary that $\Delta H_t^\infty(Ph_4As^+, W \to S) = \Delta H_t^\infty(BPh_4^-, W \to S)$ and $\Delta S_t^\infty(Ph_4As^+, W \to S) = \Delta S_t^\infty(BPh_4^-, W \to S)$ for all S must also be accepted.

The values of $\Delta H_t^\infty(i, W \to S)$ and $\Delta S_t^\infty(i, W \to S)$ calculated on this basis (on the mol dm^{-3} scale for the latter) have been compiled,[46] weighted essentially on the same basis as used for ΔG_t^∞ (p. 164), and the weighted mean taken as the selected values presented in Table 6.9. The conventions used in Table 6.8 concerning the accuracy of the data apply also to the ΔH_t^∞ data in Table 6.9. Regarding ΔS_t^∞, integral values are considered reliable to within 3 to 5 J K^{-1} mol^{-1}. There is, however, an additional criterion which must be met by these selected data, namely that $\Delta H_t^\infty - T\cdot\Delta S_t^\infty$ must equal the selected ΔG_t^∞, within the assigned uncertainties of ± 6 kJ mol^{-1}. Since the entropies of transfer are generally calculated from the Gibbs free energies and the enthalpies of transfer, which are the primary data, the entropies are quantities that are beset by the larger errors. In the few cases where the last mentioned criterion is not met, the value of ΔS_t^∞ in Table 6.9 is enclosed in brackets. For one solvent this criterion is not met systematically: N,N-dimethylformamide. An adjustment of $+(-)$ 14 J K^{-1} mol^{-1} to the cation(anion) values must be made in order to make the data conform.

The volumes of transfer of ions from the reference solvent W to solvents S have not been studied very extensively. However, not as in the cases of other thermodynamic quantities of transfer, there exists Zana and Yeager's experimental method (ultrasonic vibration potentials)[47] for the measurement of the absolute partial molar quantity for an ion in a given solvent. Therefore, the difference $\Delta V_t^\infty(\text{i, W} \to \text{S}) = \bar{V}^\infty(\text{i, S}) - \bar{V}^\infty(\text{i, W})$ between the partial molar volumes at infinite dilution is the required quantity. The main problem arises from the limited accuracy claimed at the present stage of development of the method for $\bar{V}_i$ in any solvent and at any concentration, and from the inability of the method to be applied at sufficiently low concentrations for safe extrapolation to infinite dilution, in particular if ion pairing is liable to occur.[49] It has been observed[48] that the additivity rule is not always adhered to, in the worst cases the gap amounting to 9 $cm^3\ mol^{-1}$, whereas the accuracy claimed for the method is estimated at $\pm 2\ cm^3\ mol^{-1}$. A criticism has also been raised at the ultrasonic vibration potential method, that it yields similar values of $\bar{V}_I^\infty$ in water and nonaqueous solvents for the tetraalkylammonium ions: e.g., $\bar{V}^\infty(Bu_4N^+, W) = 270.1\ cm^3\ mol^{-1}$ and $\bar{V}^\infty(Bu_4N^+$, N,N-dimethylformamide) $= 266.4\ cm^3\ mol^{-1}$. The argument forwarded against this is that in water hydrophobic interactions affect the volume, whereas in organic solvents these large ions are not solvated.[48] The same paper shows, however, that because of the compensation of the hydrophobic and electrostrictive effects, the volume of these ions in water becomes nearly identical with their intrinsic volume, hence no large difference between the volumes in water and organic solvents should be expected after all. To a first approximation, therefore, the values of ΔV_t^∞ obtained from the ultrasonic vibration potential data can act as guides for the magnitudes obtained from other methods, that conform well with the additivity rule.

Two other methods have already been discussed in Section 5.2, concerning absolute partial molar volumes of ions at infinite dilution in water. One is the extrapolation of the standard partial molar volumes of tetraaklylammonium halides to zero molar mass of the cation, to give $\bar{V}_i^\infty$ of the halide anion in the solvent employed, whether water or S. The other is the division of $\bar{V}^\infty(Ph_4AsBPh_4$, W or S) between the constituent ions according to their van de Waals volumes (i.e., not equally): $V_{vdW}(Ph_4As^+)/V_{vdW}(BPh_4^-) = 1.0337$ (see Table 6.7). The results from the two methods yield values of $\bar{V}^\infty(H^+, W)$ in good agreement: -6.2 and $-6.7\ cm^3\ mol^{-1}$, and they agree also, within the claimed accuracy, with the result from the ultrasonic vibration potentials, $-5.4\ cm^3\ mol^{-1}$. The mean value from the former two methods has been used to yield with the additivity rule the $\bar{V}^\infty(\text{i, W})$ values for many ions shown in Table 5.8. It appears, however, that these methods do not give the same value of $\bar{V}_t^\infty(X^-, \text{W} \to \text{S})$, even within wide limits. For instance, ultrasonic vibration potentials give $V_t^\infty(Br^-, \text{W} \to \text{methanol}) = -12.3\ cm^3\ mol^{-1}$, the Ph_4AsBPh_4 method gives -14, and the extrapolation method gives -44.9 for the same quantity. Another example: $\bar{V}_t^\infty(K^+, \text{W} \to \text{formamide})$ is given by these methods as $+3.9$, $+6$, and $+21.7\ cm^3\ mol^{-1}$, respectively.[46] The former two

Table 6.9 Selected data of ΔH_t^0 and ΔS_t^0: $\Delta H_t^0(\mathrm{X, W \rightarrow S})$/kJ mol^{-1} (upper values for each ion), and $\Delta S_t^0(\mathrm{X, W \rightarrow S})$/J K^{-1} mol^{-1} (mol/dm^3 scale) (*lower values*); 25 °C

Solvents with selected values for cations

Solvent	D_2O	MeOH	EtOH	PrOH	PC	NH_3*
H^+					44	
Li^+	1.9	−21.7	−20.2		2.8	−45
		−83	*−101*	*−100*	*−65*	
Na^+	2.6	−20.7	−19.4		−10.5	−36
		−95	*−108*	*−106*	*−75*	
K^+	2.8	−19.0	−19.6		−22.5	−26
		−94	*−113*	*−121*	*−89*	
Rb^+	2.9	−16.5			−24.9	−25
		−85	*−103*	*−118*	*−77*	
Cs^+	3.0	−14.1			−27.5	−33
		−76	*−92*	*−104*	*−66*	
Ag^+	2.3	−20.9			−11.1	−106
		−95			*−92*	
Ca^{2+}	5.4					−98
Sr^{2+}	5.7					−109
Ba^{2+}	6.1	−60.6				−102
NH_4^+	1.3				−19.8	
				−118		
Me_4N^+	1.8	0.3	0.2		−18	
		−18	*−36*	(*−45*)		
Et_4N^+	0.9	7.1				
		[31]				
Pr_4N^+	−0.2	15				
Bu_4N^+	−1.0	20			21	
		142				
Ph_4As^+	0.7	−1.0			−13.1	
		74		*90*	*71*	

Solvent	FA	DMF	NMPy	MeCN	DMSO	TMS†
H^+						73
Li^+	−6.0	−25.4			−27.1	
	13	*[−60] − 46*			*−43*	
Na^+	−16.5	−32.4	−41	−13.3	−29.2	−16
	−29	*[−81] − 67*		*−90*	*−56*	
K^+	−17.9	−35.7	−47	−22.9	−35.4	−26
	−41	*[−100] − 86*		*−103*	*−84*	
Rb^+	−17.8	−36.1		−24.6	−35	−28
	−41	*[−96] − 82*		*−103*	*−87*	
Cs^+	−17.7	−34.6			−33.0	−26
	−38	*[−92] − 78*		*−100*	*−74*	
Ag^+		−35		−52.7	−53.3	−14
		[−77] − 63		*−100*	*[−89]*	
Ca^{2+}						

Table 6.9—*continued*

Solvent	FA	DMF	NMPy	MeCN	DMSO	TMS†
Sr^{2+}						
Ba^{2+}					−80	
NH_4^+					−41	
Me_4N^+		−13		−15.3	−16.4	
		[*−33*] *−19*		[*−41*]	[*−24*]	
Et_4N^+						
		[*31*] *45*		*39*		
Pr_4N^+						
Bu_4N^+						
		[*151*] *165*				
Ph_4As^+	−0.5	−17.2	−17	−11.1	−10.6	−11
	80	[*68*] *82*		*75*	*85*	

* At −33 °C; † at 30 °C.

Solvents with selected values for anions

Solvent	D_2O	MeOH	EtOH	PrOH	PC	NH_3*
F^-	−2.6					
Cl^-	−0.2	8.4	10.4		26.2	6
		−17	*−33*	*−58*	*−43*	
Br^-	0.4	4.5	5.5		15.2	−16
		−23	*−43*	*−65*	*−51*	
I^-	1.0	−1.0	−0.7		−1.6	−29
		−31	*−46*	*−70*	*−64*	
N_3^-		0.5			16.7	
		−34			*−43*	
ClO_4^-	−0.2	−3.1	−2.7		−16.3	
		−26	*−46*	*−62*		
BPh_4^-	0.7	−0.9			−13.2	
		74		*90*	*71*	

Solvent	FA	DMF	NMPy	MeCN	DMSO	TMS†
F^-	19.9					
Cl^-	3.5	17.9	27	19.3	20.0	27
	−32	[*−85*] *−99*		*−74*	*−62*	
Br^-	−1.5	0.6	13	8.0	4.6	13
	−39	[*−92*] *−106*		*−76*	*−81*	
I^-	−6.8	−15.0	2	−7.6	−11.5	−8
	−47	[*−105*] *−119*		*−85*	[*−86*]	
N_3^-				8.8		15
		[*−111*] *−125*		[*−72*]	*−92*	
ClO_4^-		−23.4	−11		−18.2	
	−40	[*−86*] *−100*		*−72*	*−69*	
BPh_4^-	−0.5	−17.2	−17	−10.4	−11.0	−11
	80	[*68*] *82*		*75*	*85*	

* At −33 °C; † at 30 °C.

Table 6.10 Standard molar volumes of transfer of ions from water to nonaqueous solvents, $\Delta V_t^0(\mathrm{i}, \mathrm{W} \rightarrow \mathrm{S})/\mathrm{cm^3\ mol^{-1}}$, at 25 °C

Solvent	D_2O	MeOH	EtOH	$En(OH)_2$	EC*	PC	FA	MeCN	DMSO	HMPT
H^+		−11	−10							
Li^+		−11	−13	−3		−4	3	−15	−13	10
Na^+	0.8	−10	−3	2		1	4	−13	−3	19
K^+		−10	−4	1		−2	5	−13	−4	21
Rb^+		−10	−3	1		−3	3	−15	−4	
Cs^+		−11	−3	0		−3	3	−16	−6	
NH_4^+		−8	−3							
Me_4N^+	−0.7	−14			18	−4				
Et_4N^+	−0.7	−14	−18		22	−3	−1	−15	−14	
Pr_4N^+	−0.9	−7	−9		30	1	4	−9	−9	
Bu_4N^+	−1.1	−3	−4		36	9	12	−1	−2	
Pn_4N^+	−2.0		1				18	5	5	
Ph_4As^+		−26					8	−11		−12
F^-	−0.6	−6								
Cl^-	−0.1	−10	−11	2		−5	1	−18	2	
Br^-	0.4	−12	−15	1		−5	1	−21	−1	−32
I^-	−0.4	−14	−15	0	−19	−4	1	−17	3	−25
NO_3^-		−7	−11						3	
ClO_4^-						−3				
BPh_4^-		−26					8	−11		−12

* At 40 °C.

methods are seen to be more nearly mutually compatible, whereas the third is a long way off. No immediate reason for this is apparent, therefore it is impossible to judge the 'correctness' of any of them. Table 6.10 is an attempt at the correlation of the available data,[46] within the constraint of the additivity rule. Eventual adjustment by the algebraic addition (subtraction) of some constant for a given solvent to the cation (anion) values may be necessary.

6.5 Thermodynamics of transfer and the properties of solvents and ions

The thermodynamic quantities for the transfer of individual ions from a reference solvent to other solvents are related to the interactions of the ions with the solvents (relative to those with the reference solvent). These interactions include any effects the ion may have on the solvent molecules surrounding it (changes in the liquid structure, dielectric saturation, etc.), in addition to direct solute–solvent interactions, as detailed in Chapter 3. It is of interest to relate these interactions to the properties of the solvents and the ions taking part in them. One purpose of such a relation is the prediction of thermodynamic quantities of transfer, hence of thermodynamic properties of electrolyte solutions in nonaqueous solvents. The prediction applies to combinations of solvents and ions that have not been studied directly, provided the relevant properties are independently available. Another purpose may be the gaining of insight into the physical nature of the interactions that are important in dilute electrolyte solutions in nonaqueous solvents.

The first purpose can be achieved by two approaches:

(i) a statistical analysis of the relation between a large set of thermodynamic quantities of transfer and sets of properties of the solvents and ions involved;
(ii) the calculation of quantitative results from a detailed physical model of the interactions, with the appropriate properties of the solvents and ions used as the physical parameters of the model.

Both approaches can, in principle, be used for the predictive purpose, and the physical model approach has been used extensively for the division of thermodynamic quantities of transfer (or of solvation) into the ionic contributions, see Section 6.3. Only the first approach, however, can yield further insight, since the second involves a preselected model, and hence a biased viewpoint.

The unbiased application of the statistical approach requires a broad data base, e.g., $\Delta G_t^\infty(\mathrm{i, W \rightarrow S})$ for many ions i and solvents S, and a wide variety of mutually independent properties of the ions and the solvents. Statistical tests then determine which of these properties, or which combination of them, can account for the data in the most significant manner. An early example of this method notes that the shift of the half wave potential for the polarographic reduction of naphthaquinones in the presence of metal ions M^{z+} from that in their absence, $\Delta E_{\frac{1}{2}}$, is a measure of the solvation of the ions. The relationship between $\Delta E_{\frac{1}{2}}$ and selected properties of the solvents and the ions has been tested for four solvents (acetone, acetonitrile, dimethylsulphoxide, and N,N-dimethylformamide) and a small group of metal ions. Only one property of the metal ions: their ionic potentials $z\mathbf{e}/r_M$, and of the solvents: their donor numbers DN, have been used.[50] Within this limited program, the following relationship can be written:

$$\Delta E_{\frac{1}{2}} = a + b \cdot DN + c \cdot z\mathbf{e}/r_M + d \cdot DN \cdot z\mathbf{e}/r_M \tag{6.26}$$

where a, b, c, and d are constants, independent of the ions and the solvents. A generalization of this method involves many properties, and for the study of the transfer of ions, the solvent properties should be referred to those of the reference solvent, W. Hence the general relationship[51]

$$\Delta G_t^\infty(\mathrm{i, W \rightarrow S}) = \sum_i \sum_j A_{ij} \cdot P(\mathrm{i})_i[(P(\mathrm{S})_j - P(\mathrm{W})_j)] \tag{6.27}$$

results, where the index i refers to the ith property of an ion and the index j to the jth property of a solvent, the P's are the values of the properties (preferably normalized, dimensionless quantities), and the A's are multiple regression coefficients. For the sake of simplicity, and in view of the commonly used linear free energy relationships, linear dependencies on the properties P have been assumed (thus, dependencies on r_i^{-1} and r_i^{-2}, for example, are dependencies on two distinct properties of the ions, although these are mutually dependent). Similar equations, naturally with different coefficients A_{ij}, can be written for the enthalpy, entropy, and volume of transfer, or any other thermodynamic property, for which a sufficiently large data base is avilable.

Several statistical methods can be applied to the solution of equation 6.27, relating data bases for m ions i and n solvents S to the statistically most significant i properties P(i) and j properties $P(S)$ (with $\sum_i + \sum_j \ll \mathrm{m} \cdot \mathrm{n}$) and find the coefficients A_{ij}. Of these, two recommend themselves in particular: vector (or factor) analysis, as used, for instance, by Krygowski and Fawcett,[52] and stepwise multiple regression, as used by Glikberg and Marcus.[51] The former requires a complete matrix of $m \cdot n$ data points, and if one of these were missing, then $(m - 1)$ or $(n - 1)$ further points would have to be excluded from the statistical analysis. In the present case, the data base shown in Table 6.8 has unfortunately several gaps, and the second method is therefore preferable. It has been found convenient to carry out first a stepwise multiple regression over the solvent properties:

$$G_t^{\infty}(\mathrm{i, W \rightarrow S}) = \sum_j A_j(\mathrm{i})[(P(\mathrm{S})_j - P(\mathrm{W})_j)] \tag{6.28a}$$

for each ion separately, and then in the next stage a stepwise multiple regression over the properties of the ions:

$$A_j(\mathrm{i}) = \sum_i A_{ij} \cdot P(\mathrm{i})_i \tag{6.28b}$$

The reverse order of procedure would, of course, also be possible.

Tables 6.1 to 6.6 list many properties of solvents, which might be relevant to the present problem. Other properties could also be tested, provided they are known for all the solvents listed in Table 6.8, belonging to the data base, as can also mathematical derivatives and combinations of properties. For instance, the reciprocal of the dielectric constant, ε^{-1}, the square of the solubility parameter, δ^2, the polarization $P' = (\varepsilon - 1)(\varepsilon + 2)^{-1}$ or the bulk polarization VP', the polarizability $R' = (n_D^2 - 1)(n_D + 2)^{-1}$ or the bulk polarizability VR', the quantity $Q = 9g\varepsilon(\varepsilon - 1)^{-1}(\varepsilon + 2)^{-1}$, etc., should also be tested. On the other hand, the statistical program is able to avoid the simultaneous use of two properties which are well mutually correlated.

Particular attention must be paid to the specific interaction properties of the solvents, which are losely called 'polarity indices', acid–base properties, donor–acceptor properties, etc. In Section 6.1 it has already been pointed out that some of these are mutually well correlated, and that there are essentially only two more or less independent properties: the electron pair donating and accepting propensities of the solvents. Thus the acidity or electron pair accepting or hydrogen bond donating properties (the last one only for protic solvents), such as Y, W, Ω, α, Z, χ_B, AN, and S, can be so correlated.[53] A common scale may not work with both protic and aprotic solvents, but the E_T^N scale is well correlated with all of these for the solvents that are common to these scales, and has the advantage that it is known for over 240 solvents.[53] It must be stressed that it measures a property of the bulk solvent rather than that of an isolated solvent molecule. The basicity or electron pair donating or hydrogen bond accepting properties, such as $E_A C_A + E_B C_B$ (with A a reference acid, such as $SbCl_5$ and the base B being a solvent), B or $\Delta\nu(CH_3OD)$ or the hydrogen bond accepting β-scale are all mutually correlated, as also with the normalized donor numbers DN^N.[54] In this case, the DN are the properties of the isolated solvent molecules in a, hopefully, inert solvent, 1,2-dichloroethane, rather than those of the bulk solvent. For about ten solvents estimates of the DN in the bulk solvent are known beside the regular ones in 1,2-dichloroethane. For these the regular DN is approximating also a property of the bulk solvent.[54] Furthermore, since the logarithms of the equilibrium constants for the donor–acceptor reaction of the standard acid, $SbCl_5$, with solvents in dilute solution in 1,2-dichloroethane correlate linearly with the corresponding enthalpy values which constitute the DN scale, it is concluded that DN is an appropriate measure also for Gibbs free energies. For the normalization of the DN scale the solvent hexamethyl phosphoric triamide, HMPT, is used, rather than water, since the water DN is not unambiguously established, and HMPT has the largest DN of the solvents where this property has been determined directly in the diluent 1,2-dichloroethane. The normalized quantity is thus $DN^N = DN(S)/DN$ (HMPT).

The $\Delta G_t^\infty(i, W \rightarrow S)$ data of Table 6.8 have been treated according to the stepwise multiple regression method. The program selected three or four mutually independent variables for the fit in equation 6.28a at an 80% significance level or better (more variables were introduced with only much lower significance). These variables included E_T in all cases, DN for all the cations, and two other variables, of which δ^2 and ε^{-1} were selected in the greatest number of cases. Regressions with the four variables E_T, DN, δ^2, and ε^{-1} accounted well for all the data, even those where the latter two were not among the first choices in free selection runs. The results at this stage are four sets of $A_j(i)$ coefficients for 17 ions i. The values to be used with the normalized variables range from -113 to $+146$ for E_T^N, from -110 to $+2$ for DN^N, from -37 to $+61$ for $\delta^{2N} = \delta^2(S)/\delta^2(W)$, and from -9 to $+8$ for $\varepsilon^{-1N} = \varepsilon(W)/\varepsilon(S)$. The ranges of these coefficients indicate the relative importance of the terms, since the normalized variables are all in the range from zero to unity. The ratio of the standard error of the fit to the range of the ΔG_t^∞ values fitted was <0.13 except for five ions: H^+, Ag^+, I^-, SCN^-, and ClO_4^-, but for the latter three this range itself is small relative to the standard errors attributed to the ΔG_t^∞ data themselves. For the former two, specific interactions that are not taken into account by generalized donor–acceptor interactions may be responsible for the larger standard errors of the fits.

Each of the $A_j(i)$ coefficients was related by the stepwise multiple regression program to a large variety of properties of the ions. These included r_i^n, $(r_i + \Delta)^n$ (n ranging from -6 to $+3$), z_i or z_i^2 times these quantities (for the univalent ions included, $z_i^2 = 1$ and z_i is either $+1$ or -1, but for predictive purposes on multivalent ions both of these may be significant). The molar ionic refractivity R_i, the softness parameter[55] σ_i, ionization potentials, electron and proton affinities, as far as these are known for the 17 univalent ions tested, were also included. As mentioned above, an incomplete data base is no deterrent to the use of the stepwise multiple regression method.

The best model with the minimum number of parameters for all the ions required a modified ionic size, $r_i + \Delta r$, rather than the crystal ionic radius, r_i, *per se*, and the following functions of the variables:

$$\begin{aligned} A(E_T^N) &= 14.25[z_i(r_i + \Delta r)^{-1}] \\ A(DN^N) &= -1.15[z_i(r_i + \Delta r)^{-2}] \exp(z_i - 1) \\ A(\delta^{2N}) &= -0.56[z_i(r_i + \Delta r)^{-2}] + 5.8\sigma_i \\ A(\varepsilon^{-1N}) &= 2.87[z_i^2(r_i + \Delta r)^{-1}] - 0.360[z_i^2(r_i + \Delta r)^{-2}] \end{aligned} \tag{6.29}$$

The numerical coefficients are valid for r_i expressed in nm, and $\Delta r = 0.059 \pm 0.027$ nm. The factor $\exp(z_i - 1)$ in $A(DN^N)$ is an artificial step function that suppresses this quantity for anions but retains it for cations. There are altogether seven numerical parameters, and with their help the $A_j(\mathrm{i})$ have been fitted to within 11 to 14% of their ranges. The fit of the original 196 ΔG_t^∞ data points by these seven parameters used with equations 6.27 to 6.29 is shown in Figure 6.3. The fit of the anions is somewhat poorer than that of the cations, bad offenders

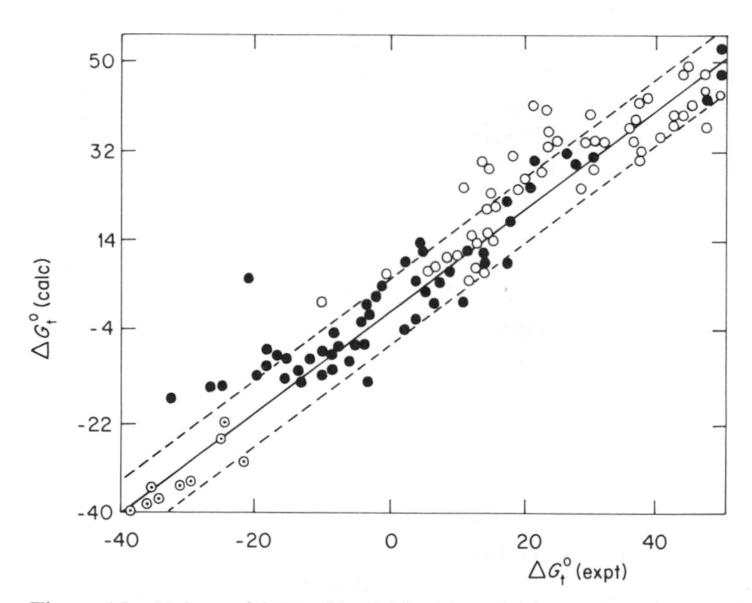

Figure 6.3 Values of ΔG_t^∞ of individual ions (cations: ●, anions: ○, tetraphenyl ions: ⊙), calculated according to equations 6.27 to 6.29, plotted against the experimental data, obtained from Table 6.8. The dashed lines are drawn at a distance of two standard deviations of the fit from the continuous line, that passes with a slope of unity through the point (0, 0). (From Ref. 51, with permission of Plenum Publishing Corp.)

being SCN^- and ClO_4^-, for which the data are not very reliable in the first place. A further difficulty arises from the fact that ΔG_t^∞ of Ph_4As^+ and BPh_4^- are necessarily equal (according to the extrathermodynamic assumption employed to obtain the individual ionic ΔG_t^∞ values of the data base), but $A(DN^N)$ discriminates between ions of practically the same size but opposite signs. For the other ions, except H^+ and Ag^+ mentioned above, the fit is within the experiment accuracy of the ΔG_t^∞ data (± 5 kJ mol^{-1}).

The dependence of ΔG_t^∞(i, W $\rightarrow$ S) on the solvent properties E_T, DN, δ^2, and ε^{-1} and on the ion properties $z_i^n(r_i + \Delta)^{-n'}$ ($n, n' = 1$ and 2) and σ_i can be rationalized in terms of the solute–solvent interactions. This rationalization is the 'insight' that has been expected from the statistical fit approach, compared with the physical model approach, that reflects only those interactions that have been intentionally included. The most important point to note is that a major fraction of ΔG_t^∞, some 70 to 80%, is accounted for by two direct interaction terms, those that involve the donor–acceptor interactions with the solvents (i.e., those that involve E_T and DN). This shows that physical models based primarily on electrostatic interactions (including ion–dipole and ion–quadrupole) interactions are not realistic, when two solvents are compared (they may have their use for studies of solvation in a single solvent).

The Gibbs free energy of transfer of anions has been found to depend primarily on the electron-pair acceptance ability of the solvents, measured by E_T, whereas that of cations on the electron donation ability, measured by DN, as could have been expected from the donor–acceptor properties of the ions. Not so clear is the dependence of ΔG_t^∞ of cations on E_T, that has also been found, even for the transfer of H^+ ions that cannot donate an electron pair to a solvent. However, this dependence is negative: the higher the E_T of a solvent, the more are cations repelled by it. This may be ascribed to the presence of a positive centre in the solvent that is responsible for its electron-pair acceptance properties, as also to its effectiveness in repulsing cations. The dependencies of the coefficients of these direct interactions on the ionic potential or the ionic field strength (or surface change density) exhibited in equation 6.29 are intuitively understood, but not theoretically sustantiated, nor is it understood why different dependencies on the ionic size are exhibited by these two short range interaction modes, electron-pair donation and acceptance.

The importance of the direct, short range interactions of the ionic solutes with the solvents in determining ΔG_t^∞(i, W $\rightarrow$ S) has been recognized by several other workers too.[50-60] In these studies, however, the electron-pair donicity and acceptance have been intentionally imposed on the regressions, rather than obtained from among many other properties by a stepwise selection procedure guided only by the statistical significance tests, as in the study described in detail here.[51] Not all authors used E_T^N and DN^N (or their non-normalized equivalents) for the electron pair acceptance and donation properties. Koppel and Palm's *B* was used instead of *DN* by Makitra and Pirig,[58] Kamlet, and Taft's β was used similarly by Wrona,[59] but these quantities are linearly related to *DN* as noted above.[54] Kamlet and Taft's α was used instead of E_T and no donicity index was employed by Kolling,[60] but his expression,

$$G_t^\infty(\mathrm{Ph_4AsBPh_4, W \rightarrow S}) = -27.34 + 3.41\pi^* + 3.58\alpha$$

cannot be right, since if S is set equal to W and since if W is water, $\pi^* = 1.090$ and $\alpha = 1.017$, -22.7 results instead of the expected zero.

Kolling[60] and Makitra and Pirig[58] added terms relating to the polarity of the solvents (Kamlet and Taft's π^* and the functions $(n_D^2 - 1)/(n_D^2 + 2)$ and $(\varepsilon - 1)/(2\varepsilon + 2)$, respectively), and Kolling,[60] Wrona,[59] and Krygowski and Fawcett[56] used a solvent-independent constant in their regressions. The correlations that the stepwise linear regression method preferred[51] included instead solvent-dependent properties that accounted for about one-quarter of the magnitude of ΔG_t^∞, i.e., the cohesive energy density, δ^2, and the reciprocal of the dielectric constant, ε^{-1}. The latter quantity, indeed, accounts for the polar properties of the solvents. The former was included by Makitra and Pirig[58] in their regressions, and some parameter that should express the resistance of the solvent to changes in structure, which must be related to δ^2, was proposed by Krygowski and Fawcett[56] as a possible improvement of their equations. (In a subsequent paper these authors dispensed with terms that depended on the above mentioned measures of the solvent polarity, but retained a constant term,[52] which might have fulfilled this function.) It is easy to accept a dependence of ΔG_t^∞ of ions on ε^{-1} as expressing the long range electrostatic effects expressed by the Born equation, which should not be allowed to carry the burden of the main part of the ion–solvent interactions, as implied in the electrostatic calculations discussed in Section 6.3. The long range aspect also rationalizes the term Δr added to the crystal ionic radius r_i. The independence of Δr from the nature of the solvents implied in equations 6.29 reflects mainly the insensitivity of the data to small changes in Δr, in view of its large standard deviation (40%). The second term in the equation for $A(\varepsilon^{-1N})$, that is proportional to the electrical field of the ion or to its surface charge density, remains unexplained.

Even more puzzling is the dependence of $A(\delta^{2N})$ on the ion size that has been found. If this term were related to the work required for the creation of a cavity in the solvent of a size needed for the accommodation of the ion, then it could be expected to increase with r_i^3. Instead, it decreases with increasing sizes of the ions, but is significant mainly for the larger ions, since the terms for direct interaction, involving E_T and DN, decrease even faster with increasing ionic sizes.

It must be admitted that the general correlations discussed above cannot account for anomalous cases, where additional interactions take place. The cases of H^+ and Ag^+ have already been briefly alluded to, The former ion hydrogen-bonds to electronegative atoms of solvents with high DN values. The latter ion, as well as Cu^+ (and to a much lesser extent also Cd^{2+}) back-donate d^{10} electrons to the π-electron system of acetonitrile (this is clearly illustrated in Figure 6.1).[61,62] Even more dramatic is the specific interaction of Ag^+ (also Cu^+ and to a much smaller extent Tl^+) with N,N-dimethylthioformamide, into which solvent ΔG_t^{∞} of the alkali metal cations from water is positive, but that of the soft cations Ag^+, Cu^+, and Tl^+ is negative, very highly so for the former one.[63]

The enthalpy and entropy of transfer of ions from water into nonaqueous solvents have not been analysed in terms of the properties of the solvents and ions in as systematic and generalizing manner as the Gibbs free energy of transfer. However, Krygowski and Fawcett[56] applied to data of Parker *et al.*[62] on the enthalpies of transfer of ions from acetonitrile to other solvents their equation

$$\Delta H_t^{\infty}(\text{i}, CH_3CN \rightarrow S) = \Delta H_t^0 + aE_T + bDN \qquad (6.30)$$

where ΔH_t^0, a and b are solvent-independent constants, characterizing the ion i. Contrary to the case of ΔG_t^{∞}, where b for anions is negligibly small,[56] this is not so for the ΔH_t^{∞} of the only anion (beside Ph_4B^-) included in the study: Cl^-. In the case of ΔH_t^{∞}, indeed, all the b's are negative and all the a's (except for Cl^-) positive, so that a certain compensation of the donor and acceptor properties of the solvents is manifested.

As Table 6.9 shows, the ΔH_t^{∞} for the monoatomic cations are generally negative (an exception is the transfer of Li^+ to propylene carbonate) and for the monoatomic anions (except for I^-) it is generally positive. The donor–acceptor interactions should dominate the enthalpy of transfer and these observations can be rationalized in their terms. The normalized donor number of water ($DN^N = 0.464$) is less than those of most of the solvents into which the cations are transferred with a negative ΔH_t^{∞}, and these solvents are preferred energetically. On the other hand, the electron pair acceptance index of water ($E_T^N = 1.000$) is higher than that of any of the solvents into which the anions are transferred with positive ΔH_t^{∞}, and water is the solvent preferred energetically. For the tetraalkylammonium cations ΔH_t^{∞} varies from small negative values for the tetramethyl one to small positive values for the tetrabutyl one. The hydrophobic bonding effect (enhancement of the water structure) is here manifested for the larger cations. Large or polyatomic anions have both positive and negative values of ΔH_t^{∞} too, depending on the solvent, but they are generally small and difficult to systematize.

The entropy of transfer, ΔS_t^{∞}, is seen in Table 6.9 to be generally (except for the tetraphenyl- and larger tetraalkyl ions) negative, for both cations and anions, small or moderately large, monovalent or divalent. Here the structural aspect seems to be of importance, since water is by far the most highly structured solvent (though not by the criteria at the basis of Table 6.6!). The next most structured solvent, formamide, does, indeed, show the least negative ΔS_t^{∞} values (and for Li^+ even a positive value). However, the dominant quantity that determines ΔS_t^{∞} is the number of solvent molecules translationally immobilized per ion in the solvent S, relative to their number in water, and the amount of entropy lost per molecule immobilized in the solvent and in water. The entropies of transfer have been used in this sense for the estimation of solvation numbers, see Table 4.6. The effect of causes other than solvent immobilization on the magnitude of the entropy of transfer must be taken into account, and this can be done in several ways. A different manner has been employed by Hedwig *et al.*[62] to give 'minimal' solvation numbers of Ba^{2+}, Zn^{2+}, and Cd^{2+}, that are not included in Table 4.6, in addition to those of Li^{+}, K^{+}, and Cs^{+}, that are.

References

1. J. A. Riddick and W. B. Bunger, *Organic Solvents*, Wiley-Interscience, New York (1970).
2. R. C. Weast, ed., *Handbook of Chemistry and Physics*, CRC Press, Boca Raton (1982/3).

3a. M. J. Kamlet, J. L. Abboud, and R. W. Taft, *J. Am. Chem. Soc.*, **99**, 6027 (1977); b. V. Bekarek, *J. Phys. Chem.*, **85**, 722 (1981); c. M. J. Kamlet, J. L. M. Abboud, M. H. Abraham, and R. W. Taft, *J. Org. Chem.*, **48**, 2877 (1983).

4. V. Gutmann and E. Wychera, *Inorg. Nucl. Chem. Lett.*, **2**, 257 (1966); V. Gutmann, *Coord. Chem. Rev.*, **18**, 225 (1976).

5a. M. J. Kamlet and R. W. Taft, *J. Am. Chem. Soc.*, **98**, 377 (1976); b. Y. Marcus, *J. Soln. Chem.*, **13**, 599 (1984).

6. K. Dimroth, C. Reichardt, T. Siepmann, and F. Bohlmann, *Liebigs Ann. Chem.*, **661**, 1 (1963); C. Reichardt, *Solvent Effects in Organic Chemistry*, Verlag Chemie, Weinheim (1979), pp. 242 ff and 2nd ed. (1985), Table 7-3.
7. R. W. Taft and M. J. Kamlet, *J. Am. Chem. Soc.*, **98**, 2886 (1976).
8. E. Kalman, I. Serke, G. Palinkas, M. D. Zeidler, F. J. Wiesmann, H. Bertagnolli, and P. Chieux, *Z. Naturforsch.*, **38a**, 231 (1983).
9. A. H. Narten, *J. Chem. Phys.*, **66**, 3117 (1977); A. H. Narten and S. I. Sandler, *ibid.*, **71**, 2069 (1979).
10. H. Michel and E. Lippert, in *Organic Liquids*, A. D. Buckingham, E. Lippert and S. Bratos, eds., Wiley, Chichester (1978), chap. 17.
11. S. I. Sandler and A. H. Narten, *Mol. Phys.*, **32**, 1543 (1976); A. H. Narten, *J. Chem. Phys.*, **65**, 573 (1976); **67**, 2102 (1977); **70**, 299 (1979).
12. G. Engel and H. G. Hertz, *Ber. Bunseng. Phys. Chem.*, **72**, 808 (1968).
13. D. Feakins and K. G. Lawrence, *J. Chem. Soc.*, **A1966**, 212.
14. H. P. Bennetto and E. F. Caldin, *J. Chem. Soc.*, **A1971**, 2191.
15. R. H. Cole, *J. Chem. Phys.*, **27**, 33 (1957).
16. M. Breant and J. L. Sue, *J. Electoanal. Chem.*, **40**, 89 (1972).
17. B. Case, N. S. Hush, R. Parsons, and M. E. Peover, *J. Electroanal. Chem.* **10**, 360 (1965).
18. B. Case and R. Parsons, *Trans. Faraday Soc.*, **63**, 1224 (1967); R. Parsons and B. T. Rubin, *J. Chem. Soc. Faraday Trans. 1*, **70**, 1636 (1974); I. Zagoska and Z. Koczorowski, *Rocz. Chem.*, **44**, 1559 (1970).
19. N. Bjerrum and E. Larsson, *Z. Phys. Chem.*, **127**, 358 (1927).
20. A. J. Parker, *J. Chem. Soc.*, **A1966**, 220; R. Alexander, A. J. Parker, J. H. Sharp, and W. E. Waghorne, *J. Am. Chem. Soc.*, **94**, 1148 (1972).
21. M. Alfenaar and C. L. DeLigny, *Rec. Trav. Chim.*, **86**, 929 (1967).
22. O. Popovych, *Crit. Rev. Anal. Chem.*, **1**, 73 (1970).
23. H.-M. Koepp, H. Wendt, and H. Strehlow, *Z. Elektrochem.*, **64**, 483 (1960); H. Strehlow, *ibid.*, **56**, 827 (1952).
24. J. F. Coetzee, J. M. Simon, and R. J. Bertozzi, *Anal. Chem.*, **41**, 766 (1969).
25. N. A. Izmailov, *Zh. Fiz. Khim.*, **34**, 2414 (1960); *Dokl. Akad. Nauks SSR*, **126**, 1033 (1959); *ibid.*, **149**, 884, 1103, 1364 (1963).
26. C. L. DeLigny, H. J. M. Denessen, and M. Alfenaar, *Rec. Trav. Chim.*, **90**, 1265 (1971); D. Bax, C. L. DeLigny, and M. Alfenaar, *ibid.*, **91**, 452 (1972).

27. V. A. Pleskov, *Usp. Khim.*, **16**, 254 (1947).
28a. S. Villermaux and J. J. Delpuech, *J. Chem. Soc. Chem. Commun.*, **1975**, 478; b. M. F. Lejaille, M. H. Livertoux, C. Guidon, and J. Bessiere, *Bull. Soc. Chim. France*, **1978**, I-373.
29. Y. Marcus and L. E. Asher, *J. Phys. Chem.*, **82**, 1246 (1978); Y. Marcus, E. Pross, and J. Hormadaly, *ibid.*, **84**, 2708 (1980).
30. C. F. Wells, *Trans. Faraday Soc.*, **61**, 2194 (1965); *J. Chem. Soc. Faraday Trans. 1*, **69**, 984 (1973); *ibid.*, **70**, 694 (1974).
31. I. V. Nelson and R. T. Iwamoto, *Anal. Chem.*, **33**, 1795 (1961).
32. G. Gritzner, *Inorg. Chim. Acta*, **24**, 5 (1977); G. Gritzner and E. Geyer, *Z. Phys. Chem.*, **125**, 7 (1981).
33. A. D. Buckingham, *Disc. Faraday Soc.*, **24**, 151 (1957).
34. J. I. Kim, *J. Phys. Chem.*, **82**, 191 (1978).
35. E. Grunwald, G. Baughman, and G. Kohnstam, *J. Am. Chem. Soc.*, **82**, 5801 (1960).
36. O. Popowych, A. Gribovsky, and D. H. Berne, *Anal. Chem.*, **44**, 81 (1972); D. H. Berne and O. Popowych, *ibid.*, **44**, 817 (1972).
37. Y. Marcus, *Rev. Anal. Chem.*, **5**, 53 (1980).
38. A. Bondi, *J. Phys. Chem.*, **68**, 441 (1964).
39. Y. Marcus, *Pure Appl. Chem.*, **55**, 977 (1983).
40. B. G. Cox and A. J. Parker, *J. Am. Chem. Soc.*, **95**, 402 (1973).
41. M. Salomon, *J. Phys. Chem.*, **73**, 3299 (1969).
42. H. F. Halliwell and S. C. Nyburg, *Trans. Faraday Soc.*, **59**, 1126 (1963).
43. G. Somsen, *Rec. Trav. Chim.*, **85**, 526 (1966); G. Somsen and L. Weeda, *J. Electroanal. Chem. Interf. Electrochem.*, **29**, 375 (1971).
44. C. M. Criss, R. P. Held, and E. Lushka, *J. Phys. Chem.*, **72**, 2970 (1968).
45. M. H. Abraham, *J. Chem. Soc. Faraday Trans. 1*, **69**, 1375 (1973).
46. Y. Marcus, *Pure Appl. Chem.*, **57**, (1984).
47. R. Zana and E. B. Yeager, in *Modern Aspects of Electrochemistry*, J. O'M. Bockris, B. E. Conway and R. E. White, eds., **14**, 1 (1982).
48. B. Krumgalz, *J. Chem. Soc. Faraday Trans. 1*, **76**, 1887 (1980).
49. Y. Marcus, N. Ben-Zwi, and I. Shiloh, *J. Soln. Chem.*, **5**, 87 (1976).
50. T. M. Krygowski, *J. Electroanal. Chem.*, **35**, 436 (1972).
51. S. Glikberg and Y. Marcus, *J. Soln. Chem.*, **12**, 255 (1983).
52. W. R. Fawcett and T. M. Krygowski, *Can. J. Chem.*, **54**, 3283 (1976).
53. C. Reichardt and E. Harbusch-Goernert, *Liebigs Ann. Chem.*, **1983,** 721.
54. Y. Marcus, *J. Soln. Chem.*, **13**, 599 (1984).
55. Y. Marcus, *Israel J. Chem.*, **10**, 952 (1972).
56. T. M. Krygowski and W. R. Fawcett, *J. Am. Chem. Soc.*, **97**, 2143 (1975); W. R. Fawcett and T. M. Krygowski, *Austr. J. Chem.*, **28**, 2115 (1975).
57. U. Mayer, *Monatsh. Chem.*, **108**, 1479 (1977).
58. R. G. Makitra and Ya. N. Pirig., *Zh. Neorg. Khim.*, **24**, 2183 (1979).
59. P. K. Wrona, *J. Electroanal. Chem.*, **108**, 153 (1980).
60. O. W. Kolling, *Anal. Chem.*, **52**, 987 (1980).
61. G. J. Janz, M. J. Tait, and J. Meier, *J. Phys. Chem.*, **71**, 963 (1971); J. F. Coetzee, D. K. McGuire, and J. L. Hedrick, *ibid.*, **67**, 1814 (1963); G. R. Hedwig, D. A. Owensby, and A. J. Parker, *J. Am. Chem. Soc.*, **97**, 3888 (1975).
62. B. G. Cox, G. R. Hedwig, A. J. Parker, and D. W. Watts, *Austr. J. Chem.*, **27**, 477 (1974).
63. R. Alexander, D. A. Owensby, A. G. Parker, and W. E. Waghorne, *Austr. J. Chem.*, **27**, 933 (1974); J. G. Clune, W. E. Waghorne, and B. G. Cox, *J. Chem. Soc. Faraday Trans. 1*, **72**, 1294 (1976).

Chapter 7

Selective solvation of ions

7.1 The relevant properties of mixed solvents

This chapter deals essentially only with binary mixtures of solvents that contain an ionic solute in dilute solution. The two solvents are supposed not to interact chemically with each other, in the sense of forming definite associated compounds, although indefinite self- and mutual association may occur in the same manner as it does in the pure individual solvents (see Section 6.1). For the present purpose the volume change on mixing of the solvents is neglected. Hence, when n_A moles of solvent A are mixed with n_B moles of solvent B, the composition of the mixture can be stated in terms of:

$$\text{mole fractions, } x_A = n_A/(n_A + n_B) \qquad x_B = 1 - x_A \tag{7.1}$$

$$\text{mass fractions, } w_A = n_A/(n_A + n_B M_B/M_A) \qquad w_B = 1 - w_A \tag{7.2}$$

$$\text{or volume fractions, } \phi_A = n_A/(n_A + n_B V_B/V_A) \qquad \phi_B = 1 - \phi_A \tag{7.3}$$

where M is the molar mass and V is the molar volume of the designated component. The interconversion formulas are:

$$\left.\begin{aligned} x_A &= w_A/(w_A + (1 - w_A)M_A/M_B) = \phi_A/(\phi_A + (1 - \phi_A)V_A/V_B) \\ w_A &= \phi_A/(\phi_A + (1 - \phi_A)d_B/d_A) = x_A/(x_A + (1 - x_A)M_B/M_A) \\ \phi_A &= w_A/(w_A + (1 - w_A)d_B/d_A) = x_A/(x_A + (1 - x_A)V_B/V_A) \end{aligned}\right\} \tag{7.4}$$

where d is the density of the designated component. Information is often presented in terms of %wt, i.e. $100w$, or percent volume, i.e. 100ϕ, of a component in a solvent mixture. There is no inherently superior manner of representation of the composition.

Of the properties of the solvents discussed and tabulated in Section 6.1, only a few are of importance for the selective solvation of ions. The mean molar mass and the mean molar volume of a binary mixed solvent are linearly related to those of the components in terms of the mole fractions:

$$\text{M} = x_A M_A + (1 - x_A)M_B \qquad V = x_A V_A + (1 - x_A)V_B \tag{7.5}$$

(for the volume, again, only if the volume change on mixing is negligible). The freezing point of the mixed solvent is generally lower, the boiling point higher, than the composition-weighted means of these quantities of the pure solvents.

Of the other bulk properties, the relative permittivity (dielectric constant) ε of the solvent mixture is of importance. Theoretical expressions relate ε of a binary mixture of solvents to the volume fractions of the components. According to Debye (see Guggenheim):[1]

$$(\varepsilon - 1)/(\varepsilon + 2) = \phi_A(\varepsilon_A - 1)/(\varepsilon_A + 2) + \phi_B(\varepsilon_B - 1)/(\varepsilon_B + 2) \qquad (7.6a)$$

or, according to Onsager:[2]

$$(\varepsilon - 1)(2\varepsilon + 1)/9\varepsilon = \phi_A(\varepsilon_A - 1)(2\varepsilon_A + 1)/9\varepsilon_A + \phi_B(\varepsilon_B - 1)(2\varepsilon_B + 1)/9\varepsilon_B \quad (7.6b)$$

Empirically, ε is found to change linearly with either x, ϕ, or w for some solvent mixtures (or even harmonically: $1/\varepsilon = \phi_A/\varepsilon_A + \phi_B/\varepsilon_B$), but then non-linearly with the other composition variables. Solutions of ionic solutes in aqueous solvent mixtures are used very widely, and the dielectric constants of aqueous mixtures with a selection of solvents from those listed in Table 6.1 are presented in Table 7.1. Nonaqueous binary mixtures have also been studied in this respect, and Table 7.2 contains references to representative reports of the dielectric constants of binary nonaqueous mixtures.

Although the polarity indices of pure individual solvents are good measures of their abilities to solvate ions by direct interactions, those of mixed solvents may not be true indicators of the ability of the solvent mixture to solvate ions, due to the possibility of selective solvation. The mean electron-pair donating and accepting properties of a solvent mixture are expected to be intermediate between those of the pure components, but not necessarily linearly dependent on the composition. Interpolated E_T data,[3] in the form of the normalized E_T^N, are shown in Table 7.3 as a function of the mole fraction x of the nonaqueous component in some aqueous solvent mixtures. Also shown are values of E_T^N calculated from acceptor number (AN) data[4] by means of the linear relationship: $E_T^N = 0.0193\ AN + 0.031$. It is seen that the two measures do not agree well, even though for (nonhydroxylic) pure solvents the linear correlation between E_T and AN is acceptable.[3] The observed dependence of E_T^N obtained from either the E_T data[3] or the AN data[4] is not linear with the composition, see Figure 7.1. These facts show that selective solvation occurs not only for ions, when dissolved in the mixed solvent, but also for the probe solutes used for the determination of the solvation indices. Assignments of effective electron pair donation and acceptance indices to each component in a mixture is not possible on the basis of the available data. Therefore the short range interactions of the solvent components in a mixture with ions cannot be evaluated by the approach that uses these indices for solutions in pure solvents.

Quantities that are important for the solvation of ions in mixed solvents but are not relevant for pure individual solvents are the excess thermodynamic quantities of mixing (denoted by superscript ^{E}). As already stated above, for the present purposes $\Delta V^E \approx 0$ has been assumed, although in fact generally $|\Delta V^E| < 1\%$ of V but is not quite zero. For some binary liquid systems, those that

Table 7.1 Dielectric constants of mixtures of water and organic solvents at 25 °C (unless otherwise specified) as a function of the mole fraction of the organic component, x_S

		x_S									
No.[a]	Solvent	0.1	0.2	0.3	0.4	0.5	0.6	0.7	0.8	0.9	1.0
8	MeOH[b]	70.8	63.9	58.0	52.7	48.1	44.0	40.3	36.9	33.9	32.7
9	EtOH[b]	65.2	55.3	47.9	42.0	37.4	33.6	30.5	27.8	25.6	24.6
10	1-PrOH[b]	58.5	46.0	38.0	32.5	28.5	25.6	23.3	21.5	20.6	20.3
11	2-PrOH[b]	57.9	45.0	36.7	30.9	26.7	23.6	21.1			19.9
15	t-BuOH[b]	50.1	35.1	26.4	20.9	17.1	14.4				(12.5)
25	TFE[m]	62.8	54.4	47.1	40.9	35.9	32.0	29.2	27.6	27.1	26.7
26	$En(OH)_2$[q]	70.8	65.6	60.9	56.7	52.9	49.6	46.8	44.4	42.5	41.0
27	Glycerol[b]	68.5	61.3	56.0	52.0	48.9	46.4	44.4			42.5
36	THF[c]	52.8	41.8	32.3	24.4	18.1	13.3	10.2	8.6	8.1	7.6
37	Dioxane[c]	46.2	30.0	20.5	14.4	9.0	6.5	4.8	3.6	2.7	2.1
41	Me_2CO[b]	62.6	51.5	43.3	37.2	32.4	28.5	25.3	22.6		20.7
60	EC[k]		82.9		82.6	86.9	88.2				
75	DMSO[dj]	76.9	73.8	70.0	65.9	62.3	59.1	56.4	53.6	49.9	46.7
76	TMS[d]	67.6	61.0	56.7	53.6	51.4	49.7	48.3	47.3		(47.2)
93	MeCN[e]	70.5	62.3	55.7	50.8	47.1	44.0	41.4	39.3	37.6	35.9
99	FA[f]	91.2	97.5	101.9	105.3	107.8	109.6	110.5	110.9	110.7	110.0
100	NMF[l]	85.4	92.2	99.8	108.3	117.7	128.0	139.1	151.1	177.7	182.4
101	DMF[j]	71.7	65.2	59.4	54.2	49.7	45.9	42.8	40.3	38.5	36.7
105	DMA[g]	67.0	60.2	54.6	50.1		44.0				37.8
107	NMPr[n]		80.8	81.7	85.6	92.7	102.8	116.0	132.3	151.7	176.0
108	NMPy[p]	68.6	61.7	55.6	50.2	45.5	41.5	38.2	35.7	33.8	32.0
109	TMU[h]	65.7	56.8	49.0	42.3	36.7	32.2	28.7	26.4	25.0	23.1
110	HMPT[i]	55.0	41.0	34.5	30.5	28.5	27.0		27.0		30.0

[a] Numbering as in Table 6.1; [b] MeOH = methanol, EtOH = ethanol, 1-PrOH = 1-propanol, 2-PrOH = 2-propanol, t-BuOH = 2-methyl-2-propanol, Me_2CO = acetone, G. Åkerlöf, *J. Am. Chem. Soc.*, **54**, 4125 (1932); [c] THF = tetrahydrofuran, G. Åkerlöf, *J. Am. Chem. Soc.*, **58**, 1241 (1936), cf. also H. S. Harned and B. B. Owen, *Physical Chemistry of Electrolyte Solutions*, Reinhold, New York, 3rd ed. (1958), p. 161; [d] DMSO = dimethylsulphoxide, TMS = tetramethylenesulphone (sulpholane), A. K. Covington and T. Dickinson, in *Physical Chemistry of Organic Solvent Systems*, Plenum, London (1973), p. 17; [e] MeCN = acetonitrile, C. Moreau and G. Douheret, *J. Chem. Thermodyn.*, **8**, 403 (1976); [f] FA = formamide, Yu. M. Kessler, V. P. Emelin, Yu. S. Tolubeev, *Zh. Strukt. Khim.*, **13**, 210 (1972); [g] DMA = N,N-dimethylacetamide, M. V. Ramanamurti and L. Bahadur, *Electrochim. Acta*, **25**, 601 (1980), at 35 °C; [h] TMU = N,N,N′,N′-tetramethylurea, C. Okpala, A. Guiseppi-Elie and D. M. Maharaj, *J. Chem. Eng. Data*, **25**, 384 (1980); [i] HMPT = hexamethylphosphoric triamide, Yu. M. Kessler, V. P. Emelin, A. L. Mishustin, *Zh. Strukt. Khim.*, **16**, 797 (1975), data read from a graph; [j] DMF = N,N-dimethylformamide, G. Douheret and M. Morenas, *Compt. Rend.*, **C264**, 729 (1967); [k] EC = ethylene carbonate, F. A. Critchfield, J. A. Gibson and J. L. Hall, *J. Am. Chem. Soc.*, **75**, 6044 (1953); [l] NMF = N-methylformamide, G. G. Karamyan and M. I. Shakhparonov, *Zh. Strukt. Khim.*, **22**, 54 (1981), interpolated to 25 °C; [m] TFE = 2,2,2-trifluoroethanol, J. Murto and E. L. Heino, *Suom. Kemist.* **B39**, 263 (1966); [n] NMPr = N-methylpropionamide, T. B. Hoover, *J. Phys. Chem.*, **73**, 57 (1969); [p] P. O. I. Virtanen, *Suom. Kemist.*, **B40**, 313 (1967), extrapolated to 25 °C; NMPy = N-methylpyrrolidinone; [q] EG = ethylene glycol, M. Morenas and G. Douheret, *Compt. Rend.*, **C270**, 2097 (1970).

follow the regular solution rules,[5] $\Delta S^E \approx 0$ too, but for solvents that are of interest for the solvation of ions, and in particular for aqueous solvent mixtures, ΔS^E is generally far from zero. In the case of aqueous mixtures ΔS^E is generally negative, which makes ΔG^E less negative or more positive than ΔH^E. The (excess) heats of mixing ΔH^E are negative for systems where the mutual

Table 7.2 Dielectric constants of binary nonaqueous mixtures at 25 °C—references to reported values

No.	Solvent + No.	3	4	5	6	7	8	9	10	11	12	13	14	15	16
1	n-Hexane		n	h			u						u		
2	Benzene	v	v	v	b		v	v			c	ev	v	m	
3	Methanol		pv	v		p	p	v	v		cq	v	v	m	g
4	Ethanol			v		p	p				bc	v	v		
5	Diethyl ether						v	v	v			v	v		
6	Tetrahydrofuran								k						
7	Dioxane						v	v		a	r		v	m	
8	Acetone							v	v	a		v	v	m	
9	Chloroform								v						
10	Carbon tetrachloride									d		v	v	m	
11	Dimethylsulphoxide													t	
12	Tetramethylene sulphone												j		
13	Nitromethane												v		
14	Acetonitrile														
15	Dimethylformamide														
16	N-Methylpyrrolidinone														

[a] E. Tommila and R. Yrjövouri, *Suom. Kemist.*, **B42**, 90 (1969); [b] E. Tommila, E. Lindell, M. L. Virtalaine and R. Laakso, *Suom. Kemist.*, **B42**, 95 (1969); [c] E. Tommila and T. Autio, *Suom. Kemist.*, **B42**, 107 (1969); [d] K. Kalliorinne, *Suom. Kemist.*, **B42**, 424 (1969); [e] K. Kalliorinne and E. Tommila, *Suom. Kemist.*, **B40**, 238 (1967); [g] V. A. Granzhan and O. G. Kirilova, *Zh. Prikl. Khim.*, **43**, 1875 (1970); [h] M. T. Rätzsch, C. Wohlfarth and M. Claudius, *J. Prakt. Chem.*, **319**, 353 (1977); [j] L. Janelli, A. Lopez and S. Saiello, *J. Chem. Eng. Data*, **25**, 259 (1980); [k] D. Guillèn, S. Otin, M. Gracia, C. G. Gutierrez-Losa, *J. Chim. Phys.*, **72**, 425 (1975); [m] J. Winkelmann and K. Quitzsch, *Z. Phys. Chem. (Leipzig)*, **250**, 355 (1972); [n] G. Oster, *J. Am. Chem. Soc.*, **68**, 2036 (1946); [p] A. V. Celliano, P. S. Gentile and M. Cefola, *J. Chem. Eng. Data*, **7**, 391 (1962); [q] A. K. Covington and T. Dickinson, *Physical Chemistry of Organic Solvent Systems*, Plenum, London (1973), p. 22; [r] L. Janelli, *Z. Naturf.*, **A30**, 87 (1975); [t] D. Giannakoudakis, G. Papanastasiou, and P. G. Mavridis, *Chem. Chron.*, **5**, 167 (1976); [u] A. Weisbecker, *J. Chim. Phys.*, **66**, 1442 (1969); [v] D. DeCroocq, *Bull. Soc. Chim. France*, **1964**, 127.

interactions of the two solvents are attractive and stronger than their self interactions. Otherwise ΔH^{E} is positive, and may lead to immiscibility. If the following inequality holds within a certain composition range $x_{\mathrm{A}}^{\mathrm{I}} \leqslant x_{\mathrm{A}} \leqslant x_{\mathrm{A}}^{\mathrm{II}}$

$$(\partial^2 \Delta G^{\mathrm{E}}(x_{\mathrm{A}})/\partial x_{\mathrm{A}}^2)_{P,T} < -\mathbf{R}T/x_{\mathrm{A}}(1 - x_{\mathrm{A}}) \quad (7.7)$$

then $x_{\mathrm{A}}^{\mathrm{I}}$ and $x_{\mathrm{A}}^{\mathrm{II}}$ denote the compositions of the two liquid phases that are immiscible and in equilibrium with each other, i.e., they are the limits of the miscibility gap of the solvents A and B.[5] The following representation may be used for the excess Gibbs free energy of mixing:

$$\Delta G^{\mathrm{E}} = x_{\mathrm{A}} x_{\mathrm{B}}[a(T) + b(T)(x_{\mathrm{A}} - x_{\mathrm{B}}) + c(T)(x_{\mathrm{A}} - x_{\mathrm{B}})^2] \quad (7.8)$$

which is a truncated form of the Redlich–Kister equation, which in its general form has added terms in $(x_{\mathrm{A}} - x_{\mathrm{B}})^k$ with $k > 2$. A combination of equations 7.7 and 7.8 leads to the following condition of immiscibility at the equimolar composition:

$$a(T, x_{\mathrm{A}} = 0.5) > 2\mathbf{R}T \quad (7.9)$$

irrespective of the values of $b(T)$ and $c(T)$. If $a(T)$ has such a value that equation 7.9

Table 7.3 Normalized E_T^N values for aqueous solvent mixtures, as a function of the mole fraction x_S of the organic solvent[3]

Solvent	x_S 0.10	0.25	0.50	0.75	0.90	0.95	1.00
Methanol	0.935	0.867	0.818	0.787	0.775	0.769	0.765
Methanol*	1.00	1.00	0.934	0.874	0.858	0.851	0.829
Ethanol	0.858	0.765	0.719	0.685	0.664	0.660	0.654
Ethanol*	1.00	0.961	0.874	0.812	0.783	0.766	0.748
2-Propanol	0.778	0.691	0.642	0.596	0.568	0.556	0.552
2-Propanol*	1.00	0.918	0.831	0.754	0.715	0.700	0.679
Dioxane	0.796	0.664	0.546	0.395	0.228	0.198	0.164
Dioxane*	0.992	0.876	0.760	0.634	0.518	0.457	0.240
Acetone	0.843	0.738	0.664	0.543	0.488	0.389	0.355
Acetone*	1.00	0.901	0.783	0.642	0.532		0.273
Pyridine	0.756	0.679	0.552	0.423	0.358	0.324	0.293
Pyridine*	0.984	0.889	0.731	0.571	0.468		0.306
Acetonitrile*	1.00	0.955	0.870	0.770	0.634		0.397
N,N-Dimethylformamide*	1.00	0.885	0.710	0.565	0.484		0.441
Dimethylsulphoxide*	1.00	0.889	0.706	0.555	0.490		0.404
Hexamethyl phosphoric triamide*	0.938	0.750	0.538	0.377	0.298	0.273	0.236

* Calculated from acceptor number *AN* data by means of a correlation function.

is no longer valid above a temperature T_c, the upper critical solution temperature, then the miscibility gap narrows down as the temperature is raised towards T_c, and complete miscibility obtains at $T > T_c$. The coexistence curve for the two liquid phases at equilibrium needs, however, not be symmetrical with respect of the composition. It turns out, in many cases, that $\Delta G^E(x)$ is not very unsymmetrical whereas $\Delta H^E(x)$ is quite skew, and for some solvent mixtures of interest in the present context even changes sign.[5]

Table 7.4 presents the excess Gibbs free energy of mixing and the enthalpy of mixing of water with representative solvents, both numerically at 0.1 unit intervals of x_A (where A is the nonaqueous solvent component of the mixture) and as the coefficients of equation 7.8 and the analogous equation for $\Delta H^E(x_A)$. Note that $x_A - x_B = 2x_A - 1$, and representations in terms of the latter variable are common, so that this is the form used in Table 7.4. The fit of the data in terms of three parameters is generally better for $\Delta G^E(x_A)$ than for $\Delta H^E(x_A)$, being within *ca.* 20 J mol^{-1} for the former and *ca.* 80 J mol^{-1} for the latter, if this changes sign and has both a maximum and a minimum (otherwise it is also within *ca.* 20 J mol^{-1}).

The thermodynamics of mixing of nonaqueous solvents with each other has been studied extensively, as summarized by several literature sources.[6,7] Representative values of $\Delta G^E(x = 0.5)$, where available, or else $\Delta H^E(x = 0.5)$, for binary mixtures of solvents are shown in Table 7.5. The solvents considered

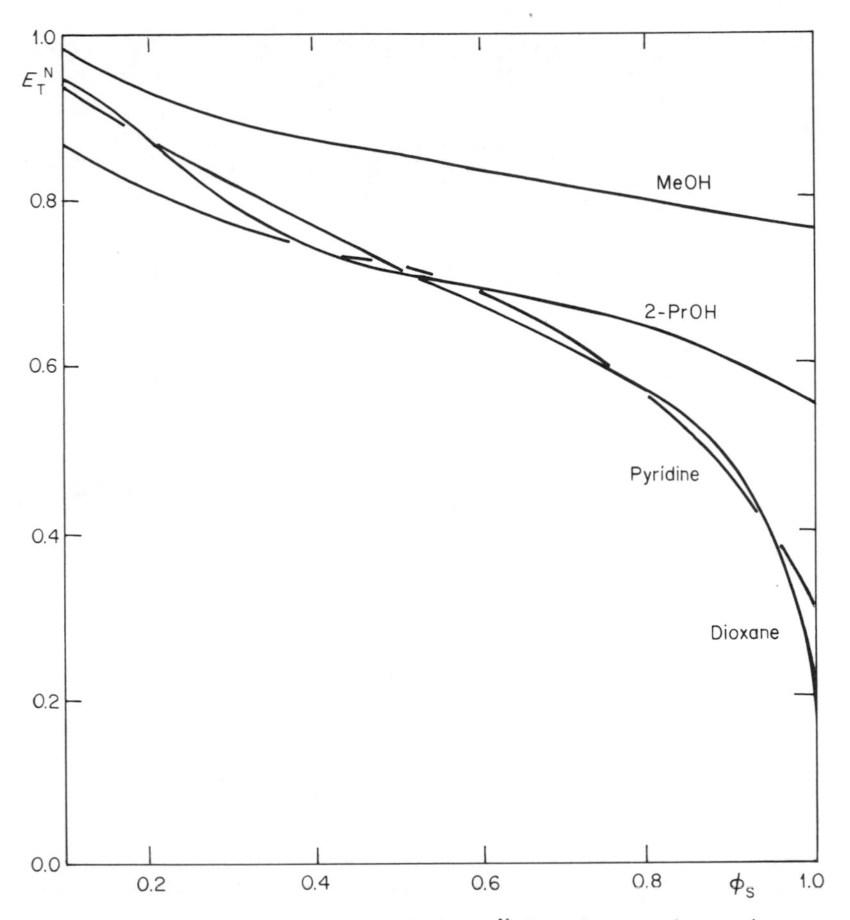

Figure 7.1 The normalized polarity index, E_T^N, for aqueous solvent mixtures, as a function of the volume fraction of the nonaqueous solvent, ϕ_S (data from Table 7.3)

in the Table are miscible in all proportions at the stated temperatures, but in a few of the systems there exist miscibility gaps at other (generally lower) temperatures. Methanol and n-hexane have an upper critical solution temperature of 34 °C, and methanol and c-hexane have one at 46 °C. Aniline and acetonitrile are not completely miscible with n-hexane below 69 and 77 °C, respectively. Miscibility gaps exist in mixtures of nitromethane with n-hexane and with c-hexane even considerably above room temperature. Of the aqueous mixtures listed in Table 7.4, only one, propylene carbonate + water, has a miscibility gap at room temperature with an upper critical solution temperature considerably above it, while another one, triethylamine + water, has a lower critical consolute temperature at 19 °C, and a miscibility gap above this temperature.

Table 7.4 Excess Gibbs free energy and enthalpy of mixing of aqueous solvents at 25 °C[a]

Solvent	0.1	0.2	0.3	0.4	0.5	0.6	0.7	0.8	0.9	a	b	c
Methanol, G^b	96	179	248	290	303	283	228	166	83	1200	−87	−330
H^b	−554	−820	−897	−864	−776	−668	−555	−428	−260	−3102	2040	−2213
Ethanol, G^b	345	572	683	738	745	669	559	400	241	2907	−777	494
H^b	−660	−837	−739	−531	−325	−188	−140	−152	−146	−1300	−3567	−4971
1-Propanol, G^b	276	545	786	986	1083	1097	1035	855	545	4230	1717	355
H^b	−348	−198	−82	48	136	191	205	171	102	709	2521	−2982
2-Propanol, G^b	400	724	910	952	952	876	717	503	276	3843	−984	−98
H^b	−712	−777	−498	−113	213	383	370	215	32	854	5167	−7243
t-Butanol, G^c	523	821	944							3319	−1979	830
Ethylene glycol, G^c	−205	−349	−437	−470	−457					−1829	689	160
H^d	−456	−697	−785	−770	−694	−585	−460	−326	−178	−2776	1933	−1172
Glycerol, G^b	−243	−440	−629									
THF, G^e	597	995	1237	1355	1371	1298	1137	881	511	5484	47	1371
H^f	−803	−991	−802	−432	−34	282	448	438	269	−136	7443	−4429
Dioxane, G^g	391	683	873	963	959	870	710	496	251	3835	−973	−421
H^h	−476	−577	−435	−157	153	419	575	576	389	611	6006	−1712
Acetone, G^i	525	851	1045	1138	1144	1069	919	692	388	4560	−163	1140
H^j	600	650	520	370	180	−50	−180	−300	−270	569	−5408	1838
PC, G^k						(1263)	976	645	307			
H^x						(1230)	1322	1146	696	3649	8130	−3788
Formic acid, H^v	−20	−80	−152	−216	−257	−264	−236	−176	−92	−1027	−500	−630
Acetic acid, G^l	313	440	546	679	803	851	759	512	189	3213	−118	−663
H^l	4	90	186	262	310	334	339	316	228	1239	1791	310

DMSO, G[k]	−598	−994	−1213	−1282	−1227	−1074	−849	−577	−286	−4909	2168	−5
H[w]	−1574	−2467	−2843	−2846	−2593	−2181	−1680	−1138	−587	−10372	6922	−2466
TMS, G[m]	460	761	937	1008	1004	933	803	611	435	3931	−1442	983
H[m]	402	703	937	1130	1247	1255	1172	962	594	4922	1988	1231
Triethylamine, G[y]	628	1013	1226	1335	1343	1230	1105	837	460	4486	−6540	7494
H[y]	−1586	−2356	−2561	−2414	−2075	−1661	−1247	−854	−460	−8294	7834	−4801
Pyridine, G[n]	424	592	644	643	612	558	481	376	231	2404	−1212	1873
H[n]	−477	−891	−1200	−1380	−1400	−1280	−1030	−695	−330	−5600	−1020	1756
Acetonitrile, G[p]	608	998	1222	1319	1316	1231	1081	822	475	5264	−639	1316
H[p]	148	403	695	964	1160	1244	1186	964	569	4640	2922	−1028
NMA, G[y]	78	96	87	75	74	86	106	118	95	296	114	1049
H[u]	−412	−851	−1235	−1536	−1811	−1975	−1975	−1564	−988	−7352	−6291	−1838
DMF, H[d]	−1293	−1893	−2133	−2107	−1920	−1613	−1267	−840	−427	−7616	7751	−1904
DMA, H[u]	−709	−1473	−2319	−3056	−3547	−3684	−3520	−2783	−1583	−14180	−6456	2282
NMPy, H[v]	−1639	−2527	−2859	−2803	−2496	−2047	−1536	−1014	−505	−9983	8917	−2496
TMU, H[r]	567	963	1152	1100	888	688	481	367	155	3736	−3133	619
HMPT, G[s]	−1059	−1301	−1291	−1238	−1172	−1083	−948	−699	−271	−4673	4185	−4270
H[t]	−2440	−3380	−3560	−3220	−2880	−2470	−2000	−1440	−940	−11367	10085	−11288

[a] ΔG^E and ΔH^E given in J(mol mixture)$^{-1}$ at evenly spaced mole fractions x of the organic solvent, 0.1 to 0.9, and the coefficients a, b, and c of their expression as $x(1-x)[a+b(2x-1)+c(2x-1)^2]$ are also given; [b] R. F. Lama and B. C. Y. Lu, *J. Chem. Eng. Data*, **10**, 216 (1965); [c] M. J. Blandamer, J. Burgess and R. I. Haines, *J. Chem. Soc. Dalton Trans.*, **1976**, 385; [d] K. Rehm and H. J. Bittrich, *Z. Phys. Chem.*, **251**, 109 (1972); [e] C. Treiner, J. F. Bocquet and M. Chemla, *J. Chim. Phys.*, **70**, 72 (1973); [f] D. N. Glew and H. Watts, *Can. J. Chem.*, **51**, 1933 (1973); [g] A. L. Vierk, *Z. Anorg. Chem.*, **261**, 286 (1950); [h] K. W. Morkom and R. W. Smith, *Trans. Faraday Soc.*, **66**, 1073 (1970); [i] R. V. Orye and J. M. Prausnitz, *Ind. Eng. Chem.*, **57** (5), 18 (1965); [j] D. O. Hanson and M. Van Winkle, *J. Chem. Eng. Data*, **5**, 30 (1960); [k] S. Y. Lam and R. L. Benoit, *Can. J. Chem.*, **52**, 718 (1974); [l] J. M. T. M. Gieskes, *Can. J. Chem.*, **43**, 2448 (1965); [m] at 30 °C, R. L. Benoit and G. Choux, *Can. J. Chem.*, **46**, 3215 (1968); [n] T. C. Chan and W. A. Van Hook, *J. Chem. Soc. Faraday Trans. 1*, **72**, 583 (1976); [p] C. Treiner, P. Tzias, N. Chemla and G. M. Poltoratskii, *J. Chem. Soc. Faraday Trans. 1*, **72**, 2007 (1976); [q] S. Murakami, R. Tanaka and R. Fujishiro, *J. Soln. Chem.*, **3**, 71 (1974); [r] K. R. Lindfors, S. H. Opperman, M. E. Glover and J. D. Seese, *J. Phys. Chem.*, **75**, 3313 (1971); [s] C. Jambon and R. Philippe, *J. Chem. Thermod.*, **7**, 479 (1975); [t] J. Jose, R. Philippe and P. Clechet, *Can. J. Chem. Eng.*, **53**, 88 (1975); [u] P. Assarson and F. R. Eirich, *Adv. Chem. Ser.*, **84**, 1 (1968); [v] A. N. Campbell and A. J. R. Campbell, *Trans. Faraday Soc.*, **29**, 1240 (1933); **30**, 1109 (1934); [w] M. F. Fox and K. P. Wittingham, *J. Chem. Soc. Faraday Trans. 1*, **71**, 1407 (1975); [x] J. Courtot-Coupez and M. L'Her, *Compt. Rend.*, **C275**, 103, 195 (1972); [y] G. L. Bertrand, J. W. Larson and L. G. Hepler, *J. Phys. Chem.*, **72**, 4194 (1968).

Table 7.5 Excess Gibbs free energy or enthalpy of mixing of equimolar binary mixtures of organic solvents (in J/mol) at indicated temperatures

	DMF	$MeNO_2$	MeCN	$PhNH_2$	Pyridine
n-Hexane				$G^{a8}_{60} = 1602$	
c-Hexane				$G^{a8}_{70} = 1456$	$G^{c1}_{45} = 854$
Benzene	$G^{a5}_{24} = 134$	$H^{a9}_{45} = 814$	$H^{b2}_{25} = 444$	$H^{b8}_{30} = 742$	$G^{c2}_{25} = 126$
Toluene	$G^{a5}_{24} = 100$		$G^{b2}_{25} = 789$		$H^{c3}_{25} = 187$
CH_2Cl_2			$H^{b2}_{25} = -303$	$H^{b9}_{25} = 412$	$G^{c3}_{30} = -313$
Chloroform			$H^{b2}_{25} = -783$		$G^{c3}_{30} = -632$
CCl_4	$G^{a6}_{30} = 554$	$H^{a9}_{45} = 1418$	$H^{b2}_{25} = 782$		$H^{c4}_{25} = -113$
Chlorobenzene	$G^{a5}_{25} = 234$	$G^{b1}_{25} = 898$	$G^{b4}_{20} = 776$	$G^{b4}_{70} = 537$	
Methanol	$H^{a4}_{20} = -107$	$G^{b1}_{25} = 1040$		$G^{b4}_{20} = 515$	$H^{c5}_{30} = -840$
Ethanol	$H^{a4}_{20} = 412$	$G^{b1}_{25} = 1167$	$G^{b5}_{20} = 845$	$G^{b4}_{40} = 645$	$H^{c6}_{25} = -162$
1-Propanol			$H^{b7}_{25} =$		$H^{c6}_{25} = -5$
2-Propanol			$H^{b7}_{25} =$		
$(CH_3)_3COH$					
Diethyl ether					
THF	$H^{a3}_{21} = 56$				
Dioxane	$G^{e1}_{30} = 270$				$H^{c7}_{25} = 52$
Acetone	$G^{a2}_{25} = 124$	$G^{b1}_{25} = -65$	$H^{b2}_{25} = -108$	$G^{b4}_{40} = -236$	$G^{c3}_{30} = 147$
Methyl acetate			$G^{b6}_{50} = 171$		
Ethyl acetate		$G^{b1}_{25} = 179$	$G^{b5}_{40} = 275$		
DMSO	$G^{i6}_{100} = 46$				
Triethylamine					
Pyridine	$G^{a1}_{25} = 167$				
Aniline	$G^{i6}_{100} = -425$		$G^{b4}_{20} = -6$		
Acetonitrile		$G^{b1}_{25} = 5$			
Nitromethane					

	Et_3N	DMSO	EtOCOMe	MeOCOMe	Acetone
n-Hexane	$H^{b2}_{25} = 82$		$H^{a7}_{25} = 1279$	$H^{a7}_{25} = 1690$	$G^{b8}_{20} = 1108$
c-Hexane	$G^{c8}_{20} = 26$		$H^{b3}_{25} = 1396$	$G^{d6}_{35} = 959$	$G^{e6}_{25} = 1132$
Benzene	$H^{b5}_{25} = 357$	$H^{a7}_{25} = 586$	$H^{e2}_{25} = 84$	$G^{c3}_{30} = 236$	$H^{b5}_{25} = 145$
Toluene					$H^{e7}_{45} = 243$
CH_2Cl_2	$G^{c9}_{10} = -142$	$G^{d6}_{25} = -331$		$G^{c3}_{30} = -374$	$G^{c3}_{30} = -404$
Chloroform	$G^{c9}_{10} = -1070$	$G^{d6}_{25} = -914$			$G^{e5}_{50} = -509$
CCl_4	$H^{d1}_{25} = -690$	$H^{a7}_{25} = 181$	$G^{e3}_{20} = 585$		$G^{e8} = 530$
Chlorobenzene	$G^{d2}_{70} = 165$		$G^{b4}_{40} = 48$		$H^{e9}_{25} = -26$
Methanol	$G^{d3}_{20} = 205$	$H^{d7}_{25} = -391$	$G^{e4}_{55} = 657$	$H^{b3}_{25} = 1009$	$G^{e5}_{50} = 431$
Ethanol	$G^{d3}_{20} = 261$	$H^{d9}_{22} = 414$	$G^{e4}_{55} = 644$	$H^{a9}_{25} = 1313$	$G^{e5}_{50} = 473$
1-Propanol	$G^{d4}_{20} = 50$	$H^{d9}_{22} = 703$	$G^{e4}_{55} = 607$		
2-Propanol	$G^{d4}_{40} = 425$		$G^{e4}_{55} = 481$		$G^{e6}_{25} = 625$
$(CH_3)_3COH$	$G^{d5}_{25} = 428$	$H^{d9}_{22} = 1025$			
Diethyl ether					$H^{i4}_{40} = -460$
THF		$G^{d8}_{25} = 775$			
Dioxane		$H^{d7}_{25} = 428$	$G^{e3}_{20} = 500$		
Acetone		$G^{d8}_{25} = 428$		$G^{c3}_{50} = 95$	
Methyl acetate					
Ethyl acetate		$G^{d8}_{25} = 757$			

Table 7.5—*continued*

	Dioxane	THF	Et_2O	Me_3COH	2-PrOH
n-Hexane	$G^{g8}_{80} = 862$	$H^{a3}_{21} = -172$		$H^{i5}_{30} = 1184$	$H^{g1} = 787$
c-Hexane	$H^{a9}_{25} = 1601$	$G^{f1}_{30} = 379$	$G^{f4}_{25} = 234$	$H^{b2}_{25} = 865$	$G^{g2}_{50} = 1163$
Benzene	$H^{a9}_{25} = -32$	$G^{f1}_{30} = -131$	$H^{i7}_{25} = -8$	$H^{f8}_{30} = 848$	$G^{g2}_{50} = 962$
Toluene	$G^{f1}_{30} = 135$	$G^{f1}_{30} = -130$			$G^{c3}_{25} = 1198$
CH_2Cl_2	$G^{c3}_{30} = -472$	$G^{c3}_{30} = -597$	$H^{f5}_{25} = -977$		
Chloroform	$G^{c3}_{30} = -763$	$G^{c3}_{30} = -1036$	$G^{f6}_{25} = -795$		
CCl_4	$H^{a7}_{25} = -250$	$G^{c3}_{30} = -138$	$H^{f5}_{25} = -487$	$H^{f8}_{30} = 941$	$H^{g3}_{25} = 628$
Chlorobenzene					
Methanol	$H^{c5}_{30} = 1040$	$G^{f3}_{25} = 549$	$G^{f7}_{25} = 816$	$G_{25} = -301$	$H^{g4}_{25} = -76$
Ethanol	$G^{e5}_{50} = 624$		$H^{i8}_{25} = 606$		
1-Propanol			$H^{i8}_{25} = 690$		$G^{g5}_{25} = -39$
2-Propanol					
$(CH_3)_3COH$	$G^{f2}_{26} = 441$				
Diethyl ether		$G^{f3}_{25} = 101$			
THF					

	1-PrOH	EtOH	MeOH	PhCl	CCl_4
n-Hexane	$G^{i3}_{25} = 1245$	$H^{c3}_{25} = 602$	$H^{c3}_{40} = 707$		$G^{i1}_{25} = 341$
c-Hexane	$G^{g6}_{55} = 1252$	$G^{g9}_{20} = 1360$	$G^{g9}_{50} = 1545$	$H^{h4}_{20} = 701$	$G^{b8}_{70} = 58$
Benzene	$H^{c3}_{25} = 949$	$G^{h1}_{30} = 1179$	$H^{c3}_{25} = 621$	$H^{h4}_{20} = 5$	$H^{a7}_{25} = 114$
Toluene	$H^{c3}_{25} = 886$	$G^{h2}_{25} = 1182$	$H^{c3}_{25} = 621$	$H^{h4}_{20} = -123$	$G^{h7}_{30} = -47$
CH_2Cl_2			$G^{h3}_{35} = 933$	$H^{h5}_{25} = -84$	$H^{h8}_{25} = -720$
Chloroform		$G^{b1}_{50} = 684$	$G^{h3}_{35} = 831$	$H^{h6}_{25} = 121$	$H^{h9}_{25} = 233$
CCl_4	$G^{g7}_{35} = 933$	$G^{e8} = 1130$	$G^{h3}_{35} = 1255$	$H^{h4}_{20} = 106$	
Chlorobenzene		$G^{h2}_{25} = 1184$	$G^{b4}_{20} = 1288$		
Methanol	$G^{b8}_{60} = 45$	$H^{b3}_{25} = 5$			
Ethanol	$H^{b3}_{25} = 19$				

	$CHCl_3$	CH_2Cl_2	PhMe	Benzene	c-Hexane
n-Hexane	$G^{i1}_{25} = 341$	$H^{i2}_{25} = 1319$		$H^{c3}_{50} = 811$	$H^{a7}_{25} = 216$
c-Hexane	$H^{f5}_{25} = 674$	$H^{f5}_{25} = 1304$	$H^{d6}_{25} = -82$	$H^{a7}_{35} = 781$	
Benzene	$H^{h5}_{25} = 423$	$H^{a7}_{25} = -82$	$H^{d6}_{25} = 68$		
Toluene	$H^{h5}_{25} = -720$	$H^{h5}_{25} = -217$			
CH_2Cl_2	$H^{h8}_{25} = 18$				

[a1] T. Ha-Phuong, P. Junghans, G. Rudakoff and R. Radeglia, *Z. Chem.*, **12**, 234 (1972); [a2] K. Quitzsch, H. P. Prinz, K. Suehnel and G. Geiseler, *J. Prakt. Chem.*, **311**, 420 (1969); [a3] K. Geier and H. J. Bittrich, *Z. Phys. Chem.*, **255**, 305 (1974); [a4] L. Grote, G. Riesel and H. J. Bittrich, *Z. Chem.*, **7**, 444 (1967); [a5] G. Rudakoff, T. Ha-Phuong, P. Junghans and R. Radeglia, *Z. Chem.*, **12**, 37 (1972); [a6] K. Quitzsch, *Z. Phys. Chem.*, **233**, 321 (1966); [a7] H. V. Kehiaian, ed., *Selected Data on Mixtures*, TRC, Texas (1974); [a8] H. Kehlen, *Z. Phys. Chem*, **249**, 41 (1972); [a9] H. V. Kehiaian, ed., *Selected Data on Mixtures*, TRC, Texas (1973); [b1] H. V. Kehiaian, ed., *Selected Data on Mixtures*, TRC, Texas (1983); [b2] *idem., ibid.* (1983); [b3] *idem., ibid.* (1973); [b4] *idem., ibid.* (1981); [b5] *idem., ibid.* (1982); [b6] H. V. Kehiaian, ed., *Selected Data on Mixtures*, TRC, Texas (1978); [b7] F. Mato and J. Coca, *An. Real. Soc. Espan. Fis. Quim.*, **65**, 1 (1969); [b8] H. V. Kehiaian, ed., *Selected Data on Mixtures*, TRC, Texas

(1979); [b9] J. L. Kapoor and P. P. Singh, *Thermochim. Acta*, **8**, 476 (1974); [c1] D. Bares, M. Soulie and J. Metzger, *J. Chim. Phys.*, **70**, 1531 (1973); [c2] P. R. Garrett, J. M. Pollock and K. W. Morcom, *J. Chem. Thermod.*, **5**, 569 (1973); [c3] H. V. Kehiaian, ed., *Selected Data on Mixtures*, TRC, Texas (1976); [c4] D. F. Gray, I. D. Watson and A. G. Williamson, *Austr. J. Chem.*, **21**, 379 (1968); [c5] P. P. Singh, D. V. Verma and P. S. Arora, *Thermochim. Acta*, **15**, 267 (1976); [c6] T. J. V. Findlay and J. L. Copp, *Trans. Faraday Soc.*, **65**, 1463 (1969); [c7] K. Amaya, *Bull. Chem. Soc. Japan*, **34**, 1278 (1961); [c8] R. Siedler and H. J. Bittrich, *J. Prakt. Chem.*, **311**, 721 (1969); [c9] Y. P. Handa, D. V. Fenby and D. E. Jones, *J. Chem. Thermod.*, **7**, 337 (1975); [d1] D. V. Fenby, *Austr. J. Chem.*, **26**, 1143 (1973); [d2] T. M. Letcher and J. W. Bayles, *J. Chem. Eng. Data*, **16**, 266 (1971); [d3] K. W. Chun and R. R. Davison, *J. Chem. Eng. Data*, **17**, 307 (1972); [d4] K. W. Chun, J. C. Drummond and R. R. Davison, *ibid.*, **19**, 143 (1974); [d5] K. W. Chun, J. C. Drummond, W. H. Smith and R. R. Davison, *ibid.*, **20**, 58 (1975); [d6] H. V. Kehiaian, ed., *Selected Data on Mixtures*, TRC, Texas (1975); [d7] K. Quitzsch, H. P. Prinz, K. Sühnel, V. S. Pham, and G. Geiseler, *Z. Phys. Chem.*, **241**, 273 (1969); [d8] Y. Sassa, R. Konishi and T. Katayama, *J. Chem. Eng. Data*, **19**, 44 (1974); [d9] J. J. Lindberg and I. Pietilä, *Suom. Kemist.*, **35B**, 30 (1962); [e1] K. Quitzsch, *J. Prakt. Chem.*, **28**, 59 (1965); [e2] J. P. Grolier, D. Ballet and A. Viallard, *J. Chem. Thermod.*, **6**, 895 (1974); [e3] M. Steinbrecher, H. J. Bittrich, *Z. Phys. Chem.*, **232**, 313 (1966); [e4] I. Nagata, T. Yamada and S. Nakagawa, *J. Chem. Eng. Data*, **20**, 271 (1975); [e5] H. V. Kehiaian, ed., *Selected Data on Mixtures*, TRC, Texas (1978); [e6] P. S. Puri, J. Polak and J. A. Reuther, *J. Chem. Eng. Data*, **19**, 87 (1974); [e7] R. W. Hanks, A. C. Gupta and J. J. Christensen, *Ind. Eng. Chem., Fund.*, **10**, 504 (1971); [e8] M. Rother and H. J. Bittrich, *Z. Phys. Chem.*, **235**, 195 (1967); [e9] D. O. Hanson and M. Van Winkle, *J. Chem. Eng. Data*, **5**, 30 (1960); [f1] D. D. Deshpande and S. L. Oswal, *J. Chem. Thermod.*, **7**, 155 (1975); [f2] M. Pedraza and K. Quitzsch, *Z. Phys. Chem.*, **255**, 1039 (1974); [f3] H. Arm, D. Bankay, R. Schaller and M. Wätli, *Helv. Chim. Acta*, **49**, 2598 (1966); [f4] H. Arm, D. Bankay, K. Strub and M. Wätli, *ibid.*, **50**, 1013 (1967); [f5] T. J. V. Findlay and P. J. Kavanagh, *J. Chem. Thermod.*, **6**, 367 (1974); [f6] F. Becker, M. Kiefer, P. Rhensius and H. D. Schafer, *Z. Phys. Chem. (NF)*, **92**, 169 (1974); [f7] H. Arm and D. Bankay, *Helv. Chim. Acta*, **51**, 1243 (1968); [f8] S. Otin, M. Gracia and C. Gutierrez Losa, *J. Chim. Phys.*, **70**, 1227 (1973); [f7] J. Polak, S. Murakami, V. T. Lam, H. D. Pflug and G. C. Benson, *Can. J. Chem.*, **48**, 2457 (1970); [g1] S. Murakami, K. Amaya and R. Fujishiro, *Bull. Chem. Soc. Japan*, **37**, 1776 (1964); [g2] I. Nagata, T. Ohta and Y. Uchiyama, *J. Chem. Eng. Data*, **18**, 54 (1973); [g3] R. F. Blanks and J. M. Prausnitz, *J. Phys. Chem.*, **67**, 1154 (1963); [g4] E. L. Taylor and G. L. Bertrand, *J. Soln. Chem.*, **3**, 479 (1974); [g5] J. Polak, S. Murakami, G. C. Benson and H. D. Pflug, *Can. J. Chem.*, **48**, 3782 (1970); [g6] K. Strubl, V. Svoboda, R. Holub and J. Pick, *Coll. Czech. Chem. Comm.*, **35**, 3004 (1970); [g7] G. C. Paraskevopoulos and R. W. Missen, *Trans. Faraday Soc.*, **58**, 869 (1962); [g8] H. V. Kehiaian, ed., *Selected Data on Mixtures*, TRC, Texas (1980); [g9] I. Klesper, *Z. Phys. Chem. (NF)*, **51**, 1 (1966); [h1] H. V. Kehiaian, ed., *Selected Data on Mixtures*, TRC, Texas (1977); [h2] W. Schulze, *Z. Phys. Chem. (NF)*, **6**, 315 (1956); [h3] E. A. Moelwyn-Hughes and R. W. Missen, *J. Phys. Chem.*, **61**, 518 (1957); [h4] B. S. Harsted and E. S. Thomsen, *J. Chem. Thermod.*, **7**, 369 (1975); [h5] R. K. Nigam and B. S. Mahl, *J. Chem. Soc. Faraday Trans. 1*, **68**, 1508 (1972); [h6] M. Tamres, *J. Am. Chem. Soc.*, **74**, 3375 (1952); [h7] R. P. Rastogi, J. Nath and J. Misra, *J. Phys. Chem.*, **71**, 1277 (1967); [h8] G. H. Cheesman and A. M. B. Whitaker, *Proc. Roy. Soc.*, **212A**, 406 (1952); [h9] L. A. Beath and A. G. Williamson, *J. Chem. Thermod.*, **1**, 51 (1969); [i1] T. G. Bissel and A. G. Williamson, *J. Chem. Thermod.*, **7**, 131 (1975); [i2] T. G. Bissel, G. E. Okafor and A. G. Williamson, *J. Chem. Thermod.*, **3**, 393 (1971); [i3] I. Brown, W. Fock and F. Smith, *J. Chem. Thermod.*, **1**, 273 (1969); [i4] B. H. Carroll and J. H. Mathews, *J. Am. Chem. Soc.*, **46**, 30 (1924); [i5] W. M. Rećko, K. W. Sadowska and M. K. Woycicka, *Bull. Acad. Polon. Sci., Ser. Sci. Chim.*, **19**, 475 (1971); [i6] J. Surový and J. Dojčanský, *Chem. Zvesti*, **28**, 313 (1974); [i7] J. E. A. Otterstedt and R. W. Missen, *J. Chem. Eng. Data*, **11**, 360 (1966); [i8] M. A. Villamañan, C. Casanova, A. H. Roux and J. P. E. Grolier, *J. Chem. Thermod.*, **14**, 251 (1982).

7.2 Solvent competition for coordination sites

In so far as ions form definite solvates when placed in a pure solvent, they may be expected to form them also in a mixture of solvents. The question then arises whether the ions are solvated selectively by one component of the mixture, or the average composition of the solvates reflects faithfully the composition of the bulk solvent mixture. The former situation is expected to obtain in cases where there exists a large discrepancy between the solvating abilities of the components of the solvent mixture, and the latter in cases where these abilities are more nearly alike. Spectroscopic methods, and in particular NMR, have been used for the study of selective solvation and the formation of mixed solvates of ions in solvent mixtures, see Figure 7.2.

In a few cases, where the solvates are kinetically inert, it is possible to isolate mixed solvates. The two solvents act as ligands and occupy coordination sites in the (first) coordination sphere of a cation such as Cr^{3+}. Salts of the series of solvated chromium(III) cations such as $Cr(H_2O)_{6-n}(C_2H_5OH)_n{}^{3+}$ ($n = 1$, 2, and 3) and $Cr(H_2O)_{6-n}(CH_3S(O)CH_3)_n{}^{3+}$ ($n = 1$ to 5) have been isolated.[8]

Low temperatures, that are attainable in certain aqueous solvent mixtures in contrast to pure aqueous solutions, may slow solvent exchange reactions down

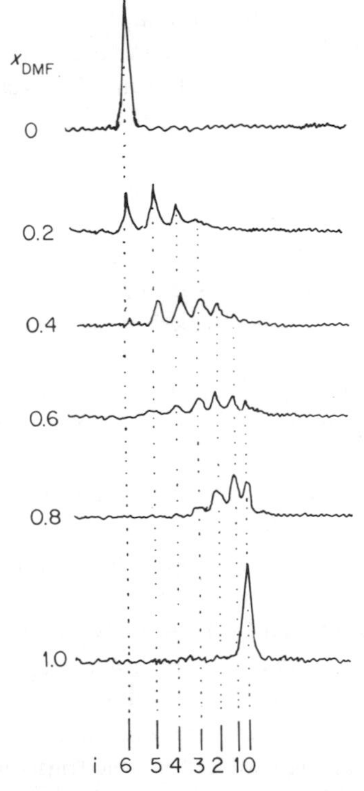

Figure 7.2 the ^{27}Al NMR spectrum of a 0.013 m solution of $Al(ClO_4)_3$ in mixtures of N,N-dimethylformamide and dimethylsulphoxide in nitromethane, at various mole fractions of the former, x_{DMF}. Note the distinct peaks obtained for the species $Al(DMSO)_i(DMF)_{6-i}^{3+}$, those for $i = 2$, 3, and 4 have contributions from the two geometrical isomers. The total spread of the peak positions from that for $i = 0$ to $i = 6$ is 2.7 ppm. (After H. Schneider, Ref. 17.)

sufficiently for the recognition of separate NMR signals from the two solvating solvent molecules in the solvation sphere as well as from those of the bulk solvents. This fact has been made use of for the determination of hydration numbers in aqueous deuterated acetone solutions (see Section 4.4). These can be cooled down to −100 °C, or even to −120 °C in the presence of dichlorodifluoromethane. Complete selective hydration obtains, the acetone being practically inert under these circumstances and does not compete with the water as long as this is sufficiently abundant relative to the amount of ions that are to be hydrated. A problem that does arise, however, is ion pairing between the cation of which the hydration is to be studied and the anions present in the solution. Even perchlorate anions, that generally are reluctant to form ion pairs, do so when the concentration of the water is not higher than required to satisfy the coordination spheres of the cations. Anions are generally preferred to acetone molecules in the coordination sphere under such circumstances. Lower limits of hydration numbers are therefore apt to be obtained by the NMR method, according to Fratiello and co-workers,[9] in solutions that are relatively poor in water. Dioxane, tetrahydrofuran and tetramethylurea behave in this respect as inert diluents in the same manner as acetone, not being able to replace water molecules from the coordination sphere of aluminium.[10] However, there are reports of solvation of cations by acetone at extremely low concentrations of water, i.e., practically anhydrous acetone solutions.[11]

In contrast to acetone, methanol, for instance, does compete with water for coordination sites around cations. A study of the solvation of magnesium ions at −75 °C demonstrates ^{1}H NMR peaks arising from both bound and free methanol hydroxylic protons and from bound and free water protons.[12] In the case of a paramagnetic ion, Co^{2+}, separate peaks are observed for each of the $Co(H_2O)_{6-n}(CH_3OH)_n$ species detected ($n = 4$, 5, and 6 for mixtures poor in water).[13]

At higher temperatures, where solvent exchange between the solvation shell and the bulk solvent is rapid, more qualitative information is obtained as a rule. The NMR signal from a metal ion, such as $^{23}Na^{+}$, exhibits a chemical shift that is a linear function of the composition if both solvent components in a mixture have essentially the same ability to solvate it. Otherwise, deviations from linearity occur, as is the case, for instance, in the dimethylsulphoxide + nitromethane system, where the former solvent is clearly preferred around the Na^{+} cations.[14] This approach can be generalized and qualitative selectivity series can be set up, where each solvent is able to displace those behind it in the series from the coordination sphere of a given ion, when present at commensurate concentrations in a binary solvent mixture. For Na^{+} this series is:

$$\text{HMPT} > \text{DMSO, DMF} > \text{pyridine, methanol} > \text{acetonitrile, acetone} > \text{nitromethane} \tag{7.10}$$

The selectivity series is, in principle, ion-specific, the position of a given solvent depending on its basicity (measured, e.g., by its donor number), but also on its bulk relative to the size of the ion and on other factors. For example, the order DMF > methanol for Na^+ is retained for Ni^{2+} but is reversed for Fe^{2+}.[15]

Attempts have been made to put such selectivity scales on a more quantitative basis, by means of the equilibrium constants (or the corresponding standard Gibbs free energies) for the solvent exchange reactions:

$$XA_n + B \rightleftharpoons XA_{n-1}B + A \qquad K_1 \tag{7.11}$$

$$XA_n + n'B \rightleftharpoons XB_{n'} + nA \qquad \bar{K}^n \tag{7.12}$$

$$XAB_{n'-1} + B \rightleftharpoons XB_{n'} + A \qquad K_{n'} \tag{7.13}$$

where X is an ion (its charge is not indicated for the sake of simplicity, but it may be a cation or an anion), and, for the sake of definiteness, the preferred solvent is designated as B, so that the equilibrium constants of reactions 7.11 to 7.13 are all larger than unity. Reaction 7.11 takes place in a dilute solution of the preferred solvent B in the less favoured solvent A. Stepwise solvent replacement generally takes place as the concentration of B in the mixture increases, until the last A molecule is displaced in reaction 7.13. This will occur as the mixture approaches pure B, if the selectivity between A and B is not very pronounced, but it may occur in solutions still relatively rich in A, if the selectivity is very strong. The solvation number n' with respect to solvent B employed in equation 7.13 need not be the same as n with respect to solvent A employed in equation 7.11. This would be the case when the bulks of the two solvents differ considerably, or when one of them has a pronounced bidentate nature, the other being monodentate (e.g., ethylene glycol and ethanol). Generally, however, $n' = n$ is either tacitly or explicitly assumed. Equations 7.11 and 7.13 focus the attention on the replacement of one solvent molecule from the coordination sphere by the other, the preferred one: the first (K_1) or the last ($K_{n'}$). In many cases only an average constant $\bar{K}$ for the solvent exchange equilibria is determinable. It relates to the total replacement of all the n A solvent molecules by n' B solvent molecules according to reaction 7.12. However, only if $n' = n$ can a definite relationship be readily established: $\bar{K} = (K_1K_2 \dots K_n)^{1/n}$, where the subscripted K's pertain to the stepwise replacement reactions. If the selectivity is relatively slight, these constants are interrelated according to the statistical probabilities of the replacement of A in species XA_iB_{n-i} by B. In this case a definite relationship exists between K_1, $\bar{K}$, and K_n, and K_i in general: $\bar{K} = K_1/n = nK_n$ and $K_i = (n - i + 1)i^{-1}\bar{K}$.

Provided that the concentration of the ion X is very small relative to the less abundant solvent in the mixture, the ratio of the activity coefficients of two solvated species participating in the replacement equilibrium is approximately unity. In this case the ratio $(n - i)/i$ of B and A in the average ion solvate is given as

$$(n - i)/i = \{[XB_n]/[XA_n]\}^{1/n} = \bar{K}a_B/a_A \tag{7.14}$$

that is, it is proportional to the ratio of the activities of the two solvents in the mixture. This ratio can be approximated by the ratio of their molal or molar concentrations or mole fractions, if the binary mixture is not too non-ideal, but this is very seldom the case. Indeed, equilibrium quotients calculated on the basis of solvent molalities in aqueous dimethylsulphoxide solutions of electrolytes containing H^+, Li^+, Na^+, or Cl^- ions have to be multiplied by a factor of *ca.* 40 to give the corresponding equilibrium constants in terms of the solvent activities.[16]

The nature and strength of the mutual interactions of the two solvents, reflected in the discrepancy between $\bar{K}$ calculated according to x_B/x_A (or m_B/m_A) from that calculated with a_B/a_A, is important also when three solvents are intercompared regarding their selective solvation of a given ion. If it is found that in equimolar mixtures of A and B solvent B is preferred and that in equimolar mixtures of B and C solvent C is preferred, it does not follow automatically that in equimolar solutions of A and C solvent C must be preferred. It will, if the selectivities in the other two mixtures are appreciable, but C may not be preferred in cases of moderate selectivities in the other two mixtures under certain circumstances.

The methodology of coordination chemistry in solution has been applied extensively to the determination of the equilibrium constants for solvent replacement in the coordination spheres of ions. The two solvents are considered as competing ligands, or one is regarded as a ligand, when the other is present in considerable excess, and is treated as the solvent. In the latter case this solvent is not generally considered explicitly in the expression for the equilibrium constant, and the latter then has the dimension of a reciprocal concentration. In order to be converted to a dimensionless numerical constant compatible with K_1 or $K_{n'}$ (or with $\bar{K}$, provided that $n' = n$) it must be multiplied with the molar concentration of this solvent.

Methods such as polarography, conductometry, solubility, and spectrophotometry have been used extensively, and have been reviewed briefly in the present context.[15,17] When NMR is used, the cases of slow and fast solvent exchange must be distinguished. In the former, separate peaks corresponding to the two bound solvents are seen, and their relative areas determine the mean solvation number by each, which can be related to the equilibrium constant for solvent replacement by means of equation 7.14. In the latter case, the dependencies of the chemical shift or the rate of relaxation on the solvent composition, the electrolyte concentration, and the temperature are interpreted in terms of these equilibrium constants. It is assumed in a widely accepted treatment due to Covington and co-workers[18] that the chemical shift of distinguishable nuclei (such as hydroxylic ^1H) originating in molecules of solvent B among the bound solvents is proportional to the fraction of B molecules in the solvate species $XA_{n-1}B_i$, i.e., that $\delta_i = (i/n)\delta_B$, where δ_B is the shift for the pure B-solvate XB_n, and that the individual displacement steps proceed statistically, i.e., $K_i = (n + 1 - i)i^{-1}\bar{K}$. The observed chemical shift δ is then

$$\delta = \delta_B\bar{K}([B]/[A])/(1 + \bar{K}[B]/[A]) \qquad (7.15)$$

which can be rearranged to give the average replacement equilibrium constant:

$$\bar{K} = (\delta_B - \delta)^{-1}[A]/[B] \tag{7.16}$$

provided that the ratio of the solvent concentrations [A]/[B] (or mole fractions x_A/x_B or molalities m_A/m_B) can replace their ratio of activities. Non-statistically proceeding replacement reactions and non-equality of n' and n in equation 7.12 cause complications that have been taken into account in modifications of this method.[19]

A collection of values of K_1 (or of $\bar{K}$ or $K_{n'}$, if K_1 values are unavailable) for the replacement of solvent A by the preferred solvent B around various ions is presented in Table 7.6. The values are generally equilibrium quotients obtained at not-infinitesimal concentrations of the electrolyte solutes, hence they are sensitive to the accompanying counter ion, as well as to the ratio of the electrolyte to the less abundant solvent. Some of the discrepancies in the values are due to this. Others may be due to the use of solvent concentrations (or mole fractions) instead of activities, and to the non-validity of some other assumptions, such as the statistical relations between the constants for successive replacement steps. The purpose of Table 7.6 is thus mainly to give a qualitative or semiquantitative picture of the selectivities of various solvents towards ions. It should be noted that in several cases the order of selectivity itself is ambiguous. Both solvents in certain mixtures have been said to be preferred over the other. In some cases this seems to be a real phenomenon: in aqueous dimethylsulphoxide Al^{3+} is preferentially hydrated at $x_{DMSO} < 0.2$ and preferentially solvated by dimethylsulphoxide at higher mole fractions of this solvent. However, when these solvents are diluted with acetone there exists a composition range in which neither of the two is preferred, their ratio in the solvation shell of the Al^{3+} ions being the same as in the bulk solvent.[10,20] A similar situation obtains in the Ag^+ + water + dimethylsulphoxide system, where hydration is the preferred mode of solvation in water-rich mixtures, whereas solvation by dimethylsulphoxide is preferred in mixtures relatively poor in water.[21] In other cases the uncertainty in which of the two solvents is the preferred one is an artifact of the assumptions by means of which the equilibrium quotients have been derived from the experimental data. This seems to be the case for the Li^+ + water + dimethylsulphoxide, Cl^- + water + dimethylsulphoxide, and NO_3^- + water + acetonitrile systems, the selectivities being small in either direction in all three.

Some general trends may be observed in Table 7.6. Cations of high charge density (e.g., Li^+, Mg^{2+}, Al^{3+}) seem to prefer the more basic solvent in a mixture (i.e., the one with a higher donicity), irrespective of whether it is protic or aprotic. Cations of low charge density (e.g., Cs^+, $(C_2H_5)_4N^+$) tend to shun water or protic solvents and favour aprotic ones. To a special category belong Cu^+ and in particular Ag^+, that have an extra strong affinity to acetonitrile, due to back-bonding from the metal d-orbitals to the π-orbitals of the solvent. Anions, except very big ones, tend to prefer protic solvents, with which they can

Table 7.6 Equilibrium constants for solvent replacement around ions

Ion	Solvent A	Solvent B	log *K*	Ref.
H^+	MeOH	Water	2.32	a
	Water	DMSO	0.40*	b
	MeCN	Water	3.48	hh
Li^+	H_2O_2	Water	$\underline{0.31}$	c
	PC	Water	0.98	d
			0.36, 0.55	e
			0.82	f
	DMSO	Water	0.32	b
			$\underline{0.10}$	g
			0.38	h
	Water	DMSO	$\sim$0.7*	d
	TMS	Water	0.85	d
	MeCN	Water	1.20	d
			0.60	b
			1.15	i
			0.82, 1.11	e
	TMU	Water	0.2, 0.36	e
	PC	MeOH	0.18	e
	MeCN	MeOH	0.76	i
			0.54	e
	TMU	MeOH	0.08	e
	MeCN	EtOH	0.77	i
	PC	DMA	3.32	ii
	DMA	DMSO	$\underline{0.08}$	g
Na^+	H_2O_2	Water	$\underline{0.01}$	c
	PC	Water	0.26	d
			0.15	f
	DMSO	Water	0.04	b
	Water	En	1.11	h
	MeCN	Water	0.40	d
			0.42	b
			0.36	i
			0.11	e
			0.30	g
			0.24	h
			1.53	j
	DMSO	MeOH	0.0	mm
	MeCN	MeOH	0.18	i
			1.32	j
	MeCN	EtOH	0.18	i
	THF	$PrNH_2$	$\underline{0.64}$	k
	Aniline	THF	$\underline{0.34}$	k
	THF	Pyridine	$\underline{0.11}$	k
	PC	DMSO	$\underline{2.67}$	ii
	PC	DMA	2.50	ii
	MeCN	DMSO	2.05	j
	MeCN	DMF	1.66	j
	MeCN	DMA	2.00	j
	MeCN	HMPT	2.74	j
K^+	Water	PC	0.40*	f
	MeCN	Water	0.11	b
			0.00*	m
Rb^+	Water	H_2O_2	$\underline{0.29}$	c
Cs^+	Water	H_2O_2	$\underline{0.45}$	c
	Water	DMSO	$\underline{0.55}$	h
	Water	TMS	$\underline{0.19}$	g
			0.78	h
	Water	MeCN	0.30*	m
	DMF	Water	$\underline{0.20}$	n
Cu^+	Water	MeCN	$\underline{3.3}$	o
Ag^+	PC	Water	0.68	d
	Water	MeCN	$\sim$1.0*	d
			2.0	o
			2.18	p
	Water	DMThF	8.5	ii
	DMSO	MeOH	0.00*	q
	MeOH	MeCN	2.4	o
	DMF	MeOH	0.36*	r
	HMPT	MeOH	0.2*	s
	Acetone	DMSO	1.2	q
	Acetone	MeCN	2.3	o
	Acetone	DMF	0.5	r
	PC	DMSO	3.3	t
			4.1	ii
	PC	MeCN	3.0	t
	PC	DMA	3.6	ii
	PC	DMThF	$\sim$10	ii
	TMS	DMSO	1.74	q
	TMS	DMF	1.3	r
	TMS	HMPT	4.04	s
	$MeNO_2$	MeCN	2.4	o
	MeCN	HMPT	0.3	s
	MeCN	DMThF	5.0	ii
Tl^+	Water	DMSO	$\underline{0.43}$	u
	water	Pyridine	$\underline{0.37}$	u
	Water	Formamide	$\underline{0.19}$	v
	Water	NMF	$\underline{0.45}$	v
	THF	Pyridine	$\underline{0.69}$	jj
	THF	DMF	$\underline{0.46}$	jj
	Acetone	Pyridine	$\underline{0.97}$	jj
	Acetone	DMF	$\underline{0.72}$	jj
	Pyridine	DMSO	$\underline{0.48}$	u
			$\underline{0.56}$	jj
	DMF	DMSO	$\underline{0.28}$	jj
	TBP	Pyridine	$\underline{0.43}$	jj
	TBP	DMF	$\underline{0.11}$	jj
	Pyridine	DMF	$\underline{0.11, 0.16}$	jj
	Pyridine	HMPT	$\underline{0.99}$	jj
	NMF	Formamide	$\underline{0.00}$	v
	DMF	HMPT	$\underline{0.33}$	jj
Et_4N^+	Water	PC	$\underline{0.52}$*	f
Bu_4N^+	Water	PC	0.52*	e
	Water	MeCN	0.15*	e
	Water	TMU	0.52*	e
	MeOH	MeCN	0.45*	e
Be^{2+}	Water	HMPT	1.92*	w
			2.09	ll
Mg^{2+}	Acetone	Water	$\underline{2.96}$	x
	MeCN	Water	$\underline{>2.5}$	i
	MeCN	MeOH	2.5	i
	MeCN	EtOH	2.15	i
Ca^{2+}	PC	Water	1.50	e
	MeCN	Water	2.06, 1.95	e
	MeCN	MeOH	1.28	e
Sr^{2+}	PC	Water	1.06	e
	MeCN	Water	1.50	e
	MeCN	MeOH	0.65	e
Ba^{2+}	MeCN	Water	1.20	i
	MeCN	MeOH	0.92	i
	MeCN	EtOH	0.77	i

Table 7.6—*continued*

Ion	Solvent A	Solvent B	log K	Ref.	Ion	Solvent A	Solvent B	log K	Ref.
Mn^{2+}	MeOH	Water	2.09	z	Cl^-	MeCN	Water	1.34, 1.28	e
	MeCN	Water	2.98	z				0.95	m
	DMF	Water	1.11	z				0.48	h
Co^{2+}	MeOH	Water	1.64	kk	Br^-	PC	Water	0.72, 0.64	e
Cu^{2+}	Water	EtOH	0.26*	aa		MeCN	Water	0.92, 1.00	e
			0.23*	bb				0.02*	h
	Water	Acetone	0.57*	aa		TMU	Water	0.30, 0.08	e
	$MeNO_2$	Water	2.90	bb		PC	MeOH	0.60	e
Cd^{2+}	PC	DMSO	$\underline{2.64}$	cc		MeCN	MeOH	0.78	e
	PC	TMP	$\underline{1.63}$	cc		TMU	MeOH	0.70	e
	PC	DMF	$\underline{1.87}$	cc	I^-	PC	Water	0.0, 0.26	e
	PC	DMA	$\underline{2.03}$	cc		Water	DMSO	1.45	h
	PC	NMPy	$\underline{2.07}$	cc		MeCN	Water	0.30, 0.48	e
UO_2^{2+}	Water	DMSO	1.01	dd				0.33*	h
Al^{3+}	DMSO	Water	1.26	ee		PC	MeOH	0.11	e
	Water	TMP	1.52*	w		MeCN	MeOH	0.18	e
Eu^{3+}	MeCN	Water	2.08	ff	NO_3^-	PC	Water	0.08	d
Cr^{3+}	MeOH	Water	2.66*	gg		Water	DMSO	$\underline{0.16}$	g
	EtOH	Water	$\underline{0.88}$	l				0.68	h
	Water	DMSO	2.34	y		Water	MeCN	0.05*	d
F^-	Water	H_2O_2	$\underline{0.38}$	c		MeCN	Water	0.43	e
Cl^-	Water	H_2O_2	$\underline{0.0}$	c				0.30	m
	PC	Water	0.79	f		MeCN	MeOH	0.53	h
			0.74, 0.90	d	ClO_4^-	Water	PC	0.7	d
			1.20, 1.00	e				0.52*	e, f
	Water	DMSO	0.22*	d		water	MeCN	0.15*	e, m
	DMSO	Water	0.38	b		water	TMU	0.52*	e
	TMS	Water	0.88	d		MeOH	MeCN	0.52*	e
	MeCN	Water	0.90	d	BF_4^-	Water	PC	0.52*	f
			1.04	b	Ph_4B^-	Water	PC	1.0*	f

* Unmarked log K values pertain to K_1, those with an asterisk to K_n, underlined ones to $\bar{K}$.

[a] Y. Kondo and N. Tokura, *Bull. Chem. Soc. Japan*, **45**, 818 (1972); [b] R. L. Benoit and C. Buisson, *Inorg. Chim. Acta*, **7**, 256 (1973); [c] A. K. Covington, T. H. Lilley, K. E. Newman and G. A. Porthouse, *J. Chem. Soc. Faraday Trans. 1*, **69**, 963 (1973); [d] R. L. Benoit and S. Y. Lam, *J. Am. Chem. Soc.*, **96**, 7385 (1974); [e] I. D. Kuntz and C. J. Cheng, *J. Am. Chem. Soc.*, **97**, 4852 (1975); [f] D. R. Cogley, D. N. Butler and E. Grunwald, *J. Phys. Chem.*, **75**, 1477 (1971); [g] A. K. Covington, I. R. Lantzke and J. M. Thain, *J. Chem. Soc. Faraday Trans. 1*, **70**, 1869 (1974); [h] A. K. Covington and J. M. Thain, *J. Chem. Soc. Faraday Trans. 1*, **70**, 1879 (1974); [i] G. W. Stockton and J. S. Martin, *J. Am. Chem. Soc.*, **94**, 6921 (1972); [j] K. Izutsu, T. Nomura, T. Nakamura, H. Kazama and S. Nakajima, *Bull. Chem. Soc. Japan*, **47**, 1657 (1974); [k] A. Delville, C. Detellier, A. Gertmans and P. Lazslo, *Helv. Chim. Acta*, **64**, 547 (1981); [l] D. W. Kemp and E. L. King, *J. Am. Chem. Soc.*, **89**, 3433 (1967); [m] M. K. Chantooni and I. M. Kohltoff, *J. Am. Chem. Soc.*, **89**, 1582 (1967); [n] H. Gustavsson, T. Ericsson and B. Lindman, *Inorg. Nucl. Chem. Lett.*, **14**, 37 (1978); [o] S. E. Manahan and R. T. Iwamoto, *J. Electroanal. Chem.*, **14**, 213 (1967); [p] H. Schneider and H. Strehlow, *Z. Phys. Chem.*, **49**, 44 (1966); [q] D. C. Luehrs, R. W. Nicholas and D. A. Hamm, *J. Electroanal. Chem.*, **29**, 417 (1971); [r] D. C. Luehrs, *J. Inorg. Nucl. Chem.*, **33**, 2701 (1971); [s] D. C. Luehrs, *J. Inorg. Nucl. Chem.*, **34**, 791 (1972); [t] B. G. Cox, A. J. Parker and W. E. Waghorne, *J. Phys. Chem.*, **78**, 1731 (1974); [u] R. W. Briggs and J. F. Hinton, *J. Soln. Chem.*, **7**, 118 (1978); [v] R. W. Briggs and J. F. Hinton, *J. Soln. Chem.*, **6**, 827 (1977); [w] J. J. Delpuech, A. Peguy and M. R. Khaddar, *J. Mag. Res.*, **6**, 325 (1972); [x] A. D. Covington and A. K. Covington, *J. Chem. Soc. Faraday Trans. 1*, **71**, 831 (1975); [y] L. P. Scott, T. I. Weeks, Jr., D. E. Bracken and E. L. King, *J. Am. Chem. Soc.*, **91**, 5219 (1969); [z] L. Burlamacchi, G. Martini and M. Romanelli, *J. Chem. Phys.*, **59**, 3008 (1973); [aa] N. J. Friedman and R. A. Plane, *Inorg. Chem.*, **2**, 11 (1963); [bb] R. C. Larson and R. T. Iwamoto, *Inorg. Chem.*, **1**, 316 (1962); [cc] J. Massaux and G. Duyckaerts, *J. Electroanal. Chem.*, **84**, 399 (1977); [dd] A. Fratiello, V. Kubo and R. E. Schuster, *Inorg. Chem.*, **10**, 744 (1971); [ee] D. R. Olander, R. S. Marianelli and R. C. Larson, *Anal. Chem.*, **41**, 1097 (1969); [ff] Y. Haas and G. Navon, *J. Phys. Chem.*, **76**, 1449 (1972); [gg] C. C. Mills and E. L. King, *J. Am. Chem. Soc.*, **92**, 3017 (1970); [hh] M. K. Chantooni and I. M. Kohltoff, *J. Am. Chem. Soc.*, **92**, 2236 (1970); [ii] G. Clune, W. E. Waghorne and B. G. Cox, *J. Chem. Soc. Faraday Trans. 1*, **72**, 1294 (1976); [jj] J. J. Dechter and J. I. Zink, *Inorg. Chem.*, **15**, 1690 (1976); [kk] Z. Luz and S. Meiboom, *J. Chem. Phys.*, **40**, 1058 (1964); [ll] H. H. Füldner and H. Strehlow, *Ber. Bunsenges. Phys. Chem.*, **86**, 68 (1982); [mm] L. Baltzer, N. A. Bergman and T. Drakenberg, *Acta Chem. Scand.*, **A35**, 759 (1981).

form hydrogen bonds, over aprotic ones of the same polarity. The bigger anions, again, tend to shun water.

Another measure of the affinity of solvents in a mixture towards ions is the average residence time in the coordination sphere of the ion. This can be determined from the NMR band widths, i.e., the rate of relaxation. The corresponding activation energy for solvent exchange can be determined from the temperature dependence of the rate of exchange. Solvents may be arranged in a series according to increasing mean residence times, and the order obtained is the reverse of the order of their affinities to the ion given in displacement series, such as equation 7.10. A set of data pertaining to Li^+ and Na^+ is shown in Table 7.7.

Table 7.7 Average lifetimes τ of solvents in the coordination spheres of Li^+ and of Na^+ mixed solvates[a]

Solvent	τ/ns	Solvent	τ/ns
Nitromethane	0.05	Formamide	4.0
Acetonitrile	0.60	N-Methylformamide	5.6
Acetone	0.80	N,N-Dimethylformamide	8.0
Tetrahydrofuran	1.00	Dimethylsulphoxide	8.0
Ethanol	1.50	Dimethylacetamide	10.0
Methanol	1.60	Hexamethyl phosphoric	
Water	3.30	triamide	15.0

[a] A. I. Mishustin, *Zh. Fiz. Khim.*, **55**, 1507 (1981).

7.3 Ion transfer into binary solvent mixtures

It is not always possible to describe the solvation of ions in terms of the formation of definite solvates, either because no coordination sphere with a definite coordination number can be assigned to the ion, or because the solvation affects further layers of solvent around the ion beyond the first. This is true not only for ions solvated by a single solvent, but also when the solvent is made up of two components. These compete for the environment of the ion, but the competition may not be forced into the narrow description of two coordinating ligands replacing each other.

In as much as an important contribution to preferential solvation arises from the electrostatic effects of ion–dipole, ion–higher multipole, and ion–induced dipole interactions, it is useful to examine the consequences of these on the solvent composition in the vicinity of the ion. From a consideration of the electrostatic work done on a fluid by the electrostatic field due to the ion,[22] the following equation has been derived by Padova:[23]

$$\ln[(x_A/x_B)/(x_A^0/x_B^0)] = \int_{r_c}^{\infty} r^2\, dr \int_0^{E_r} (\partial\mu_A/\partial \ln x_A)^{-1}(\partial\varepsilon/\partial x_A)\, d(E^2) \quad (7.17)$$

which describes the solvent sorting by the ion, due to its electric field. Here x_A and x_B are the mole fractions of the two solvents in the environment of the ion, whereas x_A^0 and x_B^0 are those in the bulk of the solvent mixture, outside the influence of the field of the ion. The solvent mixture is considered as a continuum extending from the ion outward, and the quantities r_e and E_r are the effective radius of the ion and the field at this distance from the centre of the ion. The field at any point a distance r away ($r > r_e$) is designated by E. The partial differentials are taken at constant P, T, and E, it being noted that the relative permittivity (dielectric constant) ε is dependent on E in the general case. Only in the case of an ideal solvent mixture, where $\partial\mu_A/\partial \ln x_A)_{P,T,E} = \mathbf{R}T$, and of a field-independent relative permittivity ε is it possible to evaluate the second integral in equation 7.17 analytically. The first integral evaluates to $\bar{V}/4\pi$, where $\bar{V}$ is the mean molar volume of the solvent mixture. If the ionic field is taken to be $E = z\mathbf{e}/\boldsymbol{\varepsilon}_0\varepsilon r^2$, then

$$\ln[(x_A/x_B)/(x_A^0/x_B^0)] = -z^2\mathbf{e}^2\bar{V}(4\pi\mathbf{R}T\boldsymbol{\varepsilon}_0^2\varepsilon^2r^4)^{-1}(\mathrm{d}\varepsilon/\mathrm{d}x_A) \quad (7.18)$$

where r is now the distance up to which x_A/x_B differs appreciably from x_A^0/x_B^0.

Field independence of the relative permittivity is realized, if at all, only for the environment of very large ions of low charge, hence equation 7.18 is not expected to be practically useful. It has been shown,[23] however, that retaining the assumption of ideal mixing of the solvent components, that of the field independence of the relative permittivity can be dispensed with. In fact $\int_0^{E_r}(\partial\varepsilon/\partial x_A)\,\mathrm{d}(E^2)$ represents the Gibbs free energy density of ionic solvation in the pure solvent A (E_r depends on the dielectric constant ε_A of solvent A). Hence

$$\begin{aligned}\ln[(x_A/x_B)/(x_A^0/x_B^0)] &= -(\mathbf{R}T)^{-1}(\mu^\infty_{\text{solv(X,A)}} - \mu^\infty_{\text{solv(X,B)}})\\ &= -(\mathbf{R}T)^{-1}[\Delta G_t^\infty(\text{X, W} \rightarrow \text{S}_A)\\ &\quad - \Delta G_t^\infty(\text{X, W} \rightarrow \text{S}_B)] \end{aligned} \quad (7.19)$$

provided that the electrostatic contribution predominates in the Gibbs free energy of the solvation of the ion X.

Equation 7.19 shows a definite relationship between the Gibbs free energy of transfer $\Delta G_t^\infty(\text{X, S}_A \rightarrow \text{S}_B) = \mu^\infty_{\text{solv(X,B)}} - \mu^\infty_{\text{solv(X,A)}}$ and the solvent composition near the ion. However, it does not specify the extension of the region where this composition differs from that in the bulk. This approach also views the solvent mixture as a continuum, the only characterizing property of which being its relative permittivity, that varies with the composition and the field strength. In specifying the solvent composition in the vicinity of the ion as dependent on its properties (through z, its charge, and r_e, its effective size), this approach bears some resemblence to the methods that describe the solvent sorting by means of the equilibrium constants for solvent replacement reactions, discussed in Section 7.2.

In another approach, Wells combines an electrostatic term with a solvent replacement term, and the practical limitation of the treatment given above to water-rich aqueous mixtures is retained.[24] Another limitation is the restriction to the transfer of hydrogen ions, but the transfer of other ions can be deduced from data on the transfer of complete electrolytes, obtained from e.g. potentiometric data, by means of those for the hydrogen ions. The solvated proton in water is considered to be describable by the tetrahedral species $H_3O(H_2O)_4{}^+$, the effective size of which can be approximated by a sphere having a radius equalling three times the radius of a water molecule: $r_{H^+(solv)} = 3r_{H_2O} = 0.414$ nm. The electrostatic term is given by the Born equation applied to an ion of this size: $\Delta G^{\infty}_{t,el} = \frac{1}{2}\mathbf{N}\mathbf{e}^2\boldsymbol{\varepsilon}_0^{-1}(3r_{H_2O})^{-1}(\varepsilon_{W+S}^{-1} - \varepsilon_W^{-1})$. The proton exchange reaction

$$H_3O(H_2O)_4^+ + S \rightleftharpoons HS(H_2O)_4^+ + H_2O \qquad (7.20)$$

where the proton is transferred from a purely aqueous environment to one in which the solvent S occupies one coordination site, is equivalent to the solvent exchange reaction 7.11 in the water-rich solutions. If the extra four water molecules are ignored in the notation of the species henceforth for the sake of clarity, and if the subscript $_T$ is made to denote the total molar concentration, then: $[H_3O^+] = [H^+]_T - [HS^+]$, $[S] = [S]_T - [HS^+]$, and $[H_2O]$ is approximated by $55.3 - [S]_T$, where 55.3 is the molar concentration of pure water. The equilibrium quotient for reaction 7.20 is

$$K_{exch} = [HS^+]\cdot(55.3 - [S]_T)/([H^+]_T - [HS^+])\cdot([S]_T - [HS^+]_T) \qquad (7.21)$$

The amount of proton exchange between the water and solvent species that has taken place per unit concentration of hydrogen ions is $[HS^+]$, hence the exchange contribution to the standard molar Gibbs free energy of transfer is $\Delta G^{\infty}_{t,exch} = -\mathbf{R}T[HS^+]\ln K_{exch}$. The concentration of the protonated solvent (or the hydrogen ions solvated by S), $[HS^+]$, is obtained from Hammett acidity function measurements with 4-nitroaniline as the indicator. This derivation depends on a number of assumptions, including that the ratio of the concentration of the free (i.e., unprotonated) water in the presence and the absence of the solvent S is essentially unity. This, then, severely limits the applicability to very water-rich mixtures (see also Section 6.3), although it has been applied well beyond this limit,[24,25] see Figure 7.3.

Grunwald,[26] Covington,[18,27] and Cox[28] and their respective co-workers derived equations relating the standard molar Gibbs free energy of transfer of an ion (or the sum of the ionic contributions to give the value for the complete electrolyte) from a solvent to its mixture with another to the equilibrium constants for solvent replacement. Following Cox *et al.*,[28]

$$\Delta G^{\infty}_t(X, S_A \rightarrow S_B) = -n\mathbf{R}T\ln \bar{K} \qquad (7.22)$$

(assuming $n' = n$ in equation 7.12), since on transfer from solvent A to solvent B the ion X changes from the solvate XA_n to XB_n. According to equation 7.22, $\bar{K}$ is an auxilary variable, derivable from ΔG^{∞}_t from solvent A to solvent B, provided the change in the Gibbs free energy on transfer is given completely by changes that take place in the first coordination sphere. When the transfer is from solvent

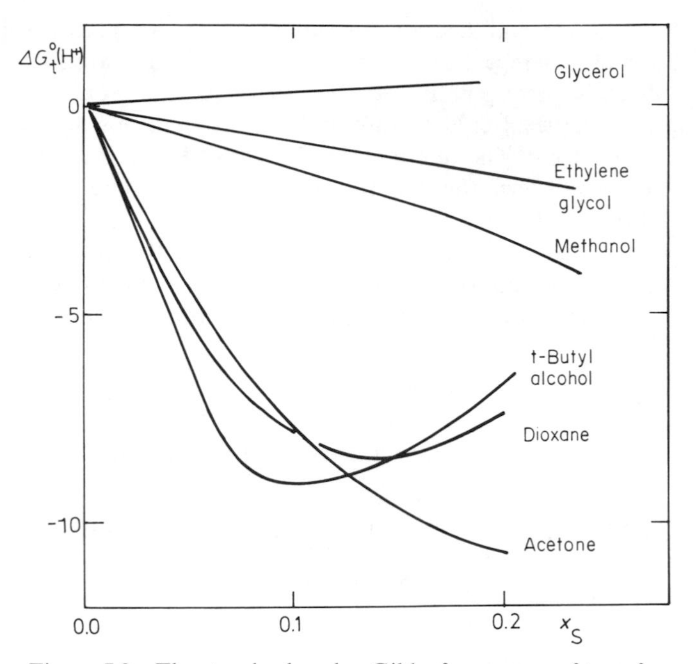

Figure 7.3 The standard molar Gibbs free energy of transfer, ΔG_t^{∞}, of hydrogen ions from water to water-rich aqueous solvents as a function of the mole fraction, x_S, of the nonaqueous component, according to the method of Wells[24,25]

A to the mixed solvents A + B, the corresponding equation is

$$\Delta G_t^{\infty}(\mathrm{X}, \mathrm{S_A} \rightarrow \mathrm{S_A} + \mathrm{S_B}) = -n\mathbf{R}T \ln \phi_A - \mathbf{R}T \ln\left[1 + \sum_{i=1}^{n}\left(\prod_{j=1}^{i} K_j\right)(\phi_B/\phi_A)^i\right] \tag{7.23}$$

where the ϕ's are the volume fractions of the solvent components. If statistical distribution of the solvated species is assumed, i.e., $K_j = (n - j + 1)j^{-1}\bar{K}$ in equation 7.23, then this equation transforms into

$$\Delta G_t^{\infty}(\mathrm{X}, \mathrm{S_A} \rightarrow \mathrm{S_A} + \mathrm{S_B}) = -\mathbf{R}T \ln \sum_{i=1}^{n} \bar{K}^i(n!/(n-i)!i!)\phi_B^i\phi_A^{n-i} \tag{7.24}$$

which apparently is a two-parameter equation in the volume fraction composition of the solvent mixture. The two parameters are the solvation number n and the average constant for solvent replacement $\bar{K}$. The latter, however, is given directly by equation 7.22, so that the standard molar Gibbs free energy of transfer to the mixed solvent is completely determined by the value for the transfer to the second component of the mixture and the single parameter n.

The equation given by Covington *et al.*[18,27] is similar to equation 7.23, except that solvent activities replace the volume fractions inside the sum in the last term on the right-hand side, and the mole fraction of solvent A replaces ϕ_A in the first term. Additionally, there is a long range electrostatic term, approximated by a difference of terms corresponding to the Born equation (equation 6.13). The radius employed in the latter is the sum of the radius of the ion and the (average) diameter of the solvent. The relative permittivity of the mixed solvent is to be used in the (positive) term for the mixture and that for pure solvent A in the (negative) term for the latter.

For situations where it is not appropriate to describe the solvation in terms of solvent replacement equilibrium constants, it is convenient to define an excess molar standard Gibbs free energy of transfer. For transfer from water into an aqueous solvent this is

$$\Delta G_t^{\infty E}(x) = \Delta G_t^{\infty}(X, W \rightarrow W + S) - x\,\Delta G_t^{\infty}(X, W \rightarrow S) \tag{7.25}$$

In the general case of transfer into a mixture of two nonaqueous solvents, the corresponding equation is

$$\Delta G_t^{\infty E}(x) = \Delta G_t^{\infty}(X, W \rightarrow S_A + S_B) - x_A\,\Delta G_t^{\infty}(X, W \rightarrow S_A) - x_B\,\Delta G_t^{\infty}(X, W \rightarrow S_B) \tag{7.26}$$

It is convenient also to define a dimensionless reduced quantity, called the *preferential solvation parameter* by Marcus[29,30]

$$g(x) = \Delta G_t^{\infty E}(x)/\mathbf{R}Tx(1 - x) \tag{7.27}$$

where $x = x_A$. A formulation of the standard molar excess Gibbs free energy of transfer in terms of volume fractions, $\Delta G_t^{\infty E}(\phi)$, and the corresponding $g(\phi)$, with ϕ replacing x, is also possible, and will be discussed more fully further below. If no preferential solvation occurs in a system, then $\Delta G_t^{\infty E} = 0$ for all values of x (or x_A) and also $g = 0$, but this is a very rare occurrence. As a crude approximation the preferential solvation parameter can be taken to be independent of x (or x_A), and it may assume either positive or negative values, the latter being the more common by far. It may change, however, as x varies, and usually does so, and it may even change its sign. The dependence of g on x is often rather symmetric, though it need not be so generally.

Values of $\Delta G_t^{\infty E}(\phi)$ can be calculated from a combination of equations 7.22 and 7.24, as shown in Figure 7.4. The resulting curves (reflected also in the behaviour of $g(\phi)$) are moderately asymmetric in ϕ, and they are invariably negative. Nonstatistical distribution of the stepwise solvent replacement equilibrium constants may aggravate the skewness of the curves. However, the case of statistical distribution of these constants permits the description of the system in terms of the single parameter n and the measured ΔG_t^{∞} into the pure second component, so that no recourse need be taken of the notion of stepwise solvent replacement at all.

Experimental data for the transfer of complete electrolytes from a reference solvent to a mixture of solvents (possibly consisting of the former and another solvent) are available, e.g., from measurements on electrochemical cells. They

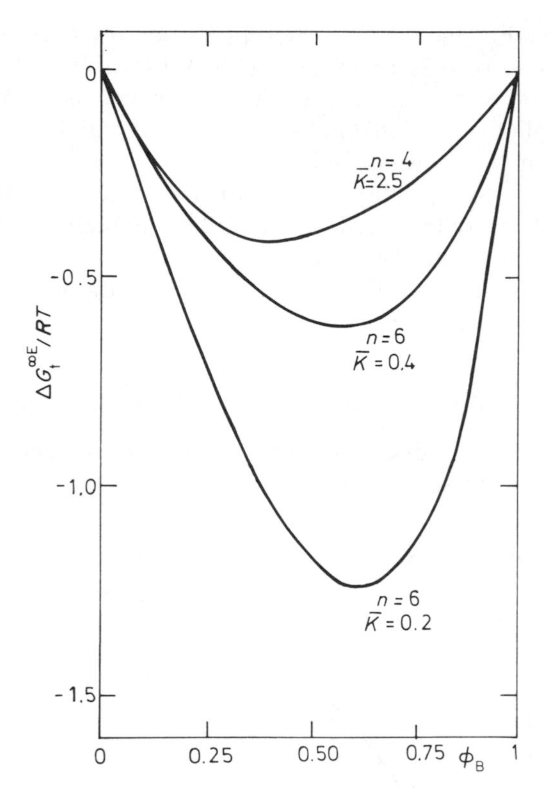

Figure 7.4 Values of the excess standard molar Gibbs free energy of transfer, $\Delta G_t^{\infty E}$ (divided by $\mathbf{R}T$) as a function of the volume fraction ϕ_B of the preferred solvent, calculated from equations 7.22 and 7.24 for various values of the parameters n (the coordination number) and $\bar{K}$ (the mean solvent replacement equilibrium constant). Note the skewness of the curves

may be split into individual ionic contributions in the same manner as for the transfer into a single solvent. This problem is discussed in detail in Section 6.3, and the arguments should not differ in principle for transfer into a mixed solvent. In particular, the reference electrolyte method has been examined by Kim as to its suitability for the transfer of individual ions to aqueous solvent mixtures (involving acetonitrile, N,N-dimethylformamide, and dimethylsulphoxide)[31] in the same manner as it has for transfer into single solvents.[32] The ratio of the standard molar Gibbs free energy of transfer for the neutral analogues Ph_4Ge (of Ph_4As^+) and Ph_4C (of BPh_4^-) is 1.09 for aqueous acetonitrile, 1.07 for aqueous N,N-dimethylformamide and 1.08 for aqueous dimethylsulphoxide mixtures, and it is 1.06 for 13 pure organic solvents, all ± 0.02.[32] This ratio is, therefore, as close to unity for the binary mixtures as for the pure solvents. For the reasons stated in Section 6.3, even splitting of ΔG_t^{∞} (Ph_4AsBPh_4, W → W + S) or of ΔG_t^{∞} (Ph_4AsBPh_4, W → $S_A + S_B$) will be employed as the most reasonable extrathermodynamic assumption.

Rather than split the standard Gibbs free energy of transfer of electrolytes into contributions from their constituent ions, 'real' Gibbs free energies of transfers of ions X from water to aqueous solvent mixtures, $\Delta\alpha_X^{\infty}$(W → W + S), can be measured directly in suitable cells, as described in Section 6.3. As discussed there, the 'real' and the extrathermodynamic Gibbs free energies of transfer differ by Faraday's constant times the charge number of the ion (in the algebraic sense) times the difference in surface potential between the solvents W and W + S:

$$\Delta G_t^{\infty}(X^z, W \rightarrow W + S) = \Delta\alpha_X^{\infty}(W \rightarrow W + S) + z\mathbf{F}\,\Delta\chi(x_S) \qquad (7.28)$$

The latter difference depends on the composition of the solvent in a complicated manner, because of the surface excess of either W or S in the mixture.

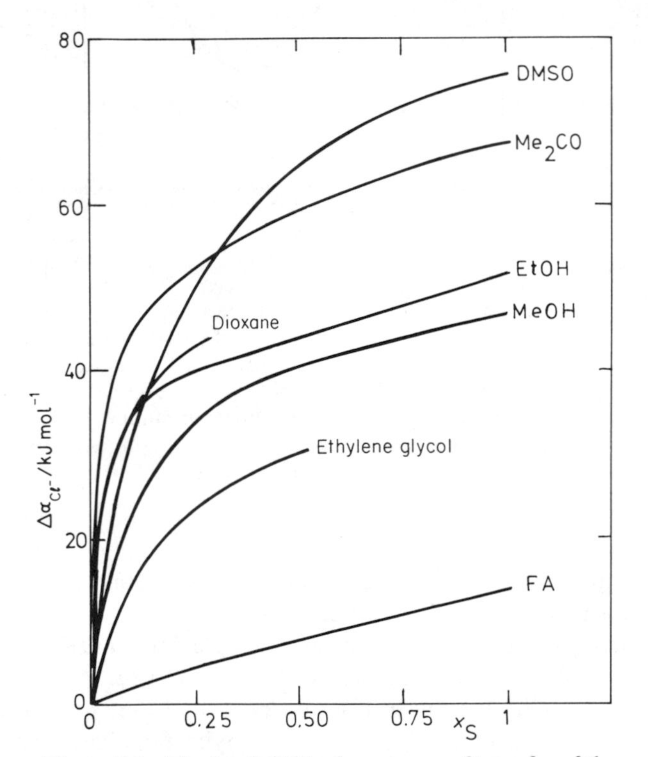

Figure 7.5 The 'real' Gibbs free energy of transfer of the chloride anion, $\alpha_{Cl^-}^{\infty}$, from water to aqueous solvent mixtures as a function of the mole fraction, x_S, of the nonaqueous component, according to Parsons and co-workers[33]. Note the steepness of the initial rise of the curves. (Reproduced by permission of the Royal Society of Chemistry)

The results for $\Delta\alpha_{Cl^-}^{\infty}$ obtained[33] for a number of aqueous solvent mixtures are shown in Figure 7.5. They all show strong positive deviations from linearity with the (mole fraction) solvent composition, and rise quite steeply with an initial small increment of the organic component (except for formamide). Indeed, $\Delta\alpha_{Cl^-}^{\infty}$ reaches 50% of its value in pure S at or below $x_S = 0.1$ for acetone, methanol, ethanol, and dimethylsulphoxide, and seems to do so also for ethylene glycol and dioxane if these curves are extrapolated to the pure solvents. This behaviour contrasts sharply with the results obtained for $\Delta G_t^{\infty}(x_S)$ according to the tetraphenylarsonium tetraphenylborate extrathermodynamic assumption, where deviations from linearity are generally negative, in particular for anions, and the steepness of the curves is much milder (see Figures 7.6 to 7.9 below).

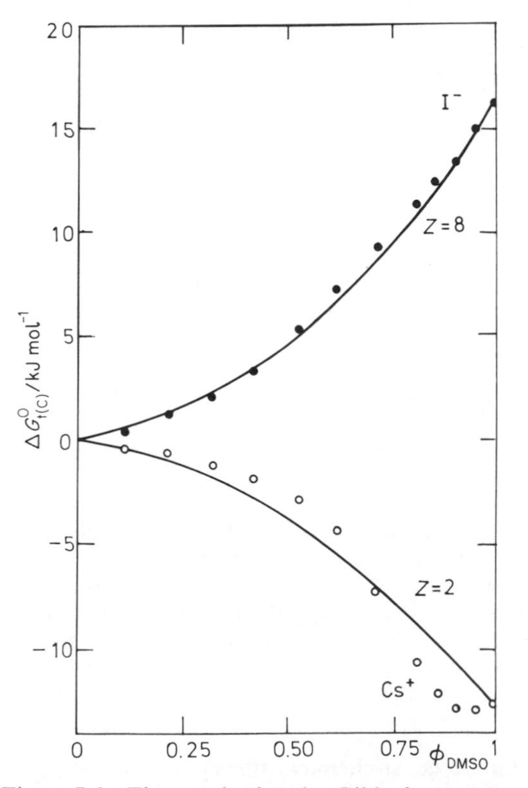

Figure 7.6 The standard molar Gibbs free energy of transfer, ΔG_t^{∞}, of caesium cations (○) and iodide anions (●) (see note (d) of Table 7.8 for source of data), according to the TATB assumption, from water to aqueous dimethylsulphoxide solutions, as a function of the volume fraction, ϕ_{DMSO}. The curves are calculated according to the quasi-lattice—quasi-chemical theory with the lattice parameters Z shown

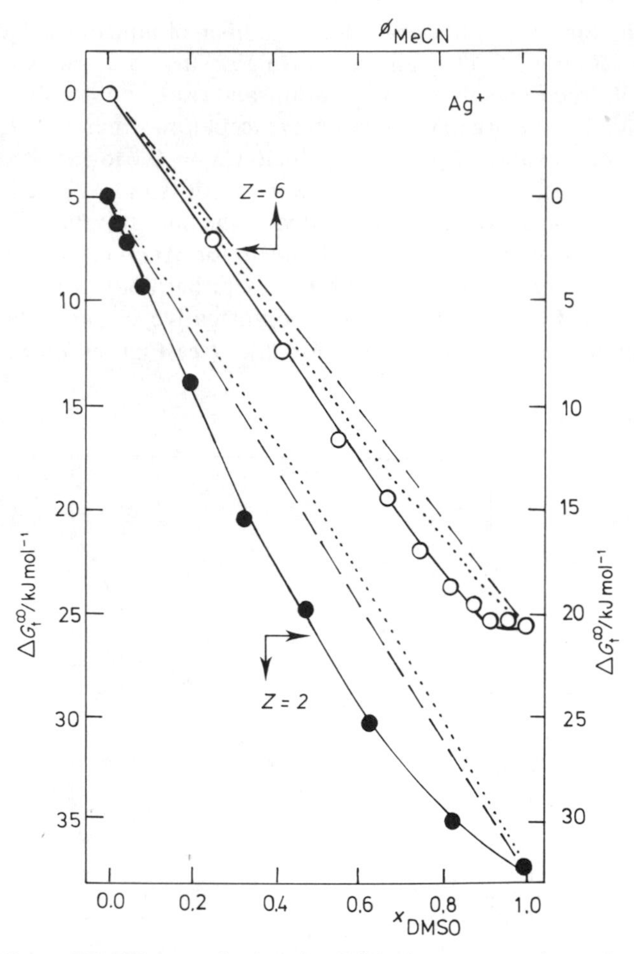

Figure 7.7 The standard molar Gibbs free energy of transfer, ΔG_t^∞, of silver cations from water into aqueous dimethylsulphoxide (●) and acetonitrile (○) (see note (b) of Table 7.8 for source of data), according to the TATB assumption, as a function of the composition of the solvent. The curves are calculated from the quasi-chemical—quasi-lattice theory with the lattice parameters Z shown

7.4 The quasi-lattice quasi-chemical theory

An alternative to the methods that use explicitly a solvent replacement equilibrium constant ($\bar{K}$ and the consecutive K_j, whether related statistically or not) employs, according to Marcus,[30] Guggenheim's quasi-lattice quasi-chemical theory.[34] The quasi-lattice theory assigns a lattice parameter Z to the system, that specifies the number of nearest neighbours each particle, whether solute ion X or solvent molecule S_A or S_B, has. The internal degrees of freedom of each particle are considered to be independent of the nature of its neighbours.

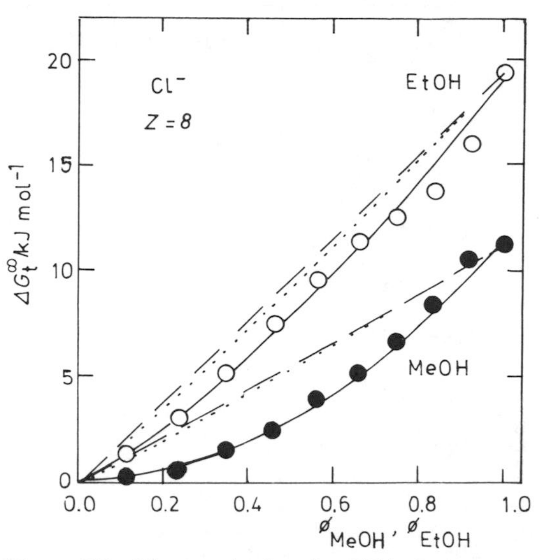

Figure 7.8 The standard molar Gibbs free energy of transfer ΔG_t^∞, of chloride anions from water to aqueous methanol (●) and ethanol (○) (see notes (f) and (g) of Table 7.8 for the source of the data), according to the TATB assumption, as a function of the composition of the solvent. The curves are calculated from the quasi-lattice–quasi-chemical theory, with the lattice parameter Z shown

The interaction energies e_{XA}, e_{XB}, and e_{AB} of neighbouring pairs of particles are considered to be independent of the other interactions in which each partner of the pair participates. The quasi-chemical theory specifies that the interactions of the particles affect the total numbers of pairs of neighbours of the various kinds, N_{ij}, in the system in the following manner:

$$N_{ij} = 2(N_{ii}N_{jj})^{1/2} \exp(\Delta e_{ij}/2\mathbf{k}T) \tag{7.29}$$

where $\Delta e_{ij} = e_{ii} + e_{jj} - 2e_{ij}$. The sum of all the pairwise interactions is $S = (Z/2)(N_A + N_B)$, at infinite dilution of the solute in a mixture consisting of N_A particles of solvent S_A and N_B particles of solvent S_B, the contribution to S from the $(Z/2)N_X$ pairwise interactions of the N_X solute particles being negligible.

The molar excess Gibbs free energy of mixing of the solvents is given by

$$G^E_{AB} = Z\mathbf{R}T[x_A \ln(N_{AA}/S)^{1/2}/x_A + x_B \ln(N_{BB}/S)^{1/2}/x_B] \tag{7.30}$$

The partial molar excess Gibbs free energy of the solute X (at infinite dilution) is obtained as follows. The total number of interactions in which the solute participates is obtained from equation 7.29:

$$ZN_X = N_{XA} + N_{XB} = 2N_{XX}^{1/2}[N_{AA}^{1/2} \exp(\Delta e_{XA}/2\mathbf{k}T) + N_{BB}^{1/2} \exp(\Delta e_{XB}/2\mathbf{k}T)] \tag{7.31}$$

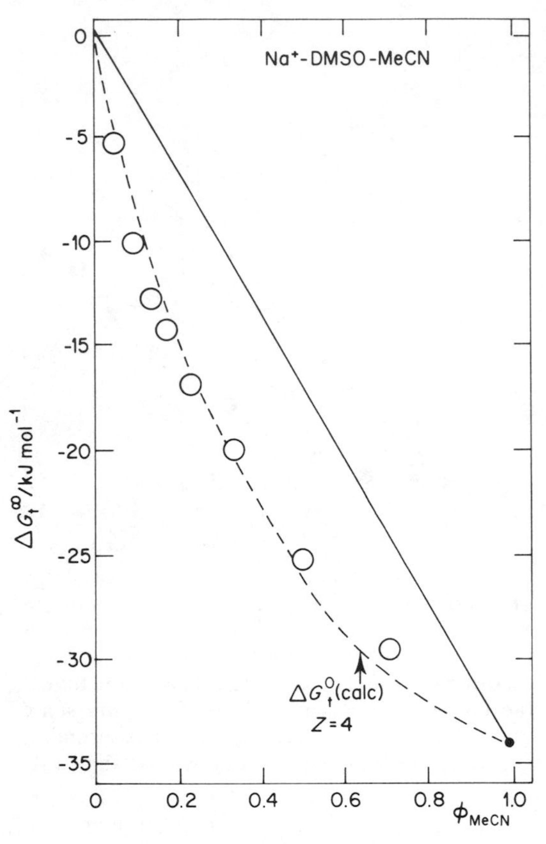

Figure 7.9 The standard molar Gibbs free energy of transfer, ΔG_t^{∞}, of sodium ions from dimethylsulphoxide into its mixtures with acetonitrile (data from B. G. Cox, A. J. Parker, and W. E. Waghorne, *J. Phys. Chem.*, **78**, 1731 (1974)), according to the TATB assumption, as a function of the volume fraction of acetonitrile, ϕ_{MeCN}. The curve is calculated from the quasi-lattice–quasi-chemical theory with the lattice parameter $Z = 4$

but this also equals $2x_X S$. The molar excess chemical potential of a component in the system is $\mu_i^E = (Z/2)\mathbf{R}T \ln(N_{ii}/x_i^2 S)$, hence from equation 7.31 and by an appropriate split into contributions from the two solvents,[30] the following expression is obtained:

$$\begin{aligned}\mu_X^{\infty E} = -Z\mathbf{R}T\{&x_A \ln(N_{AA}/S)^{1/2}/x_A + x_B \ln(N_{BB}/S)^{1/2}/x_B + x_A(\Delta e_{XA}/2\mathbf{k}T)\\ &+ x_B(\Delta e_{XB}/2\mathbf{k}T) + x_A \ln[x_A + x_B y \exp(\Delta)]\\ &+ x_B \ln[x_B + x_A y^{-1} \exp(-\Delta)]\}\end{aligned} \tag{7.32}$$

where $y = x_A x_B^{-1}(N_{BB}/N_{AA})^{1/2}$ and $\Delta = (\Delta e_{XA} - \Delta e_{XB})/\mathbf{k}T$. Inspection of equation 7.32 reveals that the first two terms in the curly brackets equal $-G_{AB}^E$ according to equation 7.30, and the next two terms are $x_A \mu_X^{\infty E}$ (in S_A) and $x_B \mu_X^{\infty E}$ (in S_B), respectively. Hence:

$$\Delta G_t^{\infty E}(X, W \rightarrow S_A + S_B) = \mu_X^{\infty E}(\text{in } S_A + S_B) - x_A \mu_X^{\infty E}(\text{in } S_A) - x_B \mu_X^{\infty E}(\text{in } S_B)$$
$$= -G_{AB}^E - \mathbf{R}T x_A x_B f(x) \qquad (7.33)$$

where

$$f(x) = Z\{x_B^{-1} \ln[x_A + x_B y \exp(\Delta)] + x_A^{-1} \ln[x_B + x_A y^{-1} \exp(-\Delta)]\} \qquad (7.34)$$

(The rational, not the molar, concentration scale must be used for $\Delta G_t^{\infty E}(X, W \rightarrow S_A + S_B)$ for equation 7.33 to be valid as written). The selective solvation parameter g is then

$$g(x) = -G_{AB}^E(x_A x_B RT)^{-1} - f(x) \qquad (7.35)$$

and is seen to be given by data independent of the transfer of the solute X to the mixed solvents: the molar excess Gibbs free energy of mixing of the two solvents, G_{AB}^E, and a quantity f that depends on Z, the single free parameter of the theory, and on the two quantities y and Δ that are evaluated, again from data independent of the transfer into the mixed solvents, as follows.

From its definition, Δ is seen to be given by the difference in the standard molar Gibbs free energies of transfer of the solute X from the reference solvent W to the two solvents S_A and S_B:

$$\Delta = (\Delta e_{XA} - \Delta e_{XB})/\mathbf{k}T = [\Delta G_t^{\infty}(X, W \rightarrow S_A) - \Delta G_t^{\infty}(X, W \rightarrow S_B)]/Z\mathbf{R}T \qquad (7.36)$$

For the evaluation of y the ratio N_{BB}/N_{AA} is required, which is obtained from the equalities $N_{AA} + N_{AB} = 2x_A S$ and $N_{BB} + N_{AB} = 2x_B S$:

$$y(x) = (x_A/x_B)[(x_B - N_{AB}/2S)/(x_A - N_{AB}/2S)]^{1/2} \qquad (7.37)$$

The application of equation 7.29 to N_{AB} yields a quadratic, from which

$$N_{AB}/2S = \{1 - [1 - 4x_A x_B(1 - \tfrac{1}{4}\exp[\Delta e_{AB}/\mathbf{k}T])]^{1/2}\}$$
$$\div 2[1 - \tfrac{1}{4}\exp(\Delta e_{AB}/\mathbf{k}T)] \qquad (7.38)$$

If now $G_{AB}^E(x_A = 0.5)$ is evaluated from equation 7.30 in terms of Δe_{AB}, the result for the latter is

$$\exp(\Delta e_{AB}/\mathbf{k}T) = 2\exp(-2G_{AB}^E(x_A = 0.5)/Z\mathbf{R}T) - 0.5 \qquad (7.39)$$

Hence a complete evaluation of y is possible, in terms of the molar excess Gibbs free energy of mixing of the two solvents at the equimolar composition and the lattice parameter Z.

The second term on the right-hand side of equations 7.33 or 7.35 is always negative, due to the inherently positive nature of the expression in the curly brackets that defines $f(x)$. The first term may be positive or negative (but for a miscible pair of solvents it may not be more positive than about $\mathbf{R}T/2$, according to equation 7.9). The quantity $f(x)$ is generally asymmetric in x, and since it dominates $\Delta G_t^{\infty E}$, the latter is also expected to be so, although the asymmetry is moderated by the factor $x_A x_B$ multiplying $f(x)$ in equation 7.33. Large negative values of G_{AB}^E can produce positive values of $\Delta G_t^{\infty E}$, which are rare, however. They signify that the mutual interactions of the two solvents are so strong that they tend to exclude the solute X, more than each of the solvents alone.

For the equimolar solvent mixture the set of equations 7.34 and 7.37 to 7.39 can be simplified, since $y(x_A = 0.5) = 1$. Insertion of this value into equation 7.34 yields

$$f(x_A = 0.5) = 4Z \ln \cosh\{[\Delta G_t^{\infty}(\mathrm{X, W \rightarrow S_A}) - \Delta G_t^{\infty}(\mathrm{X, W \rightarrow S_B})]/2Z\mathbf{R}T\} \tag{7.40}$$

or, for the case of transfer to an equimolar organic–aqueous solvent

$$f(x = 0.5) = 4Z \ln \cosh[\Delta G_t^{\infty}(\mathrm{X, W \rightarrow S})/2Z\mathbf{R}T] \tag{7.41}$$

The sum of $\mathbf{R}T/4$ times this quantity and the equimolar excess Gibbs free energy of mixing of the solvents (see Tables 7.4 and 7.5) yields the negative of ΔG_t^{∞} at the equimolar solvent composition. This is useful for the purpose of orientation. Table 7.8 presents a comparison of calculated and experimental results at the equimolar solvent composition for a number of ions and representative aqueous solvents.

Table 7.8 Standard Gibbs free energies of transfer of ions from water to equimolar aqueous solvents, compared with values calculated from the quasi-lattice quasi-chemical theory[a]

Solvent	Na^+	K^+	Ag^+	$Ph_4As^+ = BPh_4^-$	Cl^-	Br^-	I^-
MeOH	0.0[e]	0.4[f]		−23.6[e]	12.6[f]	11.0[f]	8.4[f]
	−0.7(2)	0.2(2)		−23.3(6)	12.2(8)	11.0(8)	8.8(6)
EtOH		5.3[g]	0.3[h]	−18.8[g]	12.3[g]		
		6.8(2)	2.9(2)	−19.2(12)	12.1(12)		
$En(OH)_2$	1.3[i]	1.6[i]		−13.3[i]	6.2[i]	4.6[i]	2.2[i]
	1.0[j]	1.0[j]		−14.8(12)	6.2(2)	5.5(2)	4.2(2)
DMSO	−11.7[d]	−9.2[d]	−19.7[b]	−28.5[d]	26.9[d]	19.9[d]	11.4[d]
	−11.5(6)	−9.4(6)	−28.7(12)	−33.1(12)	26.6(12)	20.3(8)	11.7(6)
MeCN	3.3[b]		−16.7[b]	−28.2[c]	7.1[b]		
	3.5(8)		−15.8(2)	−28.6(4)	5.9(10)		

[a] Molar scale, 25 °C, in kJ mol^{-1}, upper line: experimental values, lower line: calculated with the Z value given in parentheses; [b] B. G. Cox, A. J. Parker and W. E. Waghorne, *J. Phys. Chem.*, **78**, 1731 (1974); [c] J. I. Kim, A. Cecal, H. J. Born and E. A. Gomaa, *Z. Phys. Chem. (NF)*, **110**, 209 (1978); [d] J. I. Kim and E. A. Gomaa, *Bull. Soc. Chim. Belg.*, **90**, 391 (1981); [e] S. Villermaux and J. J. Delpuech, *Bull. Soc. Chim. France*, **1974**, 2534; [f] B. G. Cox and W. E. Waghorne, *Chem. Soc. Rev.*, **9**, 381 (1980); [g] O. Popovych and A. J. Dill, *Anal. Chem.*, **41**, 456 (1969); [h] J. I. Kim and H. Duschner, *J. Inorg. Nucl. Chem.*, **39**, 471 (1979); [i] A. K. Das and K. K. Kundu, *Indian J. Chem.*, **16A**, 467 (1978); [j] value indifferent to the value of Z.

The free parameter of the quasi-lattice quasi-chemical theory of selective solvation is Z, which is akin to the coordination number n for the solvent replacement theories discussed above. It differs from it in its being representative not only of the coordination of the solute but also of the 'coordination' of the solvent molecules in their mixture. It is of interest to explore[30] the possibility that the solute X has different coordination numbers in the two solvents, Z_A in S_A and Z_B in S_B, due, perhaps, to a difference in size between their molecules. The quantity used in equation 7.32 and defined in equation 7.36 should then be replaced by

$$\Delta' = [\Delta G_t^\infty(\text{X, W} \rightarrow \text{S}_A)/Z_A - \Delta G_t^\infty(\text{X, W} \rightarrow \text{S}_B)/Z_B]/\mathbf{R}T \qquad (7.42)$$

A weighting factor w is defined by $w = \exp(\Delta')x_B^r/(x_A^r + x_B^r \exp(\Delta'))$, where $r = Z_A/Z_B$. The mean lattice parameter of the system then depends on the composition according to

$$Z(x) = (1 - w)Z_A + wZ_B = (w(r - 1) + 1)Z_B \qquad (7.43)$$

Replacement of Z and Δ in equation 7.34 by $Z(x)$ and Δ' yields the value of $f(x)$ for this case of different coordination numbers. The skewness of $f(x)$ becomes more pronounced the larger the absolute value of Δ'. The use of two free parameters, Z_B and r, makes this modification more versatile, and permits a closer fitting of experimental data in some cases. However, it also entails the conceptual difficulty of having in a quasi-lattice treatment a variable lattice parameter. In particular is it difficult to accept a variable value $Z(x)$ for the mixing of the two solvents (e.g., in equation 7.30), that depends on the quantity Δ' (through the dependence of the weighting factor w on it), which is solute-specific. It is, therefore, not recommended to pursue this approach of a variable lattice parameter any further.

The difference in size between the two kinds of solvent molecules can be taken into account in another way, however, Rather than employ mole fractions x to express the solvent composition, the volume fraction scale ϕ may be used, and instead of the mole fraction x_X of the solute that defines the scale for ΔG_t^∞ that is used in equation 7.33, the molar scale c_X can be used. The standard molar Gibbs free energy of transfer on the two scales is related by

$$\Delta G_{t(c)}^\infty(\text{X, W} \rightarrow \text{S}) = \Delta G_{t(x)}^\infty(\text{X, W} \rightarrow \text{S}) + \mathbf{R}T \ln[(M_S/d_S)/(M_W/d_W)] \qquad (7.44)$$

For a mixed solvent the mean molar mass M and density d of the mixture (according to equation 7.5) should be employed. The parameter r, set equal to the ratio of the molar volumes of the two solvents, i.e. V_A/V_B for nonaqueous mixtures or $V_W/V_S = (M_S/d_S)/(M_W/d_W)$ for aqueous mixtures, that is used in the following expression, should not be confused with the r used in the expression for the variable lattice parameter above. The preferential solvation parameter on the volume fraction/molar concentration scales takes now the form:

$$g(\phi) = g(x)(rx_A + x_B)^2/r + [(rx_A + x_B - 1)(rx_A + x_B)/rx_A] \\ \times [\Delta G_{t(c)}^\infty(\text{X, W} \rightarrow \text{S}_A) - \Delta G_{t(c)}^\infty(\text{X, W} \rightarrow \text{S}_b)]/\mathbf{R}T \qquad (7.45)$$

The lattice parameter Z remains then as the single fitting parameter for the calculation of $g(x)$ from equations 7.34 and 7.35, since r is given by the independent data of the molar volumes of the solvents. The quantity $\phi_A \phi_B \mathbf{R}Tg(\phi)$ then equals $\Delta G^{\infty E}_{t(c)}(\phi)$, which, in turn, yields $\Delta G^{\infty}_{t(c)}(X, W \rightarrow S_A + S_B)$ if $\phi_A \, \Delta G^{\infty}_{t(c)}(X, W \rightarrow S_A) + \phi_B \, \Delta G^{\infty}_{t(c)}(X, W \rightarrow S_B)$ is added to it. Some instances of the application of this approach are depicted in Figures 7.6 to 7.9.

References

1. A. E. Guggenheim, *Trans. Faraday Soc.*, **45**, 714 (1949).
2. L. Onsager, *J. Am. Chem. Soc.*, **58**, 1486 (1936); J. R. Weaver and R. W. Parry, *Inorg. Chem.*, **5**, 703 (1966).
3. C. Reichardt, in *Molecular Interactions*, H. Ratajczak and W. J. Orville-Thomas, eds., Wiley, New York, Vol. 3 (1982), pp. 241–282.
4. U. Mayer, W. Gerger, and V. Gutmann, *Monatsh. Chem.*, **108**, 489 (1977).
5. Y. Marcus, *Introduction to Liquid State Chemistry*, Wiley, Chichester (1977), chap. 4 and 5.
6. J. Wisniak and A. Tamir, *Mixing and Excess Thermodynamic Properties*, Elsevier, Amsterdam (1978).
7. H. V. Kehiaian, ed., *Selected Data on Mixtures. A.*, Thermodynamic Research Center, College Station, Texas (1973–1983).
8. D. W. Kemp and E. L. King, *J. Am. Chem. Soc.*, **89**, 3433 (1967); L. P. Scott, T. J. Weeks, D. J. Bracken, and E. L. King, *ibid.*, **91**, 5219 (1969).
9. A. Fratiello, R. E. Lee, V. M. Nishida, and R. E. Schuster, *J. Chem. Phys.*, **48**, 3705 (1968); A. Fratiello, V. Kubo, S. Peak, B. Sanchez, and R. E. Schuster, *Inorg. Chem.*, **10**, 2552 (1971); A. Fratiello, V. Kubo, and G. Vidulich, *ibid.*, **12**, 2066 (1973).
10. A. Fratiello, R. E. Lee, V. M. Nishida and R. E. Schuster, *J. Chem. Phys.*, **47**, 4951 (1967).
11. A. D. Covington and A. K. Covington, *J. Chem. Soc. Faraday Trans. 1*, **71**, 831 (1975); F. Toma, M. Villemin, and J. M. Thiery, *J. Phys. Chem.*, **77**, 1294 (1973).
12. J. H. Swinehart and H. Taube, *J. Chem. Phys.*, **37**, 1579 (1962).
13. Z. Luz and S. Meiboom, *J. Chem. Phys.*, **40**, 1058 (1964).
14. R. H. Erlich, M. S. Greenberg, and A. I. Popov, *Spectrochim. Acta*, **29A**, 543 (1973).
15. J. Burgess, *Metal Ions in Solution*, Ellis Horwood, Chichester (1978), p. 169.
16. R. L. Benoit and C. Buisson, *Inorg. Chim. Acta*, **7**, 256 (1973).
17. H. Schneider, *Topics Curr. Chem.*, **68**, 103 (1976).
18. A. K. Covington, T. H. Lilley, K. E. Newman, and G. A. Porthouse, *J. Chem. Soc., Faraday Trans. 1*, **69**, 963, 973 (1973).
19. A. K. Covington, I. R. Latzke, and J. M. Thain, *J. Chem. Soc., Faraday Trans. 1*, **70**, 1869, 1879 (1974).
20. S. Thomas and W. L. Reynolds, *Inorg. Chem.*, **9**, 78 (1970); H. Chi, C.-H. Ng, and N. C. Li, *J. Inorg. Nucl. Chem.*, **38**, 529 (1976).
21. A. A. Al-Harakany and H. Schneider, *J. Electroanalyt. Chem.*, **46**, 255 (1973); A. Clausen, A. A. Al-Harakany, and H. Schneider, *Ber. Bunsenges. Phys. Chem.*, **77**, 994 (1973).
22. H. S. Frank, *J. Chem. Phys.*, **33**, 2023 (1955).
23. J. Padova, *J. Phys. Chem.*, **72**, 796 (1968); *idem.*, *J. Chem. Phys.*, **39**, 1552 (1963).
24. C. F. Wells, *Trans. Faraday Soc.*, **61**, 2194 (1965); *idem.*, *J. Chem. Soc. Faraday Trans. 1*, **69**, 984 (1973).
25. C. F. Wells, *Adv. Chem. Ser.*, **177**, 53 (1979).
26. E. Grunwald, G. Baugham, and G. Kohnstam, *J. Am. Chem. Soc.*, **82**, 5801 (1960).
27. A. K. Covington and K. E. Newman, *Adv. Chem. Ser.*, **155**, 153 (1976).

28. B. G. Cox, A. J. Parker, and W. E. Waghorne, *J. Phys. Chem.*, **78**, 1731 (1974).
29. Y. Marcus, *Rev. Anal. Chem.*, **5**, 53 (1980).
30. Y. Marcus, *Austr. J. Chem.*, **36**, 1719 (1983).
31. J. I. Kim, A. Cecal, H. J. Born, and E. A. Gomaa, *Z. Phys. Chem.*, **110**, 209 (1978); J. I. Kim and E. A. Gomaa, *Bull. Soc. Chim. Belg.*, **90**, 391 (1981).
32. J. I. Kim, *J. Phys. Chem.*, **82**, 191 (1978).
33. R. Parsons and B. T. Rubin, *J. Chem. Soc. Faraday Trans. 1*, **70**, 1636 (1974).
34. E. A. Guggenheim, *Proc. Roy. Soc.* (*London*), **A169**, 134 (1938); *idem.*, *Mixtures*, Clarendon Press, Oxford (1952).

Chapter 8

Solvation of ion pairs

8.1 The effect of ion-pairing on solvation

The previous chapters dealt with ion solvation in very dilute solutions, where no ion–ion interactions occurred. The individual ion is then entirely surrounded by solvent molecules over large distances, and the distance to the nearest ion, of either charge sign, is effectively infinite. At somewhat higher concentrations, however, the mutual distances between ions decrease sufficiently for the ion–ion interactions to become significant. The potential energy of a pair of ions is raised or lowered, relative to its value at infinite dilution that is set at zero, depending on whether the ions have the same or opposite signs. The extent of the change in the potential energy depends on the sizes of the ionic charges, on the distance to which the ions approach, and on the bulk dielectric constant (relative permittivity) of the solvent.

The distances involved in the situation just discussed are such that the solvation shells of the ions remain intact. The solvent may then be considered as a continuum characterized solely by its dielectric constant. The potential energy of the interacting pair of ions is

$$u_{12} = z_1 z_2 \mathbf{e}^2/\boldsymbol{\varepsilon}_0 \varepsilon r \tag{8.1}$$

At a given distance apart of the ions r and for ions of opposite sign, i.e., $z_1 z_2 < 0$, then as solvents with decreasing dielectric constants ε are considered, a value of ε is reached where $|u_{12}| = 2\mathbf{k}T$. At and below this value the attractive electrostatic forces between the ions produce a potential energy which can counteract successfully the kinetic energy of random thermal motion of the ions. The ions are then considered as *paired*, according to Bjerrum[1] (Figure 8.1). The coefficient 2 of **k**T arises from the position of the minimum of the distribution curve of ions of charge z_i around an ion of opposite charge z_j

$$(\rho_i(r)/\rho_i(\infty)) = \exp(-z_i z_j \mathbf{e}^2/\boldsymbol{\varepsilon}_0 \varepsilon r \mathbf{k} T) \tag{8.2}$$

where ρ is the number density of the ions. The minimum in the distribution curve occurs at $r_{min} = |z_i z_j| \mathbf{e}^2/2\boldsymbol{\varepsilon}_0 \varepsilon \mathbf{k} T$. The physical consequences of this pairing are not at all self-evident, except for the realization that in an external field the paired ions do not move individually but reorient themselves as an electrical

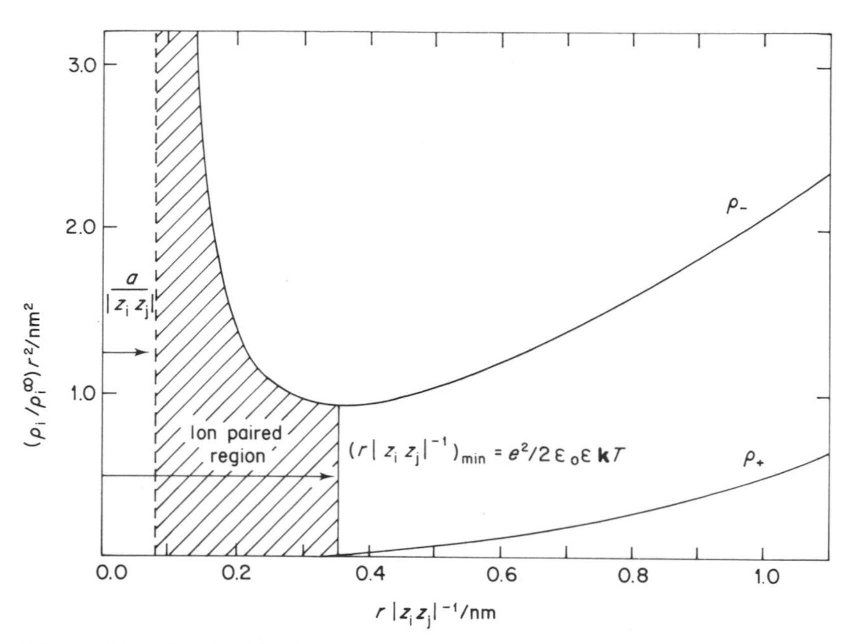

Figure 8.1 The distribution of ions i around a positive ion j, expressed as $[\rho_i(r)/\rho_i(\infty)](r/\mathrm{nm})^2$, as a function of the distance, normalized for the charges, $(r/\mathrm{nm})/|z_i z_j|$. The like-charged cations are repulsed from the ion j, but the unlike-charged anions have a region (shaded) where they are attracted to form an ion pair, according to Bjerrum's theory. (From Y. Marcus, *Introduction to Liquid State Chemistry*, Wiley-Interscience, Chichester (1977), by permission.)

dipole, i.e., their net migration is zero, if the absolute values of the charges are equal. The notions behind this concept of an ion pair have been re-examined recently[2] and found to be acceptable.

For monovalent ions at room temperature the product $r_{\min}\cdot\varepsilon = 28$ nm. For solvents of dielectric constants between 2 and, say, 28 this corresponds to values of $r_{\min}$ between 14 and 1 nm, where even the latter is sufficiently larger than the dimensions of most ions and solvent molecules of interest, for the premise that the solvent can be considered as a dielectric continuum to be valid. For more highly charged ions the product $r_{\min}\cdot\varepsilon$ is correspondingly larger, and so is the range of r values where this approach remains valid. The product εT decreases generally faster than T increases. For example, for water at 20 °C $\varepsilon T/\mathrm{K} =$ 23 560, at 50 °C it is 22 590, at 100 °C it is 20 790, at 200 °C it is 16 440, and at 350 °C it is only 8200. Hence $r_{\min}$ increases as the temperature is raised, and high temperature solvents behave in this respect as solvents of lower dielectric constants, permitting long range ion-pairing.

Under these conditions, an ion pair is defined as a pair of ions of opposite charge sign that are at a distance r apart, where $a \leqslant r \leqslant r_{\min}$ and a is their distance of closest approach. The relative concentration of ion pairs in the solution is obtained by the integration of $4\pi r^2 dr$ times the right-hand side of equation 8.2 between the limits a and r_{min}. From this the well-known Bjerrum expression for the association equilibrium constant for ion pairing is obtained:

$$K_{ip} = 4 \cdot 10^{-24} \pi \mathbf{N}(a/\text{nm})^3 b^3 Q(b) \tag{8.3}$$

Here K_{ip} is obtained on the molar concentration scale, and

$$b = |z_i z_j|(\mathbf{e}^2/4\pi\varepsilon_0\mathbf{k})T^{-1}\varepsilon^{-1}a^{-1}$$

$$= |z_i z_j| \cdot 16\,710.3(T/\text{K})^{-1}\varepsilon^{-1}(a/\text{nm})^{-1} \tag{8.4}$$

$$Q(b) = \int_2^b t^{-4} \exp(t)\, \mathrm{d}t \tag{8.5}$$

t being an auxiliary variable. The value of the integral is set by its upper limit of integration b and is readily obtained by iterative computation. The value of b, in turn, depends on the distance of closest approach of the oppositely charged ions a, on the temperature, T, and on the dielectric constant of the solvent, ε.

Ion-pairing that occurs under these conditions, by its definition, does not affect the ionic solvation at all. Thus it is not expected to affect spectral properties that depend on the solvation of the ion: the visible spectrum of transition metal ions, the UV spectrum of ions due to charge transfer, the IR and Raman spectra of the solvent molecules in the solvation shells of the ions, the NMR signals from the ions and from solvent molecules in their solvation shells, etc. All these remain the same as at infinite dilution, where no ion–ion interactions occur at all. The ion-pairing does affect the equivalent conductivity of the electrolyte solution, as mentioned above in connection with the migration of the ions in an external field. It also affects thermodynamic properties as expressed, e.g. by the activity coefficient: only unpaired, i.e. free, ions contribute to the ionic atmosphere in the Debye–Hückel expression, for instance. The effect of the ion pair on the solvent itself is also neglected: neither the field of the ions remaining separated, nor the presumably reduced field of the ion pair is considered to affect the dielectric constant of the solvent, which remains constant at ε according to this picture. Indeed, the fraction of space around an ion where dielectric saturation or the field effect of the ion is appreciable is small with respect to the space where ion-pairing is considered to take place, $(4\pi/3)r_{min}.^3$

For solvents of higher dielectric constants, including, of course, water, r_{min} becomes comensurate with the dimensions of ions and solvent molecules. The presumption that a dielectric continuum can represent the solvent is no longer tenable. The molecular nature of the solvent and specific ion–solvent interactions must be taken into account when ion-pairing is considered under these new conditions. The region in space where dielectric saturation takes place is now an appreciable fraction of the space within which ions are considered to become paired. However, the presence of another ion strongly affects the electrical field in the region between them. It is then extremely difficult to estimate the effective dielectric constant in the non-spherically symmetrical region in between the ions, so that the electrostatic calculation of the potential energy of the system is not straightforward. Non-electrostatic effects: donor–acceptor interactions, hydrogen bonding, etc., are superimposed on the coulombic interactions, as discussed in Chapter 3. Under such conditions the Bjerrum approach must be abandoned in favour of a different concept of ion-pairing.

This new concept of ion-pairing pertains to two ions that are in the close vicinity of each other, so that their (negative) pair potential is sufficiently large to hold them together at a short distance from one another for a reasonable period of time, before their thermal motion rips them apart. The time scale involved is comensurable with the time it takes for ions diffusing freely in the solvent to cover the relevant distance, but is not defined exactly. The definition of the ion pair is based on the mutual geometry of the ions and the solvent:

(i) solvent-separated ion pairs (Figure 8.2a), where both ions retain their primary solvation shells intact, these primary shells being in contact, so that some overlap of secondary and further solvation takes place;
(ii) solvent-shared ion pairs (Figure 8.2b), where the primary solvation shells of the ions interpenetrate, the two ions being held apart by one thickness of solvating solvent, i.e., one solvent molecule;
(iii) contact ion pairs (Figure 8.2c), where final penetration of the two ions through their primary solvation shells has occurred, so that no solvent molecules intervene between the ions that are in close contact. The ion pair constitutes an electrical dipole, which, in turn, is still solvated, having a primary and further solvation shells.

Ions of opposite charge where $z_+ \neq |z_-|$, i.e., ions of a non-symmetrical electrolyte, pair to form a charged ion pair, in contrast to the neutral ion pair formed in symmetrical electrolytes. The charged pair behaves as a dipolar ion, and is solvated in a manner similar to ordinary polyatomic ions having a non-spherical charge distribution. The solvation of such ions has been discussed along with that of monoatomic and symmetrical polyatomic ions in the previous chapters, and will not be further discussed here.

From this deliberation it is clear that the solvation of ion pairs that are constituted by ions in close vicinity to each other differs considerably from the

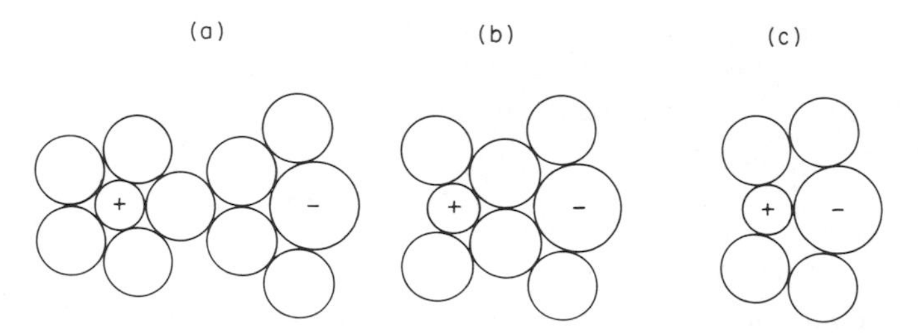

Figure 8.2 Schematic representation of (a) a solvent-separated ion pair, (b) a solvent-shared ion pair, and (c) a contact ion pair

solvation of the pair of separate ions. The concentrations of ion pairs entering categories (i), (ii), and (iii) are between that given by the Bjerrum expression, equation 8.3, and that expected from ions in contact. The latter can be evaluated from the association constant according to Fuoss[3]

$$K_{ass} = (4 \cdot 10^{-24}\mathbf{N}\pi/3)(a/\text{nm})^3 \exp(b) \tag{8.6}$$

where a is the contact distance of the ions and b is given by equation 8.4. A comparison of equation 8.6 with equation 8.3 shows the former to have the term $\exp(b)/3$ in place of $b^3Q(b)$ in the latter, which, for the region of high b where it is valid (high charges or low dielectric constants) is well approximated by $\exp(b)/b$, and since $b \gg 3$ there, $K_{ass} > K_{ip}$. However, Fuoss' theory that leads to equation 8.6 is based on electrostatic deliberations only, and does not take into account either dielectric saturation or the non-electrostatic contributions to the pair potential of the ions that have been mentioned above. The potential energy due to the electrostatic interaction between the ions can be written as

$$|u_{12}| = z_1 z_2 \mathbf{e}^2/\boldsymbol{\varepsilon}_0 \varepsilon_{eff}(a + 2n \cdot r_S) \tag{8.7}$$

where a is now defined not as the distance of closest approach of the oppositely charged ions but as the sum of the radii of the 'bare', i.e., unsolvated ions, and r_S is the radius of a solvent molecule, with $n = 2,1$, or 0 for pairs in categories (i), (ii), and (iii). The effective dielectric constant of the solvent in the space between the ions is $1 \leqslant \varepsilon_{eff} \ll \varepsilon$ (with $\varepsilon_{eff} = 1$ applying to $n = 0$, i.e., contact ion pairs), but is difficult to estimate. The attractive potential energy given by equation 8.7 now counteracts not only the kinetic energy due to thermal motion that tends to rip the ion pair apart, but also losses in potential energy due to the removal of solvent molecules from the primary solvation shells of the ions, and other changes in potential energy (that may be either positive or negative) due to the rearrangement of the remaining solvent molecules around the uncharged electrical dipole presented to them by the ion pair. No theory has so far been presented that deals with all these aspects of the energetics of the formation of ion pairs that belong to the categories (i), (ii), and (iii) of pairs of ions in close vicinity of each other.

The electrical field dependence of the dielectric constant of a solvent, that leads to dielectric saturation very near the ion and to low ε_{eff} further away, and also to electrostriction, i.e., compression, of the solvent near the ion, has been calculated for a spherically symmetrical ion only.[4,5] Equation 3.35 sets $\varepsilon(E) = \varepsilon(E = 0) + bE^2 + \cdots$ for a field strength $E = |z_i|\mathbf{e}/\boldsymbol{\varepsilon}_0\varepsilon r$, with values of the parameter b (not to be confused with the b of equation 8.4) given for several solvents on p. 41. The effect of the field of the ion on the solvent is not only to modify its dielectric constant but also to generate a pressure, P_{el}, that causes electrostriction, i.e., a compression of the solvent. The differential of the pressure is given by equation 5.14 in terms of the compressibility of the solvent and the pressure derivative of the dielectric constant. Integration yields $P_{el} = E^2(\varepsilon(E) - 1)/8\pi$, which is an implicit equation in E and in r, the distance from the centre of the ion where the values of the pressure, P_{el}, the field strength, E, and the dielectric constant, $\varepsilon(E)$, prevail. The change in the volume at a given chemical potential of the solute electrolyte, μ, is obtained as

$$(\partial \ln V/\partial E)_{\mu,T} = (E/4\pi)(\partial \varepsilon/\partial P)_{E,T} \tag{8.8}$$

The electrostrictive changes in the molar volume of water are presented in Table 5.9 as a function of r. (Water, however, is a solvent of such high dielectric constant that at room temperature for univalent ions $r_{min} = 0.35$ nm, which is less than the sum of the radii of such ions plus intervening water molecules, $a + 2nr_S$, with $r_S = 0.14$ nm. For univalent ions in water at room temperature the Bjerrum equations 8.2 to 8.5 thus do not apply, and ions do not form Bjerrum-type ion pairs.)

The distance r from the centre of an ion, up to which its electrical field causes the experimentally observable electrostriction, can be estimated from these considerations (see equation 5.12 and 5.13 and the accompanying text). In the case of a spherically symmetrical ion, E, ε, and P_{el} depend implicitly on r in a readily describable manner. This is not so in the case of the more complicated geometry that prevails when there are two ions of opposite charge in the vicinity of each other, but in the presence of solvent molecules around and in between them. It is therefore common practice to ignore changes in the electrostriction of the solvent molecules that remain in the combined field of the two ions when they form an ion pair, and focus the attention on the change in volume (from electrostricted to non-electrostricted) of those solvent molecules that are set free as the solvation shells of the pairing ions overlap each other (Figure 8.2).

Ion-pairing is generally found to cause an increase in the apparent molar volume of the electrolyte. For the situation discussed at the beginning of this Section, where the Bjerrum equation applies, this increase in the standard partial molar volume of the electrolyte, $\Delta\bar{V}^{\infty}_{ip}$, is readily estimated according to Marcus.[6] Given the experimental value of K_{ip} for the (symmetrical) electrolyte, the size of the nearest distance of approach a is fixed, hence the parameter b of equation 8.4. Given in addition the isothermal compressibility κ_T and the pressure derivative of the dielectric constant, $(\partial \ln \varepsilon/\partial P)_T$ of the solvent, equations 8.3 to 8.5 lead to

$$\Delta\bar{V}^{\infty}_{ip} = \mathbf{R}T\kappa_T\{([3 + \exp(b - \ln Q(b))/b^3]\cdot[\kappa_T^{-1}(\partial \ln \varepsilon/\partial P)_T]) - 1\} \tag{8.9}$$

The integral $Q(b)$ is obtained from the experimental K_{ip} via a combination of equations 8.3 and 8.4 as

$$Q(b) = 2.8318\cdot 10^{-14}(T/\mathrm{K})^3\varepsilon^3 K_{ip} \tag{8.10}$$

and the corresponding value of b is then obtained by iterative numerical computation. At high $Q(b)$, the approximation $Q(b) \approx \exp(b)/b^4$ holds, and can be employed for the computation of b. The values of $Q(b)$ and b are inserted into equation 8.9, together with the indicated properties of the solvent, so that $\Delta\bar{V}^{\infty}_{ip}$ can be calculated. Representative values of this quantity are[6] 16 cm^3 mol^{-1} for the Li^+I^- ion pair in acetone at 30 °C, 32 cm^3 mol^{-1} for the $Cd^{2+}SO_4^{2-}$ ion pair in 21 mole% acetonitrile in water at 25 °C, and 44 $cm^3 mol^{-1}$ for the tetrabutylammonium picrate ion pair in 50 vol.% chlorobenzene + benzene at 25 °C (but see also Table 8.1 below).

However, for situations where close ion pairs are formed, i.e., those belonging to the solvent-separated, solvent-shared, and contact categories (i), (ii), and (III), respectively, the volume change on ion-pairing must be obtained from direct experimental determinations rather than from the above kind of calculation. The measurements are most commonly carried out by a combination of measurements of ultrasonic absorption and apparent molar volumes.[7] They are based on Eigen and Tamm's concept[8] that the ions sequentially pass through stages that correspond to the categories (i), (ii), and (iii) of ion pairs, ending up with the contact pair. For a 2:2 electrolyte in water, for example (magnesium sulphate has been treated in great detail[7]), the sequence of stages is:

$$M^{2+}(aq) + A^{2-}(aq) \overset{I}{\rightleftharpoons} [MWWA] \overset{II}{\rightleftharpoons} [MWA] \overset{III}{\rightleftharpoons} [MA] \qquad (8.11)$$

where W represents a shared water molecule. For the 2:2 aqueous electrolyte r_{min} is 1.40 nm at room temperature, and since $a + 4r_S$ is *ca.* 1.0 nm only, the Bjerrum equation applies to stage I in the sequence 8.11. The estimated[7] $K_{ip} = 50$ is in reasonable agreement with $K_{ip} = 59$ obtained[8] for $MgWWSO_4$. The rate of the forward reaction of stage I is assumed to be diffusion controlled, hence extremely fast. Steps II and III are unimolecular rearrangements that are slower (see Chapter 4), for which ultrasonic relaxation measurements provide the rate and equilibrium constants, the latter being $K_{II} = 2.0$ and $K_{III} = 0.17$. The overall association constant for magnesium sulphate is

$$K_{ass} = K_{ip}(1 + K_{II} + K_{II}K_{III}) = 59(1 + 2.0 + 0.34) = 197 \qquad (8.12)$$

in good agreement with K_{ass} obtained from conductivity measurements, since all the various kinds of ion pairs do not contribute to the net migration of the ions in an external field.

The total change of the partial molar volume of the electrolyte on the formation of the contact ion pair MA from the separate ions M^{2+} and A^{2-}, $\Delta\bar{V}^{\infty}$ at infinite dilution, is given by

$$\begin{aligned}\Delta\bar{V}^{\infty} &= \bar{V}(MA) - \bar{V}^{\infty}(M^{2+}(aq) + A^{2-}(aq)) \\ &= \lim_{c\to 0}\{{}^{\phi}V - \alpha\bar{V}[M^{2+}(aq) + A^{2-}(aq)]\}(1-\alpha)^{-1} \\ &\quad - \bar{V}[M^{2+}(aq) + A^{2-}(aq)] \end{aligned} \qquad (8.13)$$

where ${}^{\phi}V$ is the measured apparent molar volume at a given concentration c of the electrolyte and

$$\alpha = 1 - K_{ass}y_{\pm}^2 c\alpha^2 \qquad (8.14)$$

is the degree of dissociation of the electrolyte at the same concentration c, with $y_{\pm}$ being the mean molar activity coefficient (on the molar scale). The total $\Delta\bar{V}^{\infty}$ for aqueous magnesium sulphate is 7.8 ± 0.6 cm^3/mol, based on the additivity of the values of the standard partial molar volumes of the individual ions $Mg^{2+}(aq)$ and $SO_4{}^{2-}(aq)$ given in Table 5.8 for 25 °C.

The total change of the partial molar volume of the electrolyte on ion-pairing may be apportioned among the stages I, II, and III of the sequence of reactions 8.11 as follows[7]:

$$\Delta\bar{V}^{\infty} = \Delta\bar{V}^{\infty}_{ip} + f(\text{MA})[\Delta\bar{V}^{\infty}{}_{III} + \Delta\bar{V}^{\infty}_{II} + \Delta\bar{V}^{\infty}_{II}/K_{III}] \tag{8.15}$$

where the first term, $\Delta\bar{V}^{\infty}_{ip}$, pertains to the formation of the solvent-shared ion pair from the free ions, and $f(\text{MA})$ is the fraction that the contact ion pair forms among all kinds of ion pairs: $f(\text{MA}) = [\text{MA}]/c(1-\alpha)$ (remember that the elimination of water from the solvent-shared ion pairs to form ultimately the contact ion pairs is a sequence of unimolecular reactions that are independent of the concentration). The equilibrium constants obtained from the ultrasonic measurements[7] yield $f(\text{MA}) = 0.10$, and $\Delta\bar{V}^{\infty}_{ip}$ can be estimated from K_{ip} according to equation 8.9. Two unknowns are still left: $\Delta\bar{V}^{\infty}_{II}$ and $\Delta\bar{V}_{III}$, to be determined from equation 8.15, and for this purpose the estimate[8] $\Delta\bar{V}^{\infty}_{II}/\Delta\bar{V}^{\infty}_{III} = 5 \pm 1$ can be invoked. The results of this analysis for magnesium sulphate ion pairing in water at 25 °C are:

$\Delta\bar{V}^{\infty}_{ip} = 4.9$ cm³/mol (from ref. 9);
$\Delta\bar{V}^{\infty}_{II} = 3.5$ cm³/mol;
$\Delta\bar{V}^{\infty}_{III} = 0.7$ cm³/mol;
$\Delta\bar{V}^{\infty} = 7.4$ cm³/mol (from refs. 7 and 9).

The alternative estimate[8] that $\Delta\bar{V}^{\infty}_{ip} = 0$ yields four times larger values for $\Delta\bar{V}^{\infty}_{II}$ and $\Delta\bar{V}^{\infty}_{III}$, but no good reason has been provided for the assumption that ion-pairing to form the solvent-shared pair MWWA should not remove a considerable part of the electrostriction. In view of the standard partial molar volumes of the free ions given in Table 5.8, the following standard partial molar volumes of the various species discussed here are:

$\bar{V}^{\infty}(Mg^{2+}(aq) + SO_4{}^{2-}(aq)) = -7.2$ cm³/mol;
$\bar{V}^{\infty}(MgWWSO_4) = -2.3$ cm³/mol;
$\bar{V}^{\infty}(MgWSO_4) = 1.2$ cm³/mol;
$V^{\infty}(MgSO_4) = 1.9$ cm³/mol.

The intrinsic molar volume of $Mg^{2+} + SO_4{}^{2-}$ is *ca.* 25 cm³/mol (this is the molar volume of a hypothetical molten salt made up of the indicated ions at room temperature, estimated from that of the anhydrous crystal, 22.4 cm³/mol, and allowing for a 10% expansion on melting). Hence, all the ion-paired species retain a considerable amount of electrostriction. The change in volume for one mole of water that is completely electrostricted has been estimated[4] at −2.1 cm³/mol, so that (1.9–25)/(−2.1) = 11 molecules of water, on the average, are still completely electrostricted around the contact ion pair $MgSO_4$. When this was formed from the solvent-shared ion pair $MgWWSO_4$ (−2.3–1.9)/(−2.1) = 2 molecules of water were released from electrostriction, and *ca.* 2.3 molecules of water were released when this solvent-shared species was formed

Table 8.1 The change of the standard partial molar volume on ion-pairing and the number of solvent molecules released from electrostriction thereby (at 25 °C)

Ion pair	Solvent	$\Delta\bar{V}^{\infty}$/cm^3 mol^{-1}	Δn_{solv}	Ref.
$LiSO_4^-$	Water	5.8	3	a
$NaSO_4^-$		7.3 (8.3)	3–4	a (b, c)
KSO_4^-		5.9	3	a (b)
$NH_4SO_4^-$		3.4	1–2	a
$RbSO_4^-$		3.3	1–2	a
$CsSO_4^-$		6.2	3	a
$MgSO_4$		7.4 (9.0, 7.8)	3–4	j (d, u)
$CaSO_4$		11.7 (10.1)	5	d (u)
$MnSO_4$		7.4 (2 and 9)	3–4	j, e (i)
$CoSO_4$		10.9 (11.5)	5–6	e (d)
$NiSO_4$		11.4 (11.6)	5–6	e (d)
$CuSO_4$		11.3 (10.0)	5	e (q)
$ZnSO_4$		10.0 (8.0)	4–5	e (n)
$CdSO_4$		3.4 (20.6)		e (m)
$LaSO_4^+$		21 to 26	10–12	f (g)
$EuSO_4^+$		25.6	12	f
$RbNO_3$		6	3	h
$TlNO_3$		15	7	h
$MgCl^+$		4.0	2	a (b)
$LaFe(CN)_6$		8.0	4	p
LiCl	Methanol	18	1–2	k
LiBr		17	1–2	k
KCl		29	3	s
LiCl	Ethanol	17	1	k
	1-Propanol	16	1	k
	2-Propanol	21.8 (19)	1	l (k)
	1-Butanol	18	1	k

from the free ions. If only water in the primary hydration shells of the ions is considered to be electrostricted, then the volume change from the intrinsic to the standard partial molar volume corresponds to 15 electrostricted water molecules, i.e., 15 water molecules residing in the primary hydration shells of the two ions (cf. Tables 4.4a and 5.12), but this number is, of course, independent of the volume change on ion-pairing.

The aqueous magnesium sulphate ion-pairing system is discussed here in considerable detail, since it seems to be the most thoroughly studied system with respect to the volume changes of the individual steps of the sequence of reactions 8.11 and the changes in the numbers of solvent molecules affected by electrostriction. Table 8.1 presents data on the partial molar volume increase on ion-pairing in several aqueous and nonaqueous systems.

For an estimation of the number of solvent molecules released from the electrostriction by the fields of the separate ions on the formation of intimate ion pairs it is necessary to have a value for the molar volume of completely electrostricted solvent, or of the change in the molar volume of the solvent on

Table 8.1—*continued*

Ion pair	Solvent	$\Delta\bar{V}^{\infty}$/cm^3 mol^{-1}	Δn_{solv}	Ref.
LiCl	1-Hexanol	20	1	k
	1-Octanol	22	1	k
NaI	2-Propanol	15	1	t
	1-Methyl-2-propanol	15	1	t
LiI	Acetone	15.9	1	m
NaI		16.8	1	m
KI		17.7	1	m
CsI		20.6	1	m
$(C_4H_9)_4NPic$*	Benzene	62 (59)		r (m)
	50 vol% chlorobenzene in benzene	51 (44)		r (m)
	Diethyl ether	115		r

* Pic = picrate.
[a] F. H. Fisher and A. P. Fox, *J. Soln. Chem.*, **7**, 561 (1978); [b] F. H. Fisher and A. P. Fox, *J. Soln. Chem.*, **6**, 641 (1977); [c] F. H. Fisher and A. P. Fox, *J. Soln. Chem.*, **4**, 225 (1975); [d] F. H. Fisher and A. P. Fox, *J. Soln. Chem.*, **8**, 309 (1979); [e] A. LoSurdo and F. J. Millero, *J. Soln. Chem.*, **9**, 163 (1980); [f] C. F. Hale and F. H. Spedding, *J. Phys. Chem.*, **76**, 2925 (1972); [g] F. H. Fisher and D. F. Davis, *J. Phys. Chem.*, **71**, 819 (1967); [h] W. L. Masterton, H. Weller, J. H. Knox, and F. J. Millero, *J. Soln. Chem.*, **3**, 91 (1974); [i] L. G. Jackopin and E. Yeager, *J. Phys. Chem.*, **74**, 3766 (1970); [j] F. H. Fisher and D. F. Davis, *J. Phys. Chem.*, **69**, 2595; [k] Y. Marcus, N. Ben-Zwi, and I. Shiloh, *J. Soln. Chem.*, **5**, 87 (1976); [l] T. Noveske, J. Stuer, and D. F. Evans, *J. Soln. Chem.*, **1**, 93 (1972); [m] Y. Marcus, *Z. Naturforsch.*, **38a**, 247 (1983); [n] Y. Taniguchi, T. Watanabe, and K. Suzuki, *Bull. Chem. Soc. Japan*, **48**, 3032 (1975); [p] S. D. Hamann, P. J. Pearce, and W. Strauss, *J. Phys. Chem.*, **68**, 375 (1964); [q] A. Dadger, D. Khorsandi, and G. Atkinson, *J. Phys. Chem.*, **86**, 3829 (1982); [r] J. Everaert and A. Persoons, *J. Phys. Chem.*, **86**, 546 (1982); [s] E. Grunwald and C. D. Brown, *J. Phys. Chem.*, **86**, 182 (1982); [t] T. Okuwa, *Acustica*, **44**, 71 (1981), T. Okuwa and K. Ohno, *Bull. Chem. Soc. Japan*, **54**, 3648 (1981); [u] A. K. Hsieh, K. P. Ang, and M. Chang, *J. Chem. Soc. Faraday Trans. 1*, **78**, 2455 (1982).

complete electrostriction. The estimate given above for the solvent water,[4] -2.1 cm^3 mol^{-1}, is based on the following expression:

$$\Delta V_{els} = -V^0\kappa^0 S_V/S_\kappa \tag{8.16}$$

where V^0 and κ^0 are the molar volume and the isothermal compressibility of the pure solvent, and S_V/S_κ is the ratio of the limiting slopes according to the Debye–Hückel theory for the molar volume and the compressibility of electrolyte solutions (i.e., the coefficients of the term in $c^{1/2}$). The numerator of this ratio involves terms in κ^0 and $(\partial \ln \varepsilon/\partial P)_T$ only, whereas the denominator involves in addition terms in the pressure derivatives of these quantities, which are in general not well known. For water, however, equation 8.16 is adequately approximated by[4,10]

$$\Delta V_{els} = -2V^0[3\kappa^{0-1}(\partial \ln \varepsilon/\partial P)_T - 1]/15[\kappa^{0-1}(\partial \ln \varepsilon/\partial P)_T]^2 \tag{8.17}$$

because of the mutual cancellation of the terms in the pressure derivatives mentioned above. If this approximation is assumed to hold also for other solvents, then the following values of ΔV_{els} are obtained from equation 8.17 and data in ref. 6 (in $cm^3\ mol^{-1}$):

methanol −11*, ethanol −17, propanol (both 1- and 2-) −21, 1-butanol −27, 1-hexanol −37, 1-octanol −47, acetone −17.

Division of $\Delta \bar{V}^{\infty}$ on ion pairing by these quantities yields the average number of solvent molecules released from electrostriction on the formation of intimate ion pairs, see Table 8.1.

The formation of an ion pair from two separate ions is expected, on first thought, to decrease the entropy of the system on account of the diminution of the number of solute particles. However, the release of solvent molecules from the solvation shells that overlap, particularly when contact ion pairs are formed, actually increases the number of particles in the system. In fact, the standard molar entropy change on the formation of an inner-sphere or contact ion pair is positive. From the observed entropy change for this process, terms pertaining to the loss of translational entropy and the long-range electrostatic contribution to the entropy should be subtracted, to obtain the entropy change of desolvation. This is divisable by the entropy of the immobilization of solvent molecules in the first solvation shell of ions to yield the number of solvent molecules released from the overlapping solvation shells.

The molar translational entropy loss for two ions forming one ion pair is

$$\Delta S_{tr} = \mathbf{R} \ln(4\pi a^3/3 \cdot 10^{-3})$$
$$= -447.6 + 24.9 \ln(a/\text{nm})\ \text{J K}^{-1}\ \text{mol}^{-1} \quad (8.18)$$

This arises from the confinement of one of the partners of the ion pair to a sphere of radius a, the contact distance, around the other partner, rather than to the whole solution volume ($10^{-3}\ m^3$ for the molar standard state used in obtaining the observed entropy change). The long-range electrostatic entropy effect is

$$\Delta S_{el} = -|z_+z_-|(\mathbf{e}^2\mathbf{N}/\boldsymbol{\varepsilon}_0)a^{-1}\varepsilon^{-1}(\text{d} \ln \varepsilon/\text{d}T) \quad (8.19)$$

where a is the contact distance of the unsolvated ions. The molar entropy change for the solvent immobilized in the primary solvation shell of an ion is approximated by the molar entropy of freezing, $\Delta S^F(T)$, extrapolated from the freezing point of the solvent, T_m, to the pertinent temperature T, see p. 80. The values of $\Delta S^F(25\ °C)$ are: water 25, methanol 15, ethanol 39, 1-propanol 46, acetone 56, propylene carbonate 17, formamide 23, N,N-dimethylformamide 76, acetonitrile 29, and dimethylsulphoxide 48, in $J\ K^{-1}\ mol^{-1}$. These values are certain to within ±10%, except the one for DMF, where the uncertainty is ±20%, due to the unavailability of data for the extrapolation from T_m to 25 °C, which had therefore to be estimated.

A positive entropy change on complex formation has long been used as a diagnostic tool to recognize the formation of inner-sphere complexes. This has

* J. Padova, *J. Chem. Phys.*, **56**, 1606 (1972), estimated $-7.5\ cm^3\ mol^{-1}$ for this quantity, but did not disclose the data on which his estimate was based.

been extended to ion-pairing, and of the many data that have been reported on this subject, only a few typical cases can be discussed here.

The average entropy of ion-pairing of divalent metal sulphates in water at 25 °C is 67 ± 7 J K^{-1} mol^{-1}, $MgSO_4$ causing the least and $CdSO_4$ the greatest change within this range.[11] This has been qualitatively interpreted as signifying the greater tendency of the larger Cd^{2+} ion to form contact ion pairs compared to the smaller Mg^{2+} ion that tends to keep its primary water of hydration more tenaciously. For a quantitative interpretation the terms for translational and electrostatic entropy changes, equations 8.18 and 8.19, must be subtracted, but to a certain extent these cancel each other out. The entropy changes ascribable to the release of the water molecules from the overlap region are 87 and 151 J K^{-1} for the $MgSO_4$ and $CdSO_4$ inner-sphere ion pairs. These numbers correspond, in turn, to the release of 3.5 and 6.0 molecules of water, respectively. It can be surmised that in both cases 3.5 of these water molecules come from the first hydration shell of the large sulphate ion, and that an extra 2.5 molecules of water, on the average, come from that of the cadmium ion in the latter case. The value obtained for $MgSO_4$ from the entropy change is in good agreement with that obtained from the volume change (Table 8.1). Since the values for the other divalent metal sulphates are in between the limits of 3.5 (for $MgSO_4$) and 6.0 (for $CdSO_4$), they are also in agreement with those in Table 8.1, where, however, discrepant values for $CdSO_4$ itself result from the discrepant volume changes.

Association to contact ion pairs in a protic nonaqueous solvent is illustrated by the data for several 1:1 electrolytes in 1-propanol.[12] The observed ΔS^0 values range between 105 and 122 J K^{-1} mol^{-1} for $NaClO_4$ and NaBr, respectively, those of ΔS_{el} range from 545 to 682 J K^{-1} mol^{-1} for RbI and NaBr, respectively, and those for ΔS_{tr} from -478 to -473 J K^{-1} mol^{-1} for NaBr and RbI, respectively. With $\Delta S^F(25\ °C) = 46$ J K^{-1} mol^{-1} for 1-propanol, the resulting numbers of solvent molecules set free from the overlapping regions of the solvation shells of the cation and the anion are 0 for NaBr, NaI, and $NaClO_4$, 0.5 for KI, and 0.8 for RbI. The total solvation numbers (Table 4.6) are in the range from 3.3 to 2.0 for these salts. It is, therefore, seen, that the ion pairs in 1-propanol retain a considerable amount of solvent around the dipolar species formed.

For the aprotic solvent acetone the data for the association of NaI and $NaClO_4$ yield the observed entropy changes of 48 and 27 J K^{-1} mol^{-1}, respectively.[13] Allowance for long range electrostatic effects and translational entropy loss on ion-pairing yields for the entropy of solvent release 105 and 113 J K^{-1} mol^{-1}, corresponding to 1.9 and 2.0 molecules of acetone released on the formation of the contact ion pairs NaI and $NaClO_4$, respectively. The solvation numbers of these salts, calculated according to equation 4.18, are 4.9 and 6.5, respectively, and it is seen that about one-third of the solvent is lost on the formation of the contact ion pairs.

The enthalpy change for ion pair formation can be interpreted in a similar fashion in terms of the differences between the enthalpies of solvation of the contact ion pair and of the ions.[14]

In addition to the electrostatic model for ion pairing that is applicable under conditions where the solvent may be regarded as a dielectric continuum, and the models that are applicable for ions in contact or sharing one or two solvent molecules, there exists a third kind of ion pairing. This is confined to ions in highly structured solvents i.e., practically to water, and to large, hydrophobic ions. This kind of ion association was called 'water-structer-enforced ion pairing'[15] and was inferred to occur not only for tetraalkylammonium bromides and iodides (possibly also chlorides[16]), but also for large alkali metal (Rb^+, Cs^+) perchlorates and other salts with large univalent anions.[15] However, this is just one example of the more general phenomenon of hydrophobic (attractive) interactions, that takes place also between uncharged particles.[17] In the case of ions, and in particular if they have long aliphatic chains (e.g., decyltrimethylammonium carboxylates, from butanoate to tetradecanoate[18]), hydrophobic association is superimposed on the electrostatic interactions. This hydrophobic interaction does differ in some way from that of aliphatic hydrocarbons in pure water, since the charges modify the water stucture in their vicinity; see Figure 5.5. As the chain becomes longer, this perturbation of the structure by the charges becomes less pronounced, and a constant increment in the Gibbs free energy of association per $-CH_2-$ group results.[17,18]

The incompatibility of the structure of the solvation environment of various parts of an ion pair is not confined to ion pairs formed by hydrophobic ions in water. The positive end of the dipole of a polar solvent molecule points towards the anion, and if the solvent is protic, then a hydrogen bond between it and the anion is formed. The negative end of the dipole points towards the cation part of the ion pair. Therefore, in the equatorial plane perpendicular to the line between the cation and the anion the orientation of the solvent molecules is undecided. This causes an increase in the partial molar volume of the solvated ion pair, over the volume that the remnant electrostriction would have caused. This increase should be larger, the bulkier the solvent, and this was indeed found for alcohols solvating lithium chloride or bromide ion pairs.[19]

8.2 Competition of solvent and counter ion

Several experimental methods, in particular Raman and NMR spectrometry and pressure-jump relaxation measurements, are sensitive to the transformation of an outer-sphere, solvent-shared, ion pair to an inner-sphere, conact, ion pair. Much use has been made of these methods in recent years in the study of this reversible transformation in aqueous, nonaqueous, and mixed solvents. The competition of the perchlorate, nitrate, or sulphate anions with a solvent molecule for a site in the first solvation shell of a variety of cations has been the main subject of research. Table 8.2 summarizes the systems where information of a quantitative nature has been obtained.

The Raman spectrum of an oxyanion is sensitive to the electrical field in which the anion is situated. Highly symmetrical oxyanions have degenerate vibrational modes in a spherical symmetric field or in the 'free' state. This degeneracy is

Table 8.2 Studies of anion/solvent competition for a site in the first solvation shell of various cations

Cation	ClO_4^-	NO_3^-	SO_4^{2-}	Other anions
Li^+	NMR: $MeNO_2$, THF Me_2CO, MeCN[a] Raman, NMR: DMF[i] NMR, IR: MeCN[j] NMR Raman, IR: Me_2CO, Et_2O[o] NMR: EtOAc, 1-PrOH, 1-BuOH[l]	Raman: H_2O[m]		NMR: $I^-/MeNO_2$; Br^-, I^-/PC, DMSO[c] Raman: Cl^-/DMF[i] NMR, IR: F^-, Cl^-/MeCN[j] IR: Cl^-, Br^-/Me_2CO[k]
Na^+	Raman: H_2O[g] NMR: EtOAc, 1-PrOH, 1-BuOH[l]	Raman: H_2O[g,m]		NMR: Br^-, I^-, SCN^-/THF, Me_2CO, TMS, DMF, TMU[c]
K^+, Rb^+		Raman: H_2O[m]		
NH_4^+			Raman: H_2O[h]	Raman: HCO_2^-, $CH_3CO_2^-/H_2O$[v]
Ag^+		Raman, IR: H_2O[q] IR: MeCN[b]		
Tl^+		Raman: H_2O[n]		
Be^{2+}		Raman: H_2O[p]	Press. jump: H_2O + FA[d], +DMSO[e], +HMPT[f]	
Mg^{2+}	NMR: EtOAc, 1-PrOH, 1-BuOH[l]			
Ca^{2+}, Sr^{2+}, Ba^{2+}		Raman: H_2O[p]		
Zn^{2+}, Cd^{2+}		Raman: H_2O[n]		
Hg^{2+}		Raman: MeCN[r] Raman, IR: H_2O[s]		
Al^{3+}		Raman: H_2O[n,p]		
In^{3+}				Raman: Cl^-, Br^-/H_2O[t,u]
La^{3+}, Th^{4+}		Raman: H_2O[n]		

[a] Y. M. Cahen, P. R. Handy, E. T. Roach, and A. I. Popov, *J. Phys. Chem.*, **79**, 80 (1975); [b] G. J. Janz, M. J. Tait, and J. Meier, *J. Phys. Chem.*, **71**, 963 (1967), C. B. Baddiel, M. J. Tait, and G. J. Janz, *ibid.*, **69**, 3634 (1965); [c] M. S. Greenberg, R. L. Bodner, and A. I. Popov, *J. Phys. Chem.*, **77**, 2449 (1973); [d] R. Lachmann, I. Wagner, D. H. Devia, and H. Strehlow, *Ber. Bunsenges. Phys. Chem.*, **82**, 492 (1978); [e] D. H. Devia and H. Strehlow, *Ber. Bunsenges. Phys. Chem.*, **83**, 627 (1979); [f] H. Strehlow, D. H. Devia, S. Dagnall, and G. Busse, *Ber. Bunsenges. Phys. Chem.*, **85**, 281 (1981); [g] R. L. Frost, D. W. James, R. Appleby, and R. E. Mayes, *J. Phys. Chem.*, **86**, 3840 (1982); [h] D. E. Irish and H. Chen, *J. Phys. Chem.*, **74**, 3797 (1970), H. Chen and D. E. Irish, *ibid.*, **75**, 2672, 2681 (1971); [i] D. W. James and R. E. Mayes, *J. Phys. Chem.*, **88**, 637 (1984); [j] J. F. Coetzee and W. R. Sharpe, *J. Soln. Chem.*, **1**, 77 (1972); [k] M. K. Wong, W. J. McKinney, and A. I. Popov, *J. Phys. Chem.*, **75**, 56 (1971); [l] H. A. Berman and T. R. Stengle, *J. Phys. Chem.*, **79**, 1001 (1975); [m] R. L. Frost and D. W. James, *J. Chem. Soc. Faraday Trans. 1*, **78**, 3223, 3235, 3249 (1982); [n] R. L. Frost and D. W. James, *J. Chem. Soc. Faraday Trans. 1*, **78**, 3263 (1982); [o] D. W. James and R. E. Mayes, *Austr. J. Chem.*, **35**, 1775, 1785 (1982); [p] D. W. James and R. L. Frost, *Austr. J. Chem.*, **35**, 1793 (1982); [q] T. C. G. Chang and D. E. Irish, *J. Soln. Chem.*, **3**, 175 (1974); [r] C. C. Addison, D. W. Amos, and D. Sutton, *J. Chem. Soc.*, **A1968**, 2285; [s] A. R. Davis and D. E. Irish, *Inorg. Chem.*, **7**, 1699 (1968); [t] M. P. Hanson and R. A. Plane, *Inorg. Chem.*, **8**, 746 (1969); [u] T. Jarv. J. T. Bulmer, and D. E. Irish, *J. Phys. Chem.*, **81**, 649 (1977); [v] L. A. Blatz and P. Waldstein, *J. Phys. Chem.*, **72**, 2614 (1968).

removed in the field produced by a cation in the immediate vicinity of the oxyanion, so that the spectral line is seen to be split. This has served for the identification of contact ion pairs, since in a solvent-shared or a solvent-separated ion pair, the polarization of the solvent produced by the cation is insufficient to cause the splitting of a degenerate spectral line of the oxyanion on the other side of the solvent molecule from the cation.[20] This is not always the case: in dilute alkali metal nitrate solution, where even outer-sphere ion pairs are absent, one of the nitrate Raman lines that is degenerate in the 'free' state of the anion is split, apparently by the water of hydration of the nitrate anion.[21]

More recent work by James and co-workers[22,23] has shown that the perturbation of the band profile of non-degenerate Raman lines, such as those corresponding to the symmetric stretching of the perchlorate or nitrate anions, can also be interpreted in terms of the anion replacing the solvent in the first solvation shell of a cation. In fact, a line corresponding to the chlorine–oxygen (in perchlorate) or nitrogen–oxygen (in nitrate) bond stretching of the anion in the contact ion pair can be resolved from the $\nu_1(A_1)$ band envelope. From its relative intensity the extent of the formation of the contact ion pair can be calculated.[22] The $\nu_1(A_1)$ frequencies of the 'free' (hydrated) perchlorate and nitrate anions are at 933.5 and 1047.6 cm^{-1}, respectively, whereas those of the contact-ion-paired anions are at 943.5 and 1052 cm^{-1}, respectively.[22,23]

Raman, infrared, and NMR spectrometry can be applied to the solvent molecules in solutions where ion-pairing occurs, whether the anion has characteristic vibrations, as in ClO_4^- or SCN^-, or not, as in Cl^- or I^-. The C–H stretching frequencies of the methyl group of acetonitrile depend on the salt concentration and the nature of the anion for a given cation, such as Li^+, Na^+, or $(C_2H_5)_4N^+$. For anions such as F^- and Cl^- with a propensity of forming contact ion pairs in an aprotic solvent that is a poor anion solvator, the frequency shift is interpreted in terms of the solvated pair

$$H_3C\text{–}CN \cdots M^+X^- \cdots H_3C\text{–}CN$$

Only this configuration can explain the sensitivity of the C–H stretching frequency to the nature of the anion.[24] In N,N-dimethylformamide, another aprotic solvent, separate lines are observed for the O–C–N deformation if the solvent is 'free' (659 cm^{-1}) or bound to Li^+ (666 and 673 cm^{-1}). The relative peak areas yield the solvation number, that decreases on the formation of contact ion pairs, at >2 M for ClO_4^- and at lower concentrations for Cl^-.[23] The association to a contact ion pair naturally depends on the ease of removing a solvent molecule from the solvation shell of the cation, and for $Li^+ClO_4^-$ it is in the order: $H_2O >$ DMF $>$ acetone.[23]

NMR spectrometry of alkali metal cations (e.g., $^7Li^+$ or $^{23}Na^+$) and of anions (e.g., $^{35}Cl^-$ or $^{35}ClO_4^-$) as studied by Popov *et al.*[25] has yielded important insight into the relative ease of the replacement of solvent molecules from the first solvation shell of a cation by an anion forming a contact ion pair with it. A combination of a low dielectric constant ε of the solvent, that promotes the formation of (outer-sphere) ion pairs in general, and a low donor number *DN* of the solvent, that causes only weak solvation of the cation, favours the formation of contact ion pairs. Their formation results in cation chemical shifts that are concentration and anion dependent.[25] For lithium salts this is evident at values of the product $\varepsilon \cdot DN < 700$, e.g., in nitromethane, acetone, tetrahydrofuran, and acetonitrile. In propylene carbonate and dimethylsulphoxide, concentration dependence is seen only for some anions but not with others, whereas for solvents with $\varepsilon \cdot DN > 800$, such as N,N-dimethylformamide and methanol, no concentration and anion dependence of the chemical shifts of the cations is discerned, and no contact ion pairs are formed. For sodium salts the border between contact ion pairing and non-formation of these is nearer $\varepsilon \cdot DN \sim 1000$.

Short relaxation times of the partners in a contact ion pair, producing line broadening of the NMR signal from nuclei having a nuclear quadrupole moment, can also serve as a diagnostic tool. This has been observed in particular for the ^{35}Cl NMR line in nonaqueous solutions of perchlorates.[23,25] In an aqueous solution a similar effect has been observed in concentrated solutions of the hydrohalic acids, and in the case of ^{81}Br, for instance, has been ascribed to the formation of the $H_3O^+Br^-$ ion pair.[26]

The separate ^1H NMR signals from the protons of 'free' water molecules and those bound to cations, in solutions containing a salt and water in acetone cooled to very low temperatures to slow down their exchange, have been used for the determination of hydration numbers, as described in Chapter 4. In the absence of contact ion-pairing in these solutions, the peak area for the signal from the water in the primary hydration shell of the cation is that expected from structure determination methods, see Table 4.4. Such an agreement is generally obtained for perchlorate salts, provided there is sufficient water in the solution to completely fill the primary hydration requirements of the cation. For other anions, however, such as chloride or nitrate, contact ion-pairing is seen to occur from the reduction of the area of the signal from the primary hydration water molecules. For instance, in the case of the uranyl(VI) cation, $UO_2(OH_2)_4{}^{2+}$ is present in perchlorate solutions, but a nitrate anion displaces two water molecules and associates in a bidentate fashion, with two of its oxygen atoms in contact with the uranium cation:

$$[ONO_2UO_2(OH_2)_2]^+$$

Chloride and bromide anions also displace water molecules from the uranyl cation, forming contact ion pairs in these low temperature (-85 °C) solutions in acetone.[27] Such observations, made by Fratiello and co-workers,[28] are quite general and pertain to many metal ion–anion systems studied in acetone at low temperatures, provided that sufficient water is present to prevent the acetone, although a poor donor solvent, from co-solvating the cation.

How the anion replaces the solvent molecule in the primary solvation shell of a cation on the formation of a contact (inner-sphere) ion pair is not generally known. In the particular case of the beryllium sulphate ion pairs, however, the reaction path has been determined by Strehlow and co-workers,[29] using the pressure jump relaxation method. Beryllium is such a small ion (its crystal ionic radius is only 0.031 nm), that it fits into the tetrahedral hole between the oxygen atoms of four water molecules arranged tetrahedrally around it. It will 'rattle', however, if these oxygen atoms belong, e.g., to four dimethylsulphoxide molecules. If a mixed solvate is formed with at least one water molecule and three other solvent molecules, the beryllium cation will be next to the water molecule.

When Be^{2+} is solvated by water, an outer-sphere ion pair with the sulphate

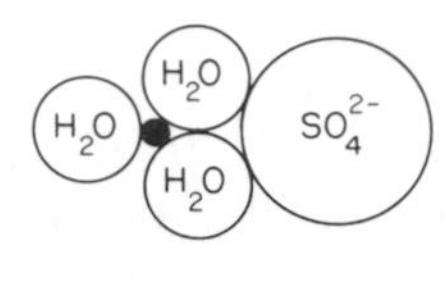

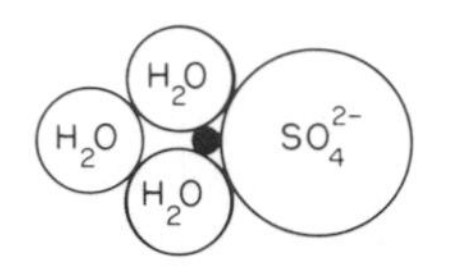

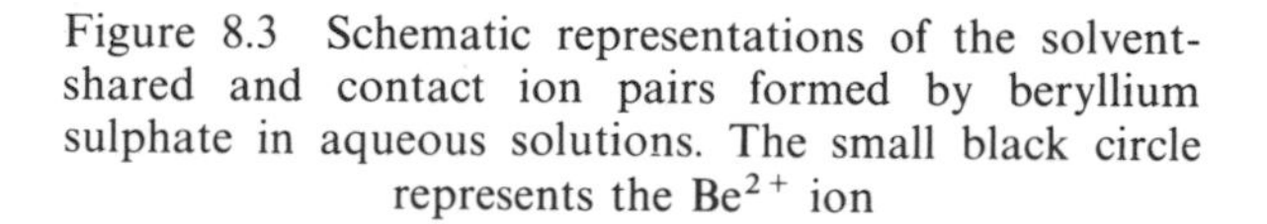

Figure 8.3 Schematic representations of the solvent-shared and contact ion pairs formed by beryllium sulphate in aqueous solutions. The small black circle represents the Be^{2+} ion

anion will have a triangle of water molecules between the cation and the anion (Figure 8.3). The electrostatic attraction between them is able to push this triangle slightly open, and cause the small Be^{2+} ion to pass through this opening into the tetrahedral hole formed by the three water molecules and one oxygen atom of the sulphate anion, and find its place next to the latter. The water molecule furthest from the sulphate anion is subsequently released. The same reaction path is followed, whether the three solvent molecules in the solvent-shared ion pair are water or formamide, since both solvents are protic, and solvate both the beryllium cation, through the lone pairs on their oxygen atoms, and the sulphate anion, through hydrogen bonds. Different relaxation rates result only on the release of the fourth solvent molecule, the one that is opposite the anion. Thus in pure water and in pure formamide solutions only one relaxation rate is observed, but in mixtures of the two solvents there are two different rates, depending on the release of a water and of a formamide molecule in the final step of the formation of the inner-sphere ion pair. In aqueous aprotic solvent mixtures, on the other hand, where the latter solvent is incapable of solvating the anion, several more relaxation steps are observed. Each of the outer-sphere ion pairs $BeS_i(OH_2)_{4-i}SO_4$, where S is the aprotic solvent (dimethylsulphoxide and hexamethylphosphoric triamide have been studied) has a different relaxation rate of releasing the fourth solvent molecule, provided that as long as $i < 4$, the water molecules present are nearest the sulphate anion, solvating it through hydrogen bonds. This fact limits the number of possible combinations of i values and releaseable solvent molecules.[29]

8.3 Molten hydrated salts

An extreme case of competition between solvation and cation–anion association arises at very high concentrations of electrolytes, such as are found in aqueous melts or molten hydrated salts. As the concentration of a salt increases, the solvation spheres of the ions approach each other, until finally a concentration is reached at which they are contiguous and beyond that concentration they must overlap. In aqueous solutions, the 'hydrated radii' of ions (see Section 5.2) range from 0.20 nm to 0.36 nm (except for highly charged, well hydrated ions that are larger: 0.46 nm for Fe^{3+}, 0.36 nm for $SO_4{}^{2-}$). The mean distance apart of the ions in a 1 M (mol dm^{-3}) uni-univalent electrolyte solution is 0.94 nm. In a 3.1 M solution the distance is only 0.64 nm, i.e., equal to the diameter of the typical larger hydrated ions; in a 13 M solution it equals the diameter of the smallest hydrated ions, 0.40 nm. Solutions containing such high concentrations of ions are encountered both in nature and in industrial and laboratory practice.

The case of brines containing the chlorides of sodium, potassium, magnesium, and calcium is a good example that has been discussed in these terms.[30]. The water of the Dead Sea contains 8.8 mol ions dm^{-3}, i.e., the ions are at a mean distance of 0.57 nm apart. Such a brine contains 28% 'free' water, calculated on the basis of the hydration numbers (primary and secondary) valid at infinite dilution. The alkali metal ions 'see' around them this free water, which separates them from the hydrated chloride ions, but the hydration shells of the divalent cations are already slightly overlapping with those of the chloride anions. In an industrial process based on the Dead Sea water, an even more concentrated so-called 'end brine' is produced, that contains 15.0 mols ions dm^{-3}, and the mean distance apart of the ions is 0.48 nm only. The solution has a deficit of −31% 'free' water, and as a consequence the hydrated alkali metal cations are in contact with the hydrated chloride anions, whereas the latter overlap with the secondary hydration shell of the magnesium ions and even penetrate the primary hydration shell of the calcium ions, see Figure 8.4. Solvent-shared ion pairing is, thus, forced on the divalent cations by the mere high concentration, with no consideration of any electrostatic interaction (which, of course, would only enhance the ion association). The conclusions from this geometrical consideration are confirmed by X-ray diffraction results. For the purpose of comparison, 5 m (mol/kg water) solutions, in which the ions are approximately 0.5 nm apart, are examined. Solutions of LiCl and NaCl do not indicate any ion–ion contacts, but CsCl solutions do. In $MgCl_2$ and $CaCl_2$ solutions an 'irregular cation hydration' has been found, that points to nonisotropic location of the water molecules and suggests some kind of ion pairing.[30]

On the other hand, extremely concentrated aqueous solutions can be considered as if they were molten salts, having hydrated ions (cations) as their constituents. The latter viewpoint permits the greater insight into the specific phenomena observable in these systems, and is supported by concrete evidence, though not for all cases so considered. This model has been proposed by Angel[31] and subsequently elaborated by Braunstein.[32] Molten hydrated salts are limited to salts of cations with a high electrostatic field strength, z/r^2, at water contents not exceeding the filling of the primary hydration shell, but also salts of low melting point (mainly salt eutectics) with some dissolved water. Molten hexa- or tetrahydrates of the chlorides and nitrates of magnesium, calcium, and zinc are typical examples, whereas molten ammonium nitrate with up to three moles of water per mole of salt is a legitimate but atypical representative of this class of systems.

If considered as molten salts, molten hydrated salts can be treated according to the quasi-lattice model, as depicted in Figure 8.5. The model involves two

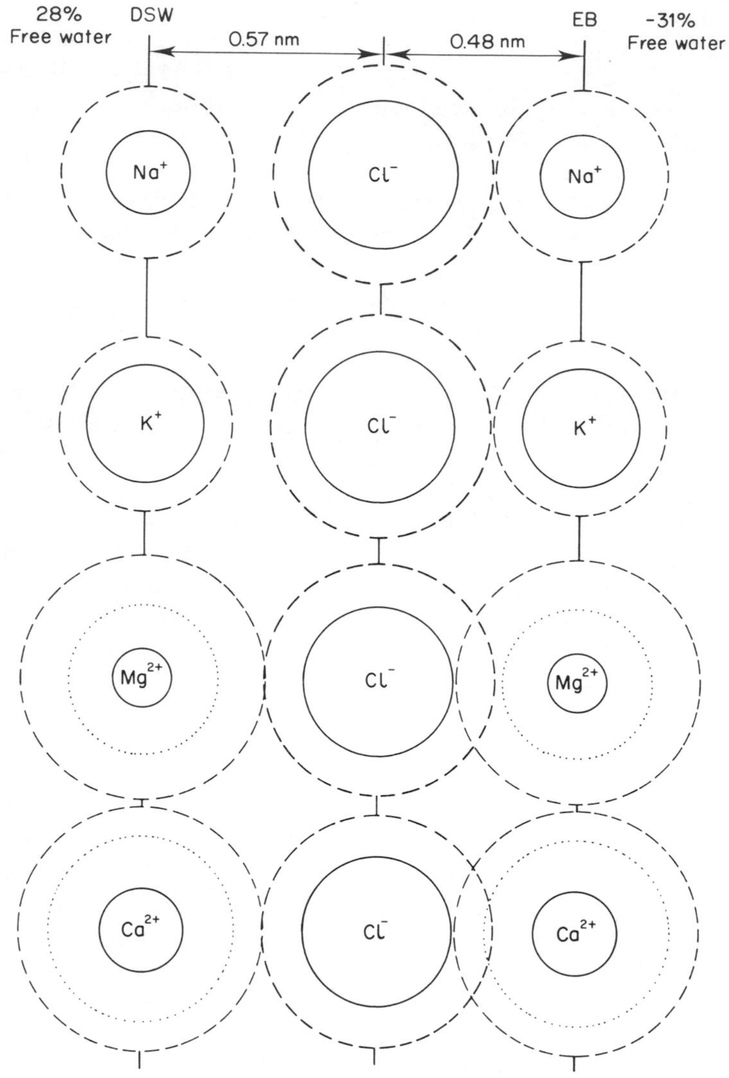

Figure 8.4 The ions in Dead Sea water (DSW, left-hand side) and in the end brine (EB, right-hand side) at the mean distances apart of the cations from the anions. The primary (dotted circles) and secondary (dashed circles) hydration shells are indicated. (From Ref. 30, by permission of Plenum Publishing Corp.)

interweaving sub-lattices of the cationic and the anionic sites, the former being occupied by the cations and the latter by the anions and the water molecules. The anions and the water molecules are therefore next-nearest neighbours to each other, and the neighbours of a cation are either anions or water molecules or both. The quasi-lattice is characterized by its coordination number Z. The

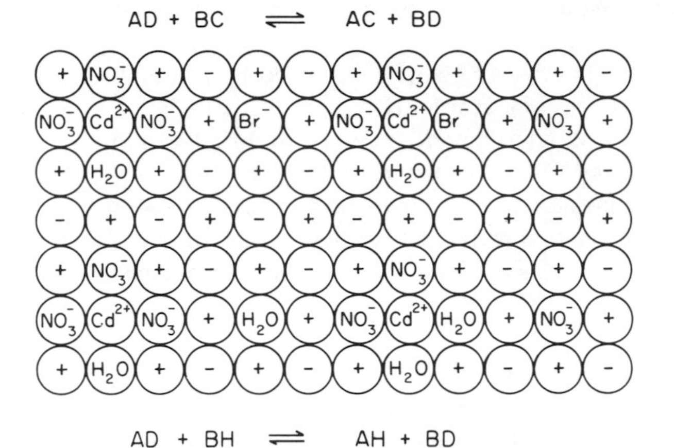

Figure 8.5 A two-dimensional representation of a hydrated molten salt solution as a quasi-lattice. The solvent salt BD (D = NO_3^-) is indicated by the circles marked with + and −. The solutes are the ions A ($=\frac{1}{2}Cd^{2+}$) and C ($=Br^-$) and the neutral species H ($=H_2O$) that occupies anionic sites in the quasi-lattice. In the upper and lower portions the exchanges of a D neighbour of an A particle for a C and H neighbour, respectively, are depicted. These exchanges involve the energies shown on the right hand side of the diagram. (After J. Braunstein, in 'Ionic Interactions', S. Petrucci, ed., Academic Press, New York, Vol. I (1971).)

model has been applied to situations where a solvent salt B^+D^- (where B^+ is typically $(Li, K)^+$ at the eutectic composition, NH_4^+, or $\frac{1}{2}Ca^{2+}$ and D^- is invariably NO_3^-) with a water (symbolized by H) content of R_W moles per mole salt contains a solute salt A^+C^- (where A^+ is Ag^+, $\frac{1}{2}Cd^{2+}$, or $\frac{1}{2}Pb^2$ and C^- is Cl^- or Br^-). The Z sites around the cation A^+ are occupied by D^- and H, but when R_W is small, then only one H per A^+ will be found. Either of the two occupants of the anionic sites can be replaced by C^-, the exchange energies being e_C and $e_C - e_H$, respectively, e_H being the energy for the replacement of D^- by H at a site near A^+, see Figure 8.5. These energies are considered to be independent of the nature of the rest of the occupants of the sites around the A^+ cations (or the B^+ cations, for that matter). EMF measurements on a cell with electrodes reversible to C^- in the presence and the absence of A^+ give the difference ΔE, from which the association quotient $Q_1 = (\mathbf{F}/\mathbf{R}T)(\partial \Delta E/\partial R_A)_{R_W}$ is obtained, where R_A is the mole ratio of A^+ to the solvent salt B^+D^-. Extrapolation of the association quotient to zero concentrations of A^+ and C^- at a series of values of R_W gives values of the association constant K_1 as a function of the latter quantity:

$$K_1 = \lim_{R_A, R_C \to 0} Q_1 = Z \exp(-e_C/\mathbf{k}T)/(1 + R_W \exp(-e_H/\mathbf{k}T)) \tag{8.20}$$

From the dependence of K_1 on R_W the values of the two energies e_C and e_H are derived.[32] Some results obtained from this approach are presented in Table 8.3.

Not in all cases are resonable values obtained by this analysis. In fact, if the affinity of B^+ to H is comparable to that of A^+ a negative intercept may be obtained in plots of Z/K_1 versus R_W, corresponding to an impossible negative value of $\exp(-e_C/\mathbf{k}T)$. This is the case for the $Zn^{2+} - Cl^-$ association in aqueous ammonium nitrate melts (see Table 8.3), and also in association systems in aqueous calcium nitrate melts at high R_W. A modified approach is then necessary, where the activity of the water in the aqueous melt, a_W, is explicitly taken into account:

$$K_1 = K_{10}/(1 + K_H a_W^n) \tag{8.21}$$

where K_{10} is the equlibrium constant for the association in the water-free molten salt solvent, K_H is the equilibrium constant for the hydration, and n is essentially a fitting parameter playing the role of the hydration number.[33] It should be noted that the latter equilibrium constants are not formulated for the exchange of the nitrate anions for the pertinent anion C^- or water molecule H, and the Gibbs free energy values obtainable from them are not comparable with

Table 8.3 Exchange energies for the replacement of a solvent anion D^- by a halide one, C^-, or by water, H, in molten salt hydrates

Solvent ions		Solute ions				
B^+	D^-	A^+	C^-	N Δe_C (kJ mol^{-1})	N Δe_H (kJ mol^{-1})	Ref.
NH_4^+	NO_3^-	Ag^+	Cl^-	−18.5	0 ± 2	a, b
				−20.5	2 ± 2	j
		Ag^+	Br^-	−26.4	0 ± 2	a, b
NH_4^+	NO_3^-	Zn^{2+}	Cl^-	−24.2	−41.8	c
NH_4^+	NO_3^-	Cd^{2+}	Cl^-	−15.1	−2.9	d
				−15.5	−0.4	e
$\frac{1}{2}Ca^{2+}$,	NO_3^-			−18.8	6.4	f
$\frac{1}{2}Ca^{2+}$, K^+	NO_3^-			−19.3	5.6	f
NH_4^+	NO_3^-		Br^-	−16.7	0.0	e
Li^+, K^+	NO_3^-			−26.5	−5.4	g
$\frac{1}{2}Ca^{2+}$	NO_3^-			−22.2	(−10.5)[i]	h
NH_4^+	NO_3^-	Hg^{2+}	Cl^-	−34.7	−3.8	c
NH_4^+	NO_3^-	Pb^{2+}	Cl^-	−10.5	1.7	e
			Br^-	−10.9	2.1	e

[a] J. Braunstein and H. Braunstein, *Inorg. Chem.*, **8**, 1558 (1969); [b] I. J. Gal, *Inorg. Chem.*, **7**, 1611 (1968); [c] R. M. Nicolic and I. J. Gal, *J. Chem. Soc. Dalton Trans.*, **1974**, 985; [d] J. M. C. Hess, J. Braunstein, and H. Braunstein, *J. Inorg. Nucl. Chem.*, **26**, 811 (1964); [e] R. M. Nicolic and I. J. Gal, *J. Chem. Soc. Dalton Trans.*, **1972**, 162; [f] I. J. Gal, R. M. Nicolic, and G. Herankovic, *J. Chem. Soc. Dalton Trans.*, **1976**, 104; [g] J. Braunstein, *J. Phys. Chem.*, **71**, 3402 (1967); [h] J. Braunstein, A. Alvarez-Funes, and H. Braunstein, *J. Phys. Chem.*, **70**, 2734 (1966); [i] H. Braunstein, J. Braunstein, and P. Hardesty, *J. Phys. Chem.*, **77**, 1907 (1973), the value for water not being the energy but the Gibbs free energy change, see the text; [j] M. Peleg, *J. Phys. Chem.*, **75**, 2066 (1971).

the energies (or to a good approximation enthalpies and even Gibbs free energies) that are estimated from the quasi-lattice model, presented in Table 8.3, that explicitly pertain to this exchange.

The competition of two cations for water in aqueous melts, in exchange for nitrate anions, is expressed by e_H. It shows, according to Table 8.3, that the water prefers as its next neighbour Cd^{2+} cations over K^+, Li^+, or NH_4^+, but that Ca^{2+} is preferred over Cd^{2+} as long as the water is scarce (not in melts of $Ca(NO_3)_2 \cdot 4H_2O$).[33] Water also prefers Zn^{2+} and Hg^{2+} over NH_4^+, is indifferent to Ag^+ versus NH_4^+, but prefers the ammonium cation over Pb^{2+}. NMR chemical shift data of Moynihan and Fratiello[34] indicate that in mixtures of $Ca(NO_3)_2 \cdot 4H_2O$ with other nitrates at 96 °C, the Ca^{2+} takes up 90% of the available water when the other cation is K^+ in an equimolar mixture, and 83% of it when it is $(CH_3)_4N^+$, which is, perhaps, surprising in view of the expected better solvation of K^+ by water than of $(CH_3)_4N^+$. However, if 20% of the calcium ions are exchanged for magnesium ions at the same total water content, the former can take up only 68% of the available water, the latter cation taking up the rest, to make its primary hydration shell of six almost 85% complete.[34]

Cations of a high field strength z/r^2 are seen to be able to dehydrate cations that have a lower field strength. This can be demonstrated visually in the case of aqueous melts of chlorides. In water-deficient $CaCl_2 \cdot 6H_2O$ melts as the solvent, Ni^{2+} retains its full primary hydration shell of six, and has, therefore, its octahedral configuration and a pale green colour. In water-defficient $MgCl_2 \cdot 6H_2O$, on the other hand, the Mg^{2+} tends more strongly than the Ni^{2+} to retain its full primary hydration, in particular when the Ni^{2+} is 'offered' Cl^- anions for coordination in recompense. At high temperatures the nickel converts to its tetrahedral configuration, forming $NiCl_4^{2-}$, which is blue.[31,35]

Spectroscopic information of another kind is provided by Raman and infrared spectral measurements on aqueous melts, in particular on aqueous nitrate melts. The studies on concentrated aqueous calcium nitrate solutions reported earlier in this chapter[21,36] have been extended to the composition of the tetrahydrate melt. The symmetry of the nitrate anion is perturbed at this composition not only by the water (that hydrates it at lower concentrations[21]) and by the solvent-shared ion pairs $Ca–OH_2–NO_3^+$, but also by direct contact with the Ca^{2+} ions. This does not involve *isolated* contact ion pairs, as surmised for lower concentrations,[36] but a polarization of the nitrate anion by low-field $Ca(H_2O)_4^{2+}$ cations that surround it, in addition to hydrogen-bonding with those water molecules that belong to the primary hydration shells of the calcium ions.

The Raman spectrum of the magnesium nitrate + water system has been investigated over the entire composition range by Peleg.[37] As long as $R_W \geqslant 6$, the primary hydration shell of the cation remains intact, and the nitrate anions are not in contact with bare magnesium cations, but rather form solvent-shared ion pairs with the hydrated cations, just as described above for the calcium system. For $R_W < 6$ penetration of the anion into the first coordination shell of the magnesium cation occurs. When R_W is lowered below about 2.5, a specific rearrangement of the coordination in the system takes place, resulting in a perturbed quasi-lattice having two different sites for the nitrate anions. Finally, when all the water is removed, a quasi-lattice molten salt results, where all the nitrate anions are again equivalent.[37]

The Raman spectrum of molten zinc nitrate hydrate also exhibits features indicative of close association of the nitrate anions with the cations at the expense of water molecules. No 'free' nitrate anions remain when the water content is lowered from $R_W = 6$ to $R_W = 1.4$, all having then entered the first coordination shell of the zinc cation.[38] The Raman spectral data for cadmium nitrate tetrahydrate melts, on the other hand, indicate that some 'free' nitrate is still present, although at least one nitrate anion occupies a site in the first coordination shell of the cation.[39] The difference between molten zinc and cadmium nitrate hydrates is understandable in terms of the lower field strength of the larger cadmium cation.

Few other molten salt hydrates have been studied by Raman spectroscopy, beside the nitrates. A concentrated aqueous solution that can be construed as molten ammonium hydrogen sulphate monohydrate has been subjected to Raman spectral measurements at 93 °C. The spectrum shows that the HSO_4^- anions have not transferred their protons to the water molecules, and that they have NH_4^+ cations as their next neighbours. The whereabouts of the water molecules in this system have not been specified, however.[40] Sodium acetate trihydrate melts at 58 °C to form a 10.7 M aqueous melt. A 10.2 M aqueous solution of ammonium acetate (that crystallizes without water) at room temperature may, therefore, be treated as essentially a molten salt hydrate. Similarly, a 10.7 M aqueous ammonium formate solution may be so treated, too. Contrary to the case of aqueous nitrate melts, the Raman spectral data show in the present systems the *anions* to be primarily hydrated, by hydrogen-bonding presumably.[41]

Beside the NMR results quoted above,[34] a few others may also be noted. The ^{1}H–NMR signals from molten magnesium, calcium, and zinc nitrate hydrates and from magnesium chlorate and acetate hydrates at 95 °C are found to be shifted upfield from that in pure water. The data have been interpreted as showing that most of the water is bound in the primary coordination shells of the cations, but that some of it is displaced by the anions, and finds itself in sites next to the latter, possibly hydrogen bonded to the latter. This interpretation[42] is not consistent with the model of the hydrated cation–anhydrous anion quasi-sublattices discussed above,[34] but is in line with the results for the carboxylate salts studied Raman-spectroscopically.[41]

Lithium chloride and bromide crystallize with one and two water molecules, respectively. A very concentrated aqueous solution of one of these salts has at 18 to 28 m concentration only three to two water molecules per mole of salt, and may, therefore, be considered as an aqueous melt or molten hydrated salt. The ^{7}Li–NMR spectral data show no preference of the cations for the anion or the water as neighbours, these being found in the vicinity of the Li^+ cation at the same ratio as given by the bulk composition. As the water content increases beyond $R_W = 2$, however, only two of the water molecules are exhangeable for chloride or bromide anions.[43]

Besides the models and approaches discussed above, in which the competition of the water and the anions for coordination sites in the quasi-lattice around the cation is emphasized, other approaches are concerned only with the change of the hydration (number) of the salt as a whole, as obtained from thermodynamic data. The hydration number of the salt at infinite dilution is obtainable from the limiting compressibility of the solution according to equation 5.9. The considerations upon which this equation is based can be extended to finite concentrations, as has been shown by Padova.[44] The apparent molar compressibility of the electrolyte, ${}^{\phi}\kappa = -(\partial^{\phi} \ln V/\partial P)_T$, where ${}^{\phi}V$ is the apparent molar volume, is known to be concentration dependent according to a relationship analogous to Root's equation for ${}^{\phi}V$: ${}^{\phi}\kappa = {}^{\phi}\kappa^{\infty} + S_{\kappa}c^{1/2}$. This dependence yields, together with equation 5.9, the following relationship between the hydration number, h, of the salt at concentration c, to that at infinite dilution, h^{∞}:

$$h/h^{\infty} = 1 - S_{\kappa}(h^{\infty}\kappa_{TH_2O}V_{H_2O})^{-1}c^{1/2} \tag{8.22}$$

which, with the data for water at 25 °C, becomes approximately $1\text{–}1.85h^{\infty\,-1}c^{1/2}$. The approximation arises by the substitution of the limiting slope of the Debye–Hückel theory for the actual, electrolyte-specific, experimental slope S_{κ}. This approach predicts the expected decrease of the hydration number with increasing concentrations, the larger, the less hydrated the electrolyte is already at infinite dilution.

According to Robinson and Stokes,[45] the osmotic coefficients of concentrated electrolyte solutions can be fitted in terms of a concentration-dependent hydration number h, that is a weighted mean of the extent of hydration of several 'hydrated species' of the electrolyte that are at equilibrium in the solution. The following function of the water activity a_W is obtained:

$$h = \sum_0^n i(K/Y)^i k^{i(i-1)/2} a_W^i / \sum_0^n (K/Y)^i k^{i(i-1)/2} a_W^i \tag{8.23}$$

with the three parameters: n (a 'maximal' hydration number), K and k, and the variable $Y = \exp(c_S(\bar{V}_S + (h - v)\bar{V}_W))$, that is a function of the molar concentration of the electrolyte, c_S, and the partial molar volumes (in dm^3/mol) of the electrolyte and the water, v being the stoichiometric coefficient of the electrolyte. Values of n, K, and k have been presented for a number of salts, acids and bases at high concentrations. For lithium chloride, for instance, h decreases from 3.8 at 10 m to 2.5 at 20 m, where di- and trihydrates are at equal concentrations, with small amounts of the mono- and tetrahydrates also present.[45]

This use of the thermodynamic data (essentially of densities and water vapour pressures) is a development of an earlier approach of the authors,[46] that regarded the hydration of the ions in concentrated aqueous solutions as a multilayer adsorption process, according to the BET model. The water can be considered as condensing on to a site near the ions (there being h such sites per formula of electrolyte) or on to the bulk of the water, the interaction energies being (in the absolute sense) E_{ad} and ΔE^V, respectively. An elaboration of this approach gave the following equation[47]:

$$a_W(1 - x_W)/(1 - a_W)x_W = (1/h)\exp[(E_{ad} - \Delta E^V)/\mathbf{R}T] \times \{1 + \exp[-(E_{ad} - \Delta E^V)/\mathbf{R}T]a_W - a_W\} \tag{8.24}$$

where x_W is the mole fraction of water in the solution. The difference $E_{ad} - \Delta E^V$ was found to range from -2.5 kJ mol^{-1} for (Cs, Li)NO_3 through -6.3 kJ mol^{-1} for $Ca(NO_3)_2$ to -10.5 kJ mol^{-1} for LiBr, all at 100 °C in aqueous melts containing up to 5 moles of water per mole salt (i.e. at $x_W \leqslant 0.83$). The corresponding numbers of sites h (i.e., hydration numbers of the entire electrolytes) are 1.7, 3.6, and 3.4 respectively for the salts named (also 3.3 for LiCl and 2.7 for $LiNO_3$) under these conditions.[47] For the salts $AgNO_3$, $TlNO_3$, and $Cd(NO_3)_2$ similar data were obtained from measurements on binary or ternary salt mixtures. The values of h for these three salts, obtained by extrapolation of the results for the mixtures to the pure hydrated salts, are 0.8, 0.3, and 3.7, respectively. The corresponding values of $E_{ad} - E^V$ are 1.5, 0.6, and 4.0 kJ mol^{-1}, respectively.[48]

Some information on the hydration of ions and their pairing in concentrated solutions has been obtained from diffraction methods applied to them, as presented in Table 5.12. This information is limited to the formation of solvent-shared ion pairs in certain chloride solutions of multivalent ions, the chloride ions being found to occupy sites in the second coordination shell of the cations.

In analogy with molten hydrated salts, molten solvated salts may also be studied in terms of the competition of solvent and an ion of one kind of sign for coordination sites around the ion of the other kind of sign. The different solvation abilities of protic solvents, such as water, and aprotic ones, in particular with respect to anions must be taken into account, however. In fact, very few investigations have applied themselves towards this problem. One such study is that of the silver nitrate–acetonitrile system, already mentioned above. The study includes melts of the mono- and disolvates, in which the ions of opposite charge are in contact, but the silver ions are also solvated by the acetonitrile molecules present, which do not solvate the nitrate ions at all. This behaviour is contrary to the analgous aqueous system, where the water solvates the nitrate anions too.

Molten tetra-n-pentylammonium thiocyanate, to which p-xylene[49] or nitrobenzene[50] has been added, shows a peculiar behaviour of the conductivity, or rather of the Walden product (viscosity times equivalent conductivity). The pure molten quaternary salt seems to be completely dissociated to ions, but the addition of p-xylene to the melt at 90 °C causes it to associate partly to non-conducting ion pairs, so that the Walden product decreases monotonously as xylene is added. Nitrobenzene, like xylene, lowers the viscosity, but due to its high dipole moment does not cause the ions to associate, and indeed a maximum in the Walden product is reached when 2 moles of nitrobenzene are added per mole of salt. Solvation is hinted at,[50] but not explicitly invoked, in the attempted explanation of the observed phenomena.

Molten calcium nitrate, containing 4 or more moles of dimethylsulphoxide per mole of salt, has been treated by Sacchetto and co-workers[51] according to equation 8.23, modified for adsorption in layers beyond the first. A value of $E_{ad} - \Delta E^{V} = -15.2$ kJ mol^{-1} and a solvation number of 3.1 have been found,[51] the energy difference being much higher than for the aqueous system discussed above.[47] The dimethylsulphoxide does not effectively solvate the nitrate anions, and the latter are more effectively associated with the solvated cation. Lithium nitrate with two moles of dimethylsulphoxide per mole of salt behaves similarly, in that only the cations are solvated: $h = 1.8$ and $E_{ad} - \Delta E^{V} = -10.1$ kJ mol^{-1}, and for ammonium nitrate $h = 1.3$ and $E_{ad} - \Delta \mathbf{E}^{V} = -5.6$ kJ mol^{-1}, the values for mixtures of these two salts being in between those for the single ones.[51]

References

1. N. Bjerrum, *Kgl. Dan. Vidensk. Selsk. Mat.-Fys. Medd.*, **7**, No. 9 (1926).
2. J. C. Justice, *J. Phys. Chem.*, **79**, 454 (1975); J. C. Justice, M. C. Justice, and C. Micheletti, *Pure Appl. Chem.*, **53**, 1291 (1981).
3. R. M. Fuoss, *J. Am. Chem. Soc.*, **80**, 5059 (1958); *ibid.*, **100**, 5576 (1978).
4. J. Padova, *J. Chem. Phys.*, **39**, 1552, 2599 (1963); **40**, 691 (1964).
5. J. E. Desnoyers, R. E. Verall, and B. E. Conway, *J. Chem. Phys.*, **43**, 243 (1965).
6. Y. Marcus, *Z. Naturforsch.*, **38A**, 247 (1983).
7. F. J. Millero and W. L. Masterton, *J. Phys. Chem.*, **78**, 1287 (1974); G. Atkinson and S. Petrucci, *ibid.*, **70**, 3122 (1966).
8. M. Eigen and K. Tamm, *Z. Elektrochem.*, **66**, 93, 107 (1962).
9. P. Hemmes, *J. Phys. Chem.*, **76**, 895 (1972).
10. F. T. Gicker, *Chem. Rev.*, **13**, 111 (1933).
11. J. W. Larson, *J. Phys. Chem.*, **74**, 3392 (1970).
12. R. Wachter and K. Riederer, *Pure Appl. Chem.*, **53**, 1301 (1981).
13. B. S. Krumgalz and Yu. I. Gerzhberg, *Zh. Obshch, Khim.*, **43**, 462 (1973); V. M. Tsentovskii, V. P. Barabanov, R. B. Bairamov, and B. D. Chernokalskii, *ibid.*, **44**, 2379 (1974).
14. P. L. Huyskens, *Bull. Soc. Chim. Belg.*, **89**, 937 (1980).
15. R. M. Diamond, *J. Phys. Chem.*, **67**, 2513 (1963).
16. S. Lindenbaum, *J. Phys. Chem.*, **70**, 814 (1966).
17. A. Ben-Naim, *Hydrophobic Interactions*, Plenum Press, New York (1980), p. 92.
18. D. G. Oakenfull and D. E. Fenwick, *J. Phys. Chem.*, **78**, 1759 (1974).
19. Y. Marcus, N. Ben-Zwi, and I. Shiloh, *J. Soln. Chem.*, **5**, 87 (1976).
20. G. J. Janz, M. J. Tait, and J. Meier, *J. Phys. Chem.*, **71**, 963 (1967); C. B. Baddiel, M. J. Tait, and G. J. Janz, *ibid.*, **69**, 3634 (1965).
21. D. E. Irish, A. R. Davies, and R. A. Plane, *J. Chem. Phys.*, **50**, 2262 (1969); D. E. Irish and A. R. Davies, *Can. J. Chem.*, **46**, 943 (1968).
22. R. L. Frost and D. W. James, *J. Chem. Soc. Faraday Trans. 1*, **78**, 3223, 3235, 4249 (1982); R. L. Frost, D. W. James, R. Appleby, and R. E. Mayes, *J. Phys. Chem.*, **86**, 3840 (1982).
23. D. W. James and R. E. Mayes, *J. Phys. Chem.*, **88**, 637 (1984).
24. J. F. Coetzee and W. R. Sharpe, *J. Soln. Chem.*, **1**, 77 (1972).

25. M. K. Wong, W. J. McKinney, and A. I. Popov, *J. Phys. Chem.*, **75**, 56 (1971); M. S. Greenberg, R. L. Bodner, and A. I. Popov, *ibid.*, **77**, 2449 (1973); J. M. Cahen, P. R. Handy, E. T. Roach, and A. I. Popov, *ibid.*, **79**, 80 (1975); H. A. Berman and T. R. Stengle, *ibid.*, **79**, 1001 (1975).
26. N. Soffer and Y. Marcus, *Ber. Bunsenges. Phys. Chem.*, **86**, 72 (1981).
27. A. Fratiello, V. Kubo, R. E. Lee, and R. E. Schuster, *J. Phys. Chem.*, **74**, 3726 (1970).
28. A. Fratiello, V. Kubo, and G. A. Vidulich, *Inorg. Chem.*, **12**, 2066 (1973).
29. R. Lachmann, I. Wagner, D. H. Devia, and H. Strehlow, *Ber. Bunsenges. Phys. Chem.*, **82,** 492 (1978); D. H. Devia and H. Strehlow, *ibid.*, **83**, 627 (1979); H. Strehlow and D. H. Devia, *ibid.*, **85**, 281 (1981).
30. Y. Marcus, in *Ionic Liquids*, D. Inman and D. G. Lovering, eds., Plenum Press, London (1981), p. 97.
31. C. A. Angell, *J. Phys. Chem.*, **69**, 2137 (1965).
32. J. Braunstein, *J. Phys. Chem.*, **71**, 3402 (1967); *Inorg. Chim. Acta Rev.*, **2**, 19 (1968); in *Ionic Interactions*, S. Petrucci, ed., Academic Press, New York (1971), Vol. I, p. 180.
33. H. Braunstein, J. Braunstein, and P. Hardesty, *J. Phys. Chem.*, **77**, 1907 (1973); H. Braunstein, J. Braunstein, A. S. Minano, and R. C. Hagman, *Inorg. Chem.*, **12**, 1407 (1973).
34. C. T. Moynihan and A. Fratiello, *J. Am. Chem. Soc.*, **89**, 5546 (1967).
35. C. A. Angell and D. M. Gruen, *J. Am. Chem. Soc.*, **88**, 5192 (1966).
36. R. E. Hester and R. A. Plane, *J. Chem. Phys.*, **40**, 411 (1964); **45**, 4588 (1966).
37. M. Peleg, *J. Phys. Chem.*, **76**, 1025 (1972).
38. R. E. Hester and C. W. J. Scaife, *J. Chem. Phys.*, **47**, 5253 (1967).
39. R. Caminiti, P. Cucca, and T. Radnai, *J. Phys. Chem.*, **88**, 2382 (1984).
40. D. E. Irish and H. Chen., *J. Phys. Chem.*, **74**, 3796 (1970).
41. L. A. Blatz and P. Waldstein, *J. Phys. Chem.*, **72**, 2614 (1968).
42. V. S. Ellis and R. E. Hester, *J. Chem. Soc.*, **A1969**, 607.
43. J. W. Akitt and A. J. Downs, *Chem. Comm.*, **1966**, 222.
44. J. Padova, *J. Chem. Phys.*, **40**, 691 (1964).
45. R. H. Stokes and R. A. Robinson, *J. Soln. Chem.*, **2**, 173 (1973).
46. R. H. Stokes and R. A. Robinson, *J. Am. Chem. Soc.*, **70**, 1870 (1948).
47. H. Braunstein and J. Braunstein, *J. Soln. Chem.*, **3**, 419 (1971).
48. M. C. Abraham, M. Abraham, and J. Sangster, *J. Chim. Phys.*, **76**, 125 (1979); J. M. Sangster, and M. C. Abraham, and M. Abraham, *Can. J. Chem.*, **56**, 348 (1978).
49. L. C. Kenausis, E. C. Evers, and C. A. Kraus, *Proc. U.S. Acad. Sci.*, **48**, 121 (1962); **49**, 141 (1963).
50. F. R. Longo, J. D. Kerstetter, T. F. Kumosniki, and E. C. Evers, *J. Phys. Chem.*, **70**, 431 (1966); F. R. Longo, P. H. Daum, R. Chapman, and W. G. Thomas, *ibid.*, **71**, 2756 (1967).
51. G. A. Sacchetto and Z. Kodjes, *J. Chem. Thermod.*, **15**, 457 (1983); **16**, 15 (1984); G. A. Sacchetto, G. G. Bombi, and C. Mocca, *J. Electroanal. Chem.*, **50**, 300 (1984).

Chapter 9

Some applications of ion solvation

9.1 Introduction

The concepts of ion solvation include on the one hand qualitative indications of ion–solvent interactions, including the effects of the ions on structural and other properties of the solvent. On the other hand they include quantitative data, such as (primary, secondary, etc.) solvation numbers, the sizes of solvated ions, the strengths of the ion–solvent bonds, thermodynamic quantities pertaining to the solvation process, and other measures of the ion–solvent interactions. In the cases where there is more than one solvent in the system, both the qualitative and the quantitative aspects of preferential solvation or solvent-sorting by the ions come into play. In solutions sufficiently concentrated for ion–ion interactions to take place, the competition between these interactions (ion-pairing to form solvent-separated, solvent-shared, or contact ion pairs) and ion–solvent interactions become of importance.

All these interactions leave their mark on the gross properties and the behaviour of electrolyte solutions, be they aqueous, nonaqueous, or mixed. In fact, ion solvation is an essential requirement for the mere existence of electrolyte solutions, since the non-solvative interactions (the entropy of solution due to the volume effect or translation and free rotation and the dispersive forces not counted ordinarily as involving solvation) are not sufficiently strong to overcome the electrostatic forces that hold together the lattice of solid electrolytes. Thus the solubilities of electrolytes are strongly affected by the solvation of their constituent ions. The properties of the solvated ions, either measured directly, such as their equivalent conductivities, or estimated as their contributions to partial molar quantities of the electrolytes or to their transport properties, are very different from those that may be ascribed to the 'bare' ions, if the latter exist in the solution. (Some apparently non-solvated ions do exist in solution, e.g., symmetrical large quaternary ammonium ions in polar solvents in which they are neither ion-paired nor aggregated). The solvation of the ions is, therefore, of importance in chemical processes involving salts, acids and bases that are capable of ionization in the solvents employed, or even processes involving neutral molecules that are capable of being protonated in the course of such processes. Since the effects of ion solvation on chemical processes of practical or theoretical importance are legion, both in respect to the variety of the effects and the systems that have been studied or can potentially be affected, only a limited selection can be dealt with in this Chapter.

Applications in four fields of chemistry have been chosen for discussion. In each of them some features of ion solvation play the major role that make the specific application feasible. Applications in electrochemistry, for instance, devolve around several topics. These include measures of the acidity of solutions, i.e., of their pH on a suitable solvent-independent scale, and measures of electrode potentials, again on a solvent-independent scale, as examples of theoretical applications of ion solvation concepts. Practical applications include high power density batteries based on certain electrode materials and electrolytes in nonaqueous solvents. Applications in extractive metallurgy include electrorefining and other treatments of ore solutions in aqueous as well as in mixed solvents, and above all applications of selective solvent extraction, e.g., in the nuclear fuel industry. Solvent extraction is an example for the application of ion solvation not only in the field of extractive metallurgy, but also in other aspects of separation science, for example in drug purification. Ion exchange techniques also rely heavily on ion solvation, both for large scale and for analytical separations. Finally, organic reaction chemistry does involve ion solvation in some of its implementations. Rates of reactions involving ions as reactants or as intermediates are strongly affected by their solvation. Phase transfer catalysis is a field where ion solvation is put to practical use. It must be emphasized, however, that as already mentioned above, the applications discussed in this chapter are very far from being able to exhaust the many instances where ion solvation is of great importance, and where judicious manipulation of the free variables involved produces useful results.

9.2 Applications in electrochemistry

The acidity of solutes, solvents, and solutions is a central concept in chemistry. A widely used measure of the acidity of a solution is its pH, defined operationally by the manner it is measured by in an electrochemical cell. The cell consists of an indicating electrode sensitive to hydrogen ions, a reference electrode not sensitive to them, and the electrolyte solution, the pH of which is to be measured. The open-cell electromotive force E of the cell is compared with that, E_s, measured with the same electrodes and a standard buffer as the electrolyte solution, having an internationally agreed[1] value $pH_s(T)$ at the temperature of measurement, T. The pH of the solution that is being measured is

$$pH = pH_s + (E - E_s)(\mathbf{F}/\mathbf{R}T \ln 10) \tag{9.1}$$

This equation is valid for aqueous solutions having an ionic strength I not exceeding a few tenths molal and a pH within the range from 2 to 12. One implementation of the electrochemical cell for pH measurements is

$$Pt(s), H_2(g)/H^+(aq, m_H), Cl^-(aq, m_{Cl}), \text{other ions (aq)}/AgCl(s), Ag(s) \tag{9.2}$$

Such a cell, indeed, is used for the assignment of the pH_s value to the specified standard buffer solution, so that

$$pH_s = \lim_{m_{Cl} \to 0} [(E - E^0(AgCl, Ag))(\mathbf{F}/\mathbf{R}T \ln 10) + \log m_{Cl}] + \log \gamma_{Cl} \tag{9.3}$$

where $\log \gamma_{Cl}$ is assigned a value according to the Bates–Guggenheim convention

$$\log \gamma_{Cl} = -A(T)I^{1/2}(1 + 1.5I^{1/2})^{-1} \tag{9.4}$$

The coefficient $A(T)$ is given a value according to the Debye–Hückel theory, but the coefficient 1.5 in the denominator is the basis of the conventional value. Implicitly it involves the hydration of the ions, because it:

(a) takes care of the ion size parameter a of the Debye–Hückel theory;
(b) covers up for the lack of a term linear in I.

The Debye–Hückel theory has in the denominator on the right-hand side of equation 9.4 the term $B(T)aI^{1/2}$, where a is the mean distance of closest approach of the ions as they exist in the solution, i.e., of the hydrated ions. Its replacement by the temperature-independent quantity $1.5I^{1/2}$ makes the ion size parameter a temperature-dependent in a reciprocal manner to the dependence of B. The linear term in I is required in $\log \gamma_{Cl}$ at the ionic strengths involved, of the order of 0.1 m or more, since individual properties of electrolytes are manifested already at these relatively low concentrations. One formulation of this term relates it to the hydration of the ions directly: $[2(M_W/\ln 10)(h - \nu)/\nu(\nu - 1)]I$, where M_W is the molar mass of water (in kg mol^{-1}), h is the mean hydration number of the electrolyte, and ν is the number of ions into which one formula unit of the electrolyte dissociates. All these aspects of ion hydration that appear implicitly in the fixed value of 1.5 in the denominator of equation 9.4, hence also implicitly in the aqueous pH-scale, are minor, however.

The cell (9.2) is often replaced for practical purposes by the cell

$$GE/H^+(aq, m_H), \text{ other ions (aq)}/KCl(aq, sat)/SCE \tag{9.5}$$

where GE stands for a glass-membrane electrode sensitive to hydrogen ions and SCE for the aqueous saturated calomel electrode, i.e., saturated with potassium chloride, that also constitutes the salt bridge with the solution, the pH of which is to be measured. The provisions concerning the ionic strength and the pH of this solution specified above must still be met, in order for cell (9.5) to be useful for pH measurements with the electrodes calibrated with the same standard buffers, and the same equation 9.1 to be used for the calculation of the pH. When cell (9.5) is used, another aspect of ion hydration comes into play. Both E and E_s now involve a term in the liquid junction potential E_j between the saturated KCl salt bridge of the SCE and the solution to be measured and the standard buffer solution, respectively. The two values of E_j depend on the mobilities of the ions constituting the salt bridge, K^+ and Cl^-, and on their activities in the region where the liquid junction is formed. These quantities, in turn, depend on the hydration of these ions. It is assumed, however, that if the provisions specified above are met, the two E_j values are the same, hence cancel out in the difference $E - E_s$. Hence, again, ion hydration effects in the determination of aqueous pH values turn out to be minor.

This is not the case, however, when pH values are to be determined in nonaqueous or mixed organic–aqueous solutions. In such cases, the difference between the hydration and the solvation of the ions may play a major role. If the water in cell (9.2) is gradually replaced by a nonaqueous constituent, the changes in the measured E of the cell will be accompanied by changes in E^0(AgCl, Ag) and in $\log \gamma_{Cl}$. In a given solvent mixture, the value of E^0(AgCl, Ag) can be obtained in cells containing only HCl and no other ions, on extrapolation to zero HCl molality. A value of $\log \gamma_{Cl}$ may be assigned conventionally for this particular solvent mixture. Then values of $pH_s(S)$ can be assigned to specified standard buffers in this solvent mixture S, and finally pH(S, S) values can be measured in S and referred to the standards assigned in S. For water-rich solvents involving organic components such conventional assignments have been made by Mussini and co-workers[2]. It must be realized that the set of standard buffers with $pH_s(S)$ assigned to them is potentially infinite, since there is an infinity of solvents and solvent compositions that may be studied, with no interrelation among them. In the general case, therefore, no relationship exists between the pH(S, S) measured for a given acid concentration in one solvent or at one composition of a mixed solvent, and the pH(S, S) measured in the next solvent or composition for the same acid concentration.

It is expedient, therefore, to seek an alternative to the measurement of unrelated pH(S, S) values, and define pH(S) = pH(S, W) values, referred to a single pH scale, namely the aqueous one, with its well established $pH_s(W)$ values. The pH(S) is measured in the cell

$$\mathrm{Pt(s), H_2(g)/H^+(S, \mathit{m}_H), Cl^-(S, \mathit{m}_{Cl}), other\ ions(S)/AgCl(s), Ag(s)} \qquad (9.6)$$

but the calibration of the electrodes is carried out with the aqueous standard buffer. This, in effect, produces a value of

$$\mathrm{pH(S)} = (E - E^0(\mathrm{AgCl, Ag, aq}))(\mathbf{F}/\mathbf{R}T \ln 10) + \log m_{Cl}(\mathrm{S}) + \log \gamma_{Cl}(\mathrm{S}) + \log \gamma_t^0(\mathrm{Cl^-, W \rightarrow S}) \qquad (9.7)$$

The value of $\log \gamma_{Cl}(S)$ may be approximated according to the Debye–Hückel theory for the solvent or mixture S. The quantity $\log \gamma_t^0(Cl^-, W \rightarrow S)$, the standard transfer activity coefficient of chloride ions from water to the solvent or mixture S, equals $(1/\mathbf{R}T \ln 10)\Delta G_t^0(Cl^-, W \rightarrow S)$, values of which for transfer into pure solvents having been presented in Table 6.8.

Practical measurements of pH(S) are generally carried out not with cell (9.6) but with a cell analogous to (9.5):

$$\mathrm{GE/H^+(S, \mathit{m}_H), other\ ions(S)/KCl(aq, sat)/SCE} \qquad (9.8)$$

Now the liquid junction potential $E_j(S)$, obtained with the solution the pH(S) of which is to be measured, differs from the liquid junction potential $E_j(W)$, obtained with the aqueous solution used for the calibration of the electrodes. There are some indications that $E_j(S) - E_j(W)$ is independent of pH(S), within the specified limits of acidity, and depends only on the solvent S. A comparison of the results obtained with cells (9.6) and (9.8) in which the same solution was employed permitted the estimation of these differences in E_j (at 25 °C), being -0.035 V for S = methanol,[3] -0.071 V for S = ethanol,[4] and 0.25 V for S = acetonitrile.[5]

The values of pH(S) obtained from cell (9.6) or from cell (9.8), corrected by subtraction of $(\mathbf{F}/\mathbf{R}T \ln 10)(E_j(S) - E_j(W))$, are related to the values obtained in aqueous solutions of the same acidity by

$$\mathrm{pH(S)} = \mathrm{pH(W)} + \log \gamma_t^0(\mathrm{H^+}, \mathrm{W} \rightarrow \mathrm{S}) \tag{9.9}$$

This expression may also be construed as a definition of pH(S) = pH(S, W), i.e., the pH measured in the solvent S but referred to the aqueous pH scale. Expression 9.9 has the disadvantage that values of $\log \gamma_t^0(\mathrm{H^+}, \mathrm{W} \rightarrow \mathrm{S}) = (1/\mathbf{R}T \ln 10)\Delta G_t^0(\mathrm{H^+}, \mathrm{W} \rightarrow \mathrm{S})$ are inaccurate at best and lacking for most pure solvents of interest in the present context. They are available for some mixed aqueous–organic solvents, in particular those involving alcohols, but not generally for mixtures. Relationship 9.9 may, however, be rewritten as

$$\mathrm{pH(S)} = \mathrm{pH(W)} + \log \gamma_t^0(\mathrm{HCl}, \mathrm{W} \rightarrow \mathrm{S}) - \log \gamma_t^0(\mathrm{Cl^-}, \mathrm{W} \rightarrow \mathrm{S}) \tag{9.10}$$

Values of the standard activity coefficient of transfer of the complete electrolyte hydrogen chloride are available for a great many pure solvents and their aqueous mixtures. Values of

$$\Delta G_t^0(\mathrm{Cl^-}, \mathrm{W} \rightarrow \mathrm{S}) = \exp[\mathbf{R}T \cdot \ln 10 \cdot \log \gamma_t^0(\mathrm{Cl^-}, \mathrm{W} \rightarrow \mathrm{S})]$$

are available for these, too, see Table 6.8 for transfer into pure solvents. Furthermore, the correlations and theoretical expressions developed for the transfer of ions in Sections 6.5 and 7.4 have been found to be very successful for the transfer of chloride ions, more so, perhaps, than for several other ions, and in particular hydrogen ions. Hence equation 9.10 can be used for the estimation of pH(S) with advantage.

Values of $\log \gamma_t^0(\mathrm{H^+}, \mathrm{W} \rightarrow \mathrm{S})$ for some pure solvents and some of their mixtures with water are shown in Table 9.1. Unfortunately, not all of the values are based on a sound extrathermodynamic assumption, and the original literature should be consulted for details. To supplement this information, references to the literature where values of $\log \gamma_t^0(\mathrm{HCl}, \mathrm{W} \rightarrow \mathrm{S})$ are reported for aqueous–organic solvent mixtures are presented in Table 9.2. Values of $E^0(\mathrm{S}) - E^0(\mathrm{W})$ for the AgCl/Ag electrode determined in cell (9.6), in the absence of ions other than those resulting from the ionic dissociation of hydrogen chloride, are equivalent to $(\mathbf{R}T \ln 10)/\mathbf{F}$ times $\log \gamma_t^0(\mathrm{HCl}, \mathrm{W} \rightarrow \mathrm{S})$.

Table 9.1 Values of pH(S) − pH(W) = $\log \gamma_t^0(H^+, W \rightarrow S)$ at 25 °C. (For the extra-thermodynamic assumptions employed in the derivation of the individual values, the original references should be consulted.) The second component in mixtures is water

Solvent(S)	$\log \gamma_t^0$	Ref.	Solvent(S)	$\log \gamma_t^0$	Ref.
Methanol	1.82	b	Acetic acid	7.2	i
80 wt.% methanol	−0.99	a	60 wt.% acetic acid	0.15	g
60 wt.% methanol	−1.34	a	40 wt.% acetic acid	−0.09	g
40 wt.% methanol	−1.15	a	20 wt.% acetic acid	−0.19	g
20 wt.% methanol	−0.72	a	Acetone	0.52	j
Ethanol	1.94	b	Propylene carbonate	8.8	b
90 wt.% ethanol	−0.46	c	Dimethylsulphoxide	−3.40	b
80 wt.% ethanol	−0.67	c	80 wt.% DMSO	−4.20	a
60 wt.% ethanol	−0.80	c	60 wt.% DMSO	−2.35	a
40 wt.% ethanol	−0.32	c	40 wt.% DMSO	−1.10	a
20 wt.% ethanol	0.06	c	20 wt.% DMSO	−0.09	a
1-Propanol	1.6	b	Tetramethylenesulphone	−12.1	j
2-Propanol	0.5	d	Nitrobenzene	5.8	b
2-Methyl-2-propanol	1.1	d	Acetonitrile	8.13	b
1-Hexanol	2.3	d	60 wt.% acetonitrile	2.05	k
Ethylene glycol	0.9	b	40 wt.% acetonitrile	1.49	k
90 wt.% ethylene glycol	−0.12	e	20 wt.% acetonitrile	0.68	k
70 wt.% ethylene glycol	−0.39	e	N,N-Dimethylformamide	−3.2	b
50 wt.% ethylene glycol	−0.17	e	80 wt.% DMF	−3.22	l
30 wt.% ethylene glycol	−0.21	e	60 wt.% DMF	−2.79	l
50 wt.% dimethoxyethane	−1.82	f	40 wt.% DMF	−1.59	l
30 wt.% dimethoxyethane	−1.10	f	20 wt.% DMF	−0.51	l
10 wt.% dimethoxyethane	−0.35	f	N-Methylpyrrolidinone	−4.4	b
45 wt.% dioxane	−4.18	g, h	Hexamethylphosphoric triamide	−7.2	b
20 wt.% dioxane	−1.76	g, h			

[a] S. Villermaux and J. J. Delpuech, *Bull. Soc. Chim. France*, **1974**, 2534; [b] Y. Marcus, *Pure Appl. Chem.*, **55**, 977 (1983); [c] O. Popovych, A. Gibovsky, and D. H. Berne, *Anal. Chem.*, **44**, 811 (1972); [d] M. K. Chantooni and I. M. Kolthoff, *J. Phys. Chem.*, **82**, 994 (1978); [e] A. K. Das and K. K. Kundu, *Indian J. Chem.*, **16A**, 467 (1978); [f] A. Bhattacharya, A. K. Das, and K. K. Kundu, *Indian J. Chem.*, **20A**, 353 (1981); [g] H. P. Bennetto, D. Feakins, and D. J. Turner, *J. Chem. Soc.*, **1966**, 1211; [h] B. K. Das and P. K. Das, *Experientia*, **35**, 372 (1979); [i] J. C. Touler, quoted by D. Bauer and M. Breant in *Electroanalytical Chemistry*, A. J. Bard, ed., M. Dekker, New York, Vol. **8**, p. 251 (1975); [j] Y. Marcus, *Rev. Anal. Chem.*, **5**, 53 (1980); [k] C. Barraque, J. Vedel, and B. Tremillon, *Bull. Soc. Chim. France*, **1968,** 3421; [l] K. Das, K. Bose, and K. K. Kundu, *Electrochim. Acta*, **26**, 479 (1981).

The transfer of hydrogen ions from water to nonaqueous solvents or to aqeous solvent mixtures is intimately connected with the relative basicities (and acidities) of these solvents or mixtures and water. One measure of these quantities is the gas-phase proton affinities shown in Table 2.3. The transfer of interest in the present context, however, takes place in condensed phases, hence other measures are more relevant.

If the comparison is made in the presence of a large excess of the reference solvent water, then the ordinarily tabulated values of pK_a of acids and pK_b of bases should be consulted. (Note that in the following, 'S' and 'HS' denote solvent molecules and not sulphur and its monohydride!) These constants pertain to the reactions $HS + H_2O \rightleftharpoons H_3O^+ + S^-$ for the former and $S + H_2O \rightleftharpoons HS^+ + OH^-$ for the latter, all species being hydrated and, except for

Table 9.2 Some references to sources of data on log γ_t^0(HCl, W → S) for the transfer of hydrogen chloride to nonaqueous and mixed solvents (see also the references to Table 9.1)

Solvent	Composition range	Temp. range	Ref.
Methanol	$0.0 \leqslant x_S \leqslant 1.0$	25	a
Ethanol	$0.0 \leqslant x_S \leqslant 1.0$	25	b
1-Propanol	$0.0 \leqslant x_S \leqslant 1.0$	5–45	c
2-Propanol	$0.8 \leqslant x_S \leqslant 1.0$	5–25	c
2-Methyl-2-propanol	$x_S \leqslant 0.36$	5–45	c
Glycerol	$0.79 \leqslant x_S$	5–45	c
Diethylene glycol	$x_S \leqslant 0.66$	5–35	d
1,2-Dimethoxyethane	$x_S \leqslant 0.61$	15–35	c
Dioxane	$x_S \leqslant 0.71$	25	b
Tetrahydrofuran	$0.0 \leqslant x_S \leqslant 0.79$	35	e
Acetone	$x_S \leqslant 0.17$	25	c
Acetic acid	$x_S \leqslant 0.31$	25	g
Propylene carbonate	$x_S \leqslant 0.04$	5–45	c
Dimethylsulphoxide	$x_S \leqslant 0.05$	5–45	c
N-Methylformamide	$0.0 \leqslant x_S \leqslant 1.0$	18–30	f
N,N-Dimethylformamide	$x_S \leqslant 0.03$	5–45	c

[a] D. Feakins and P. J. Voice, *J. Chem. Soc. Faraday Trans. 1*, **68**, 1390 (1972); [b] H. Strehlow, in *The Chemistry of Non-Aqueous Solvents*, J. J. Lagowsky, ed., Academic Press, New York, Vol. 1, Ch. 4 (1966); [c] B. Sen, R. N. Roy, J. J. Gibbons, D. A. Johnson, and L. H. Adcock, *Adv. Chem. Ser.*, **177**, 215 (1979); [d] C. Kallidas and V. S. Rao, *J. Chem. Eng. Data*, **24**, 255 (1979); [e] M. M. Elsemongy, *Electrochim. Acta*, **23**, 881 (1978); [f] Z. Kozlowski, C. Kinart, and W. Kinart, *Pol. J. Chem.*, **53**, 2621 (1979); [g] H. P. Bennetto, D. Feakins, and D. J. Turner, *J. Chem. Soc.*, **1966**, 1211.

the solvent H_2O, at infinite dilution. The molar concentration scale is generally employed, and the concentration of the solvent is not explicitly included in the expressions for the equilibrium constants. Values of these quantities for some of the solvents listed in Chapter 6 are shown in Table 9.3. Several protic solvents are *amphiprotic*, i.e., react with water, used as the solvent (or with other suitable solvents) both by dissociation of HS to form S^- and by its association with hydrogen ions from the solvent to form H_2S^+.

The comparison of the acidities and basicities of solvents is less easy when it is attempted not in some reference solvent but in a more general manner, e.g., in the solvents themselves. The quantity of interest in this case is the ionic product for auto-ionization, pK_{ai}. For an amphiprotic solvent this proceeds according to $2HS \rightleftharpoons H_2S^+ + S^-$ (called autoprotolysis), but auto-ionization may occur also in aprotic solvents. It then involves ionic fragments of the molecule, but it is not always known which these are. The information on auto-ionization is generally obtained from measurements of the specific conductivity of the carefully purified solvent, compared with the equivalent conductivity of the presumed ionic fragments, available from measurements on the conductivity and transport numbers of solutions of electrolytes that contain these ions. The values that are available in the literature are presented in Table 9.3. Except for formic acid and ethanolamine, most solvents are less prone to auto-ionization than water, and

Table 9.3 The acid and base dissociation constants, pK_a and pK_b, in water, and the auto-ionization ionic product, pK_{ai}, all at 25 °C[a]

No.[b]	Solvent	pK_a	pK_b	pK_{ai}(cation)[c]
7	Water			14.00 (H_3O^+)
8	Methanol	15.5	11.97[j]	16.7, 17.20[i] ($CH_3OH_2^+$)
9	Ethanol	15.9	11.54[j]	19.1, 18,88[i] ($C_2H_5OH_2^+$)
10	1-Propanol			19.43[i] ($C_3H_7OH_2^+$)
11	2-Propanol	19.43[g]	11.96[j]	20.8[g] ($(CH_3)_2CHOH_2^+$)
12	1-Butanol		11.48[j]	21.56[i] ($C_4H_9OH_2^+$)
15	2-Methyl-2-propanol			19, 26.8[i] ($(CH_3)_3COH_2^+$)
21	Phenol	10.02		
22	2-Methylphenol	10.29		
24	2-Chloroethanol	14.31		
25	Trifluoroethanol	12.37[d]		
26	Ethylene glycol	14.24		
27	Glycerol	14.40		
28	2-Methoxyethanol	14.8		
41	Acetone		12.93[j]	32.5[i] (?)
42	2-Butanone			25.94[i] (?)
46	Formic acid	3.75		6.2 ($HCO_2H_2^+$)
47	Acetic acid	4.76		14.45 ($CH_3CO_2H_2^+$)
48	Propanoic acid	4.87		
49	Butanoic acid	4.82		
52	Trifluoroacetic acid	0.23		
53	Acetic anhydride			14.5[i] (CH_3CO^+)
55	Ethyl acetate	24.5		22.83 ($C_2H_5OC(OH)CH_3^+$)
75	Dimethylsulphoxide		13.20[j]	17.3, 33.3[i] ($(CH_3)_2SOH^+$)
76	Tetramethylene sulphone			25.45[i] (?)
77	Ammonia		4.75	30.9,[f] 32.5[i] (NH_4^+)
79	Cyclohexylamine		3.36	
80	Di-1-propylamine		3.00	
81	Triethylamine		3.29	
82	Aniline		4.61	
83	Pyridine		5.25	
85	2-Chloroaniline		11.36	
86	Ethylene diamine		4.07	15.3[d] ($H_2NC_2H_4NH_3^+$)
87	Ethanolamine		4.50	5.1,[e] 5.7[i] ($HOC_2H_4NH_3^+$)
88	Diethanolamine		5.12	
89	Triethanolamine		6.24	
90	Morpholine		5.51	
91	Nitromethane	10.74		
93	Acetonitrile		10.3[j]	26.5, 33.3[i] (CH_3CNH^+)
99	Formamide	14.48[d]		
100	N-Methylformamide	14.04[d]		16.8[i] (?)
101	N,N-Dimethylformamide	14.01[d]		18.0,[g] 29.4[i] (?)
105	N,N-Dimethylacetamide	14.9[d]		23.95[i] ($CH_3C(OH)N(CH_3)_2^+$)
108	N-Methylpyrrolidinone			24.15[i] (?)
110	Hexamethyl phosphoric triamide			20.56[i] (?)

[a] From ref. 1; [b] see listing in Table 6.1, for comparison with other properties of the solvents; [c] the cation resulting from the auto-ionization; [d] at 20 °C; [e] R. G. Bates, in *Solute-Solvent Interactions*, J. F. Coetzee and C. D. Ritchie, eds., Dekker, New York, p. 52 (1969); [f] P. M. Laughton and R. E. Robertson, see note e, p. 412; [g] R. H. Boyd, see note e, p. 199, 209; [h] from the standard potentials of $NH_4^+/NH_3, \frac{1}{2}H_2$ and $NH_3/NH_2^-, \frac{1}{2}H_2$ electrodes, W. L. Jolly and C. J. Hallada, in *Non-aqueous Solvent Systems*, T. C. Waddington, ed., Academic Press, London, p. 22 (1965); [i] Ch. Reichardt, *Solvent Effects in Organic Chemistry*, Verlag Chemie, Weinheim, p. 283 (1979) (at 20 °C for acetic anhydride and formamide, at −33 °C for ammonia); [j] E. J. King, *Acid–Base Equilibria*, Pergamon, Oxford, p. 294 (1965).

amphiprotic solvents are more prone to autoprotolysis than aprotic solvents are to auto-ionization.

Even more difficult is the comparison of the affinities of the components of mixtures to hydrogen ions. As is seen in Table 9.1, the transfer activity coefficient of hydrogen ions is a non-monotonous function of the composition. For instance, 60 wt% methanol in water constitutes a solvent that is more basic than water, even though methanol itself is less basic. Similarly, although dimethylsulphoxide is more basic than water, a mixture that contains 80 wt% of this solvent is even more basic. These facts, by themselves, do not reveal whether the methanol or the dimethylsulphoxide on the one hand or the water on the other is protonated to the greater extent in the mixtures. Such information is not available from thermodynamics, i.e., from the kind of measurements that lead to the transfer activity coefficients. It can, however, be obtained from spectroscopic measurements, e.g., from NMR or infrared spectra.

A problem cognate to the measurement of the pH in nonaqueous or mixed aqueous–organic solutions is the measurement of the standard potentials of various electrodes. Again, it is possible to measure the electromotive force of a cell that has the standard hydrogen electrode, $\mathrm{SHE} \equiv \mathrm{Pt(s)}, \mathrm{H_2(g)/H^+(S}, a_{\mathrm{H}}=1)$, as one of the electrodes, and obtain formally the standard potential of the other electrode with the convention that $E^0(\mathrm{SHE, S}) = 0$.

The stipulation that $a_{\mathrm{H}} = 1$ and the adjustment required for obtaining the *standard* electrode potential are achieved by the extrapolation of the electrolyte concentration to zero. Thus, the measurement of the EMF of cell (9.6), in the absence of ions other than H^+ and Cl^- that are derived from the hydrogen chloride electrolyte, leads on extrapolation to zero HCl molality to the standard potential of the $\mathrm{Cl^-(S)/AgCl(s), Ag(s)}$ electrode. As an aid to the extrapolation, a suitable expression of the activity coefficient of HCl can be used, so as to obtain as nearly a linear extrapolation curve as possible:

$$E^0(\mathrm{Cl^-(S)/AgCl(s), Ag(s)}) = \lim_{m_{\mathrm{HCl}} \to 0} [E(\text{cell 9.6}) - (\mathbf{R}T/\mathbf{F}) \ln \mathrm{f}(\gamma_{\mathrm{HCl}})] \qquad (9.11)$$

The true value of the activity coefficient of hydrogen chloride in the solvent S is, of course, obtained once E^0 has been determined: $\gamma_{\mathrm{HCl}}(\mathrm{S}) = \exp[(E - E^0)(2\mathbf{F}/\mathbf{R}T)]/m_{\mathrm{HCl}}(\mathrm{S})$. Once this standard electrode potential of the chloride/silver chloride electrode is known for any solvent or solvent mixture S, it can be used as a secondary reference electrode, at the side of the primary SHE(S), for the determination of standard electrode potentials.

The establishment of standard electrode potential tables with reference to the convention that $E^0(\mathrm{SHE, S}) = 0$ produces an infinity of potential scales (as many as there are solvents and solvent mixtures, S), with no direct relationships among them. It is expedient to relate the potential scales to a single conventional one, that based on $E^0(\mathrm{SHE, W, W}) = 0$ for aqueous solutions. In the same manner as for the case of the pH discussed above

$$\begin{aligned} E^0(\mathrm{SHE, S, W}) &= E^0(\mathrm{SHE, W, W}) + (1/\mathbf{F})\,\Delta G_t^0(\mathrm{H^+, W \to S}) \\ &= \Delta G_t^0(\mathrm{H^+, W \to S})/\mathbf{F} \end{aligned} \qquad (9.12)$$

where the first solvent given in the parentheses is that in which the measurement is carried out, and the second is the solvent to the scale of which the potentials are referred. For any other electrode, designated symbolically by (X, S, the reference solvent S or W)

$$E^0(\mathrm{X, S, W}) = E^0(\mathrm{X, S, S}) + E^0(\mathrm{SHE, S, W})\cdot n \tag{9.13}$$

where n is the number of electrons involved in the electrode reaction. For the reasons presented above, more accurate results for more solvents or solvent mixtures can be obtained when the expression

$$E^0(\mathrm{X, S, W}) = E^0(\mathrm{X, S, S}) - [E^0(\mathrm{HCl, S, S}) - E^0(\mathrm{HCl. W, W})]/n + \Delta G_t^0(\mathrm{Cl^-, W \rightarrow S})/\mathbf{F} \tag{9.14}$$

is employed, where E^0(HCl) is the conventional standard electrode potential of the Cl^-/AgCl(s), Ag(s) electrode in the designated solvent.

A more direct way to obtain standard electrode potentials in nonaqueous and mixed solvents S referred to the aqueous scale is to add algebraically the term for the Gibbs free energy of transfer of the ion involved in the electrode reaction to the aqueous standard potential E^0(W).[6] For metal electrodes M(s) dipping into solutions containing the cations M^{z+} the standard electrode potential is

$$E^0(\mathrm{M^{z+}/M, S, W}) = \mathrm{E}^0(\mathrm{M^{z+}/M, W, W}) + \Delta G_t^0(\mathrm{M^{z+}, W \rightarrow S})/z\mathbf{F} \tag{9.15}$$

and for anions X^{z-} in solutions saturated with the silver salt $Ag_{|z|}X$, into which a silver electrode dips

$$E^0(\mathrm{X^{z-}/Ag_{|z|}X, Ag, S, W}) = E^0(\mathrm{X^{z-}/Ag_{|z|}X, Ag, W, W}) - \Delta G_t^0(\mathrm{X^{z-}, W \rightarrow S})/|z|\mathbf{F} \tag{9.16}$$

The number of electrodes for which standard potentials E^0(S, W) are available is limited mainly to monovalent ions, the data being presented in Table 9.4.

A graphical presentation of these standard potentials of certain electrodes is shown in Figure 9.1. It should be noted that the ordinate of the graph is an arbitrarily spaced sequence of solvents, chosen so that for one of the electrodes, the K^+/K one, a monotonous curve is produced. The consequence of this is that for some other electrodes, e.g., the Ag^+/Ag or the Cl^-/AgCl, Ag ones, curves with sharp peaks and valleys are obtained, because the relative solvation of the ions by the various solvents differs considerably from what they are for the potassium ion.

An important practical electrochemical application of ion solvation takes place in high-energy-density batteries. These are commonly based on lithium metal as the anode, although sodium, magnesium, calcium and aluminium may also serve, a transition metal salt such as a halide or an oxide as the cathode material, and a lithium salt in a nonaqueous solvent as the electrolyte. The main reason for the use of a nonaqueous solvent is that the low-equivalent-weight metals suitable for the use as anodes are very reactive, and an aqueous solvent,

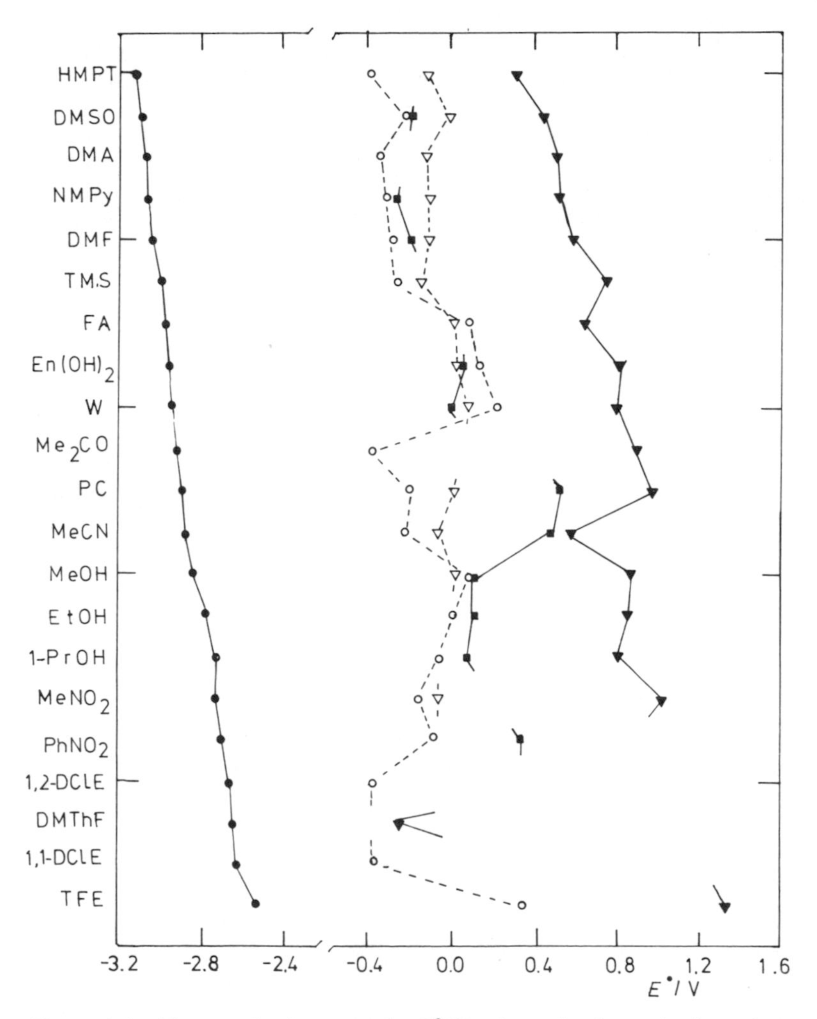

Figure 9.1 The standard potentials, E^0/V, of certain electrodes in various solvents versus the aqueous standard hydrogen electrode, SHE(W), at 25 °C. The arrangement of the ordinate is arbitrarily selected so that the resulting curve for the K^+/K electrode (●) will be monotonous, but the distances along the ordinate and the connecting lines have no physical meanings. The other electrodes are: $H^+/\frac{1}{2}H_2$(Pt)(■), Ag^+/Ag(▼), Cl^-/AgCl(s)/Ag(○), SCN^-/AgSCN/Ag(▽)

and even a nonaqueous protic solvent, would not produce a stable anode/electrode system.

The following criteria have been found to be essential for the proper choice of a solvent for the electrolyte in a high-energy-density battery:

(a) A long liquid range, extending at least from −50 °C to +50 °C; some of the applications focus on the low temperature part of this range.

Table 9.4 Standard potentials of electrodes in non-aqueous solvents at 25 °C against the aqueous standard hydrogen electrode (in V), for cations

Solvent	Electrode								
	$H^+(1/2)H_2$	Li^+/Li	Na^+/Na	K^+/K	Rb^+/Rb	Cs^+/Cs	Ag^+/Ag	Tl^+/Tl	Cu^{2+}/Cu
Water	0.000	−3.040	−2.714	−2.936	−2.943	−3.027	0.799	−0.336	0.339
MeOH	0.10	−2.99	−2.63	−2.84	−2.84	−2.94	0.87	−0.29	0.47
EtOH	0.12	−2.93	−2.57	−2.77	−2.78	−2.87	0.85	−0.26	0.58
PrOH	0.09	−2.93	−2.54	−2.75	−2.75	−2.85	0.81		0.56
TFE				−2.53			1.32		
$En(OH)_2$	0.05	−3.04	−2.74	−2.96			0.81		
Me_2CO				−2.90	−2.90	−2.99	0.89		
PC	0.52	−2.79	−2.56	−2.88	−2.95	−3.10	0.99	−0.22	0.73
FA		−3.14	−2.80	−2.93	−3.00	−3.09	0.64		
DMF	−0.19	−3.14	−2.81	−3.04	−3.04	−3.14	0.58	−0.46	0.25
DMA			−2.84	−3.06	−3.03	−3.20	0.50		
DMThF		−2.47	−2.31	−2.66		−2.88	−0.26	−0.50	
NMPy	−0.26	−3.40	−2.87	−3.06	−3.03	−3.13	0.53	−0.49	
MeCN	0.48	−2.73	−2.56	−2.88	−2.88	−2.97	0.56	−0.25	0.65
$MeNO_2$		−2.54	−2.45	−2.74	−2.92	−3.02	1.02		
$PhNO_2$	0.34	−2.65	−2.54	−2.74	−2.75	−2.87		−0.18	
DMSO	−0.20	−3.20	−2.85	−3.07	−3.05	−3.16	0.44	−0.56	0.09
TMS			−2.75	−2.93	−3.04	−3.13	0.76		0.71
HMPT				−3.10			0.32		
1,1DClE			−2.41	−2.63	−2.64	−2.74			
1,2DClE			−2.46	−2.67	−2.68	−2.78			

Table 9.4—*continued, for anions*

Solvent	Electrode					
	Cl^-, AgCl/Ag	Br^-, AgBr/Ag	I^-, AgI/Ag	CN^-, AgCN/Ag	SCN^-, AgSCN/Ag	N_3^-, AgN_3/Ag
Water	0.222	0.073	−0.152	−0.161	0.090	0.290
MeOH	0.09	−0.04	−0.23	−0.25	0.03	0.20
EtOH	0.01	−0.12	−0.29	−0.23		0.12
PrOH	−0.05	−0.16	−0.35			
TFE	0.33	0.15	−0.07			
$En(OH)_2$	0.13	0.00	−0.18		0.04	0.22
Me_2CO	−0.37	−0.42	−0.41	−0.66		0.15
PC	−0.19	−0.24	−0.29	−0.53	0.02	0.01
FA	0.08	−0.04	−0.23	−0.30	0.02	0.18
DMF	−0.28	−0.38	−0.36	−0.58	−0.10	−0.02
DMA	−0.34		−0.37		−0.13	−0.12
DMThF						
NMPy	−0.31	−0.31	−0.35		−0.10	−0.18
MeCN	−0.21	−0.25	−0.33	−0.52	−0.06	−0.09
$MeNO_2$	−0.16	−0.24	−0.33		−0.07	0.00
$PhNO_2$	−0.14	−0.23	−0.34			
DMSO	−0.20	−0.21	−0.26	−0.52	−0.01	0.02
TMS	−0.26	−0.29	−0.37		−0.14	−0.13
HMPT	−0.38	−0.41	−0.46		−0.12	−0.22
1,1DClE	−0.38	−0.37	−0.47			
1.2DClE	−0.38	−0.33	−0.42			

(b) A low vapour pressure up to the maximal temperature of use, in order to avoid loss and danger of explosion in unvented systems.
(c) A low viscosity over the temperature range of interest, in particular at its lower end. This feature helps to produce adequate mobilities of the ions in the electrolyte.
(d) A sufficiently high relative permittivity (dielectric constant), that ensures complete ionic dissociation of the electrolyte and the avoidance of ion-aggregation.
(e) A sufficiently high solvating power for both the cation and the anion of the electrolyte, for the achievement of adequate electrolyte solubility. For small ions this is produced by good electron pair donicity or acceptance ability, for large organic ones by the applicable dispersion forces.
(f) Stability against attack by the electrode materials and possibly present depolarizing materials, at least in the kinetic sense, which may involve passivation.
(g) Ready availability, ease of purification, low cost, and non-toxicity.

Of these criteria, several do not pertain directly to ion solvation, but are indirectly connected with it: the liquid range, the heat of vapourization, that together with the normal boiling point indicates the vapour pressure, the viscosity and its temperature coefficient, and the relative permittivity and its temperature coefficient. The values of these properties are presented in Tables 6.1 to 6.3 (except for the temperature coefficient of the viscosity) for many of the solvents of interest. Some inorganic solvents that have been suggested for use in high-energy-density batteries and that are not included in that Table have their relevant properties listed in Table 9.5.

Ion solvation does pertain, directly or indirectly, to some of these criteria. Electrolyte solubility of >0.3 mol dm^{-3} at all temperatures is required for the presence of a sufficient number of charge carriers per unit volume. This can be achieved for a small ion, such as lithium, by solvents of high donor strength, *DN*, see Table 4.5. However, the anion must be large, e.g., ClO_4^-, BF_4^-, AsF_6^-, in order for the lattice energy of the salt not to be excessive for good solubility. Since protic solvents are excluded on account of their reactivity towards the lithium anode, aprotic solvents that are adequate electron-pair acceptors besides having good donicities are preferred. These, by solvating both the cation and the anion, provide a distance of closest approach of the oppositely charged ions that is larger than the distance that corresponds to the formation of solvent-shared ion pairs at the prevailing relative permittivity. Solvents with at least a moderately high relative permittivity ($\varepsilon \geqslant 30$) are therefore the ones of choice in this respect. Popular solvents are above all propylene carbonate and also dimethylsulphoxide, acetonitrile, and γ-butyrolactone.

However, for the assurance of an adequate mobility of the ions, the size of the solvated charge carriers should not be too large and the viscosity of the solvent should be low. These requirements may be contradictory to those of adequate

Table 9.5 Some physical properties of inorganic solvents of potential use in high-energy-density batteries

Property	SO_2[a]	$SOCl_2$[b]	$POCl_3$[a]	$SeOCl_2$[a]
Molar mass, M/g mol^{-1}	64.1	119.0	153.4	165.9
Melting point, t_m/°C	−75.5	−105	1	10.9
Normal boiling point, t_b/°C	−10.2	79	108	176
Density, d/g cm^{-3} (at t °C)	1.46(0)	1.64(25)[d]	1.71(0)	2.42(22)
Vapour pressure, p/kPa (at t °C)	155(0)	5.3(2)	5.3(27)[b]	0.1(35)[b]
Enthalpy of vapourization, H^V/kJ mol^{-1}	24.9	28.6[f]	35(105)[c]	31.8[f]
Viscosity, η/mPa · s (at t °C)	0.43(0)	0.63(25)[d]	1.15(25)	
Relative permittivity (at t °C)	15.6(0)	9.25(20)	13.9(22)	46.2(20)[b]
Donicity, DN/kcal mol^{-1}		0.4[e]	11.7[e]	12.2[e]
Specific conductivity, κ/Ω^{-1} cm^{-1} (at 20 °C)	$3 \cdot 10^{-8}$[g]	$3 \cdot 10^{-9}$[f]	$2 \cdot 10^{-8}$[f]	$2 \cdot 10^{-5}$[f]

[a] J. Jander and Ch. Lafrez, *Ionizing Solvents*, Wiley, New York (1970); [b] R. C. Weast, ed., *Handbook of Chemistry and Physics*, Chemical Rubber Co., Cleveland, 47th ed. (1966/7); [c] I. Barin and O. Knacke, *Thermochemical Properties of Inorganic Substances*, Springer, Berlin (1973); [d] H. V. Venkatasetty and D. J. Saathoff, *J. Electrochem. Soc.*, **128**, 773 (1981); [e] V. Gutmann, A. Steininger, and E. Wychera, *Monatsh. Chem.*, **97**, 460 (1966); [f] D. W. Meek, in *The Chemistry of Non-Aqueous Solvents*, J. J. Lagowski, ed., Academic Press, New York, Vol. I, p. 12 (1966); [g] D. F. Burow, note f, Vol. III, p. 138.

solvating power, and a compromise must be resorted to. This often takes the form of the employment of a mixture of solvents: one that provides the solvation, the other that provides the low viscosity. For the latter purpose various ethers, preferably cyclic ones such as tetrahydrofuran or 1,3-dioxolane, but also the non-cyclic dimethoxyethane are employed, in spite of their low relative permittivities and poor solvating power for ions. For instance, a mixture that comprises 25 wt.% of propylene carbonate and 75 wt.% of dimethoxyethane has the following values at the two temperatures (−45 °C, +25 °C): viscosity (2.3, 0.7) mPa · s, relative permittivity (27, 19), and maximal specific conductivity of lithium perchlorate $(3.4, 14.2) \cdot 10^{-3}\ \Omega^{-1}\ \mathrm{cm}^{-1}$, which are in the range of desirable properties.[7]

Commercial implementations of these high-energy-density batteries to date have been quite impressive.[7,8] An energy density of 0.3 W · h g^{-1} or up to 0.5 W · h cm^{-3}, coupled with the very low self-discharge rate of <2% per year and open-circuit tensions of up to 3 V have been achieved in primary batteries. These have been produced by the hundreds of thousands for such uses as cardiac pace-makers and for the powering of electronic devices in remote locations (including missiles and other military applications). Less successful so far have been applications as secondary (i.e., rechargeable) batteries for vehicle propulsion or load levelling, i.e., off-peak-hours power storage. The cycling efficiency of the lithium anodes has proved inadequate.

9.3 Applications in hydrometallurgy

There are many applications of ion solvation in the field that is commonly called hydrometallurgy, but should, perhaps, be called *solvometallurgy* instead, since it may not necessarily employ exclusively aqueous solutions for its processes. If not completely water-free solvents, then at least mixed aqueous–organic solvents are used, since on the whole it is difficult to exclude water from the process solutions. Otherwise, a two-phase system may be employed, with one phase rich in water and the other in an organic solvent. Liquid–liquid distribution or solvent extraction is then practised with such systems. The solvation of metal and other ions is found to play an important and sometimes a decisive role in the processes.

One of the more striking examples of the use of nonaqueous solvents in hydrometallurgy is the employment of acetonitrile in the recovery of copper and silver from ores or scrap metal. The process is based on the fact that whereas Cu^{2+} is better solvated by water than by acetonitrile, the opposite is the case for Cu^{+} and Ag^{+}, due to the back-bonding from the π-electrons of the cyano-group in acetonitrile to the d^{10} monovalent metal ions. These selective solvation phenomena are based on thermodynamic data and confirmed by spectroscopic data, e.g., from NMR, and are manifested also in mixed aqueous–acetonitrile solutions, as shown in Table 9.6. In the mixtures, the monovalent ions are selectively solvated by the acetonitrile component, and the divalent one by the aqueous component.[9]

Table 9.6 Thermodynamic quantities for the transfer of Cu^{+}, Cu^{2+}, and Ag^{+} in the water + acetonitrile system at 25 °C

Quantity	x_{MeCN}	Cu^{+}	Cu^{2+}	Ag^{+}
$\Delta G_t^0(W \rightarrow W + S)$/kJ mol^{-1}	0.057	-27.8^c	13.0^d	-7.3^c
ΔH_t^0		-71.6^c		-24.9^c
$T\Delta S_t^0$		-43.8^c	$12.1^{c,e}$	-17.6
$\Delta G_t^0(W \rightarrow W + S)$/kJ mol^{-1}	0.247	-37^d	14.0^d	$-12^{c,d}$
$\Delta G_t^0(W \rightarrow W \rightarrow S)$/kJ mol^{-1}	0.510	-48^d	7.2^d	
$\Delta G_t^0(W \rightarrow S)$/kJ mol^{-1}	1.000	-48^b	59^b	-23.2^a
ΔH_t^0				-52.7^a
$T\Delta S_t^0$				-29.5^a

From the data in water and in aqueous acetonitrile ($x_{MeCN} = 0.25$) the following differences in the standard Gibbs free energies of reactions are obtained[c]

$\Delta\Delta G^0(Cu^{2+} + Cu^0 = 2Cu^{+})/\text{kJ mol}^{-1} = -89.0$
$\Delta\Delta G^0(Cu^{2+} + Ag^0 = Cu^{+} + Ag^{+})/\text{kJ mol}^{-1} = -63.3$
$\Delta\Delta G^0(Cu^{2+} + Fe^{2+} = Cu^{+} + Fe^{3+})/\text{kJ mol}^{-1} = -50.2$
$\Delta\Delta G^0(Cu^{2+} + 2Cu_2S = CuS + 2Cu^{+})/\text{kJ mol}^{-1} = -88.5$

[a] Y. Marcus, *Pure Appl. Chem.*, **55**, 977 (1983) and further data to be published there (1985); [b] G. F. Coetzee and W. K. Istone, *Anal. Chem.*, **52**, 53 (1980); [c] A. J. Parker, *Pure Appl. Chem.*, **53**, 1437 (1981); [d] A. J. Parker, D. A. Clarke, R. A. Couche, G. Miller, R. I. Tilley, and W. E. Waghorne, *Austr. J. Chem.*, **30**, 1661 (1977).

When slightly acidified (pH = 2) solutions of copper(II) sulphate in 6 M aqueous acetonitrile are applied to copper scrap or to copper or silver sulphide ores, dissolution takes place according to Parker and co-workers to form Cu_2SO_4 and Ag_2SO_4.[9,10] The selectivity of this dissolution is very high, and hardly any base metals accompany the copper and the silver into the solution. High concentrations (>1 M) of Cu_2SO_4 in the 6 M aqueous acetonitrile are achieved in this process. When low-grade steam is subsequently used for the distillation of the acetonitrile out of the solution, the equilibria (see Table 9.6) are reversed, and metallic copper and silver of high purity are precipitated. The stripped vapour, consisting of equal weight fractions of water and acetonitrile, is fractionated to give a ≈85 wt.% acetonitrile azeotrope, which is recycled, as is the copper(II) sulphate solution formed on the disproportionation of the Cu_2SO_4 when the metallic copper precipitates. The copper(II) in aqueous acetonitrile can replace the iron(III) commonly used as an oxidant for the dissolution of scrap copper, with the advantage that no foreign metal ions are introduced. The Cu^{2+}/Cu potential in 6 M aqueous acetonitrile in the presence of sulphate ions is 0.5 V, that of the Cu^{+}/Cu electrode is only 0.12 V, hence the powerful oxidizing properties of Cu(II) in these solutions.

When chloride ions replace the sulphate ones another interesting phenomenon takes place. Sulphate ions do not complex the ions in question appreciably, but chloride ions do, in particular the Cu^{+} and Ag^{+} ions, and they do so much more strongly in aprotic dipolar solvents than in water. This is due to the weak solvation of anions in the aprotic solvents, so that their thermodynamic activity is raised manyfold. In dry dimethylsulphoxide, $CuCl_2$ oxidizes in the presence of NaCl the metals copper, silver, and even gold (and the sulphides of the former two), and dissolves cement silver, silver halide residues, and other scrap (such as circuit boards). It does so by producing the anionic complexes $CuCl_2^-$, $AgCl_2^-$, and $AuCl_2^-$ (or, perhaps, $AuCl_4^-$). The standard Gibbs free energy of transfer of $AgCl_2^-$ from water to dimethylsulphoxide is $+4\ kJ\ mol^{-1}$, that of the other two anionic complexes is presumably the same,[9] compared with the much more positive value for chloride anions, $+40\ kJ\ mol^{-1}$. Because of the good solubility of the products (the Na^+ salts of the chlorocomplexes) in dimethylsulphoxide, the dissolution is very rapid. Metallic gold and pure AgCl are precipitated from the solution on the addition of water, since the thermodynamic activity of the chloride ions is drastically reduced on their hydration, so that the complexes become destabilized. The water is subsequently readily removed and the solvent dried by distillation.[11]

A process opposite to that of metal dissolution by an oxidant in a nonaqueous or a mixed solvent is the electrodeposition of a metal as an adherent coating on a suitable substrate. Several metals cannot be deposited satisfactorily from aqueous solutions, including highly reactive metals such as magnesium and aluminium, and those that are readily hydrolysed at aqueous solution acidities below 0.1 M, such as tantalum, zirconium, etc. The solvation of these multiply charged cations in organic solvents has so far not been studied systematically, hence only empirical rules can be given for the choice of the electrolytes and solvents for the purpose of the electrodeposition of these metals. One such rule is that oxygen and nitrogen atoms should be avoided in the electrolytes of the plating baths, since anions containing these atoms decompose to yield oxides and nitrides rather than metallic deposits.

The electroplating of aluminium can be effected from solutions containing $AlCl_3$ and AlH_3 as the electrolyte, and ethers and aromatic hydrocarbons as the solvent, in spite of their low ionic dissociating power. As an example, the NBS process[12] involves 3 M $AlCl_3$ and 0.3 M $LiAlH_4$ in diethyl ether, to which some methyl borate has been added. The replacement of the diethyl ether by the less volatile tetrahydrofuran and the addition of benzene to the latter have been found to be advantageous.[13] Specific conductances of a few times $10^{-3}\,\Omega^{-1}\,cm^{-1}$ are attainable in these solutions. The ionic species present have been claimed[13] to be Li^+ (presumably solvated) and $AlHCl_3^-$ or $AlH_2Cl_2^-$, but the possibility of the existence of solvated aluminium chloride cations: $AlCl_2(THF)_2^+$, and of $AlCl_4^-$ and $Al_2Cl_7^-$ anions in the solutions containing a large excess of $AlCl_3$ over $LiAlH_4$ should not be overlooked. A system offering similar high specific conductances is that based on aluminium bromide in toluene (e.g., 1 M Al_2Br_6 + 0.8 M KBr), where the ions $K_2Al_2Br_7^+$ and $K(Al_2Br_7)_2^-$ have been suggested to exist at the side of ion pairs, such as $KAlBr_4$ and KAl_2Br_7.[14] The function of the toluene in these solutions is not clear under the circumstances described, since its solvating power for ions is admittedly small, as is its relative permittivity.

A further review of the electrodeposition of metal coatings from nonaqueous solvents is found in ref. 7.

A final example of the application of ion solvation to hydrometallurgy is the reprocessing of spent nuclear fuels by solvent extraction. The most widely used commercial process for this purpose is the Purex process that employs tri-n-butyl phosphate (TBP) as the extractant. The problem consists of the recovery of the unused (generally isotopically enriched, hence very valuable) uranium and of the plutonium produced as a by-product while the nuclear power reactor is run, and their separation for re-use from the highly radioactive fission products that have to be treated as waste. The latter include neutron-absorbing 'poisons' and nuclides that emit very high intensitities of gamma-rays. Both features of this nuclear waste make a very thorough decontamination of the uranium from the fission products mandatory.

The spent nuclear fuel, consisting mainly of UO_2 and the fission products, after being removed from the reactor and a period of 'cooling', during which a major fraction of a gamma activity has decayed, is treated by a 'head-end' process for the removal of the cladding. It is then dissolved in aqueous nitric acid, to produce a solution containing mainly UO_2^{2+}(aq) and NO_3^-(aq) ions, and that may also contain a salting agent (e.g., aluminium nitrate or excess nitric acid) in variants of the process, and of course the fission product ions and ions of Pu(IV). This aqueous solution is then contacted with TBP, normally diluted with a hydrocarbon diluent, such as kerosene. The U(VI) and Pu(IV) are efficiently extracted into the organic phase, whereas the fission products remain behind in the aqueous phase, that is directed to the waste treatment plant. The organic phase is washed with aqueous nitric acid, and the plutonium is then stripped, after its selective reduction to Pu(III), by moderately concentrated aqueous nitric acid, which leaves the U(VI) behind in the organic phase. The latter is finally stripped by very dilute aqueous nitric acid, and is sent to the 'tail-end' processing plant for recovery.

The main reaction of interest for the extraction of uranium in this process is

$$UO_2^{2+}(aq) + 2NO_3^-(aq) + 2TBP(org) \rightleftharpoons UO_2(NO_3)_2(TBP)_2(org) \quad (9.17)$$

An analysis of the energetics of this process has been made by Marcus[15,16] in terms of the following hypothetical steps. The uranium with two nitrate anions is removed from the aqueous phase into the gas phase, with the investment of energy. Much less energy has to be invested in the removal of the TBP from its solution in the diluent into the gas phase. The gaseous uranyl and nitrate ions then associated electrostatically, and the resulting moiety is solvated by two gaseous TBP molecules. These two reactions return most of the invested energy. The solvate is then condensed into the diluent, with a further small gain in energy. The word 'energy' has been used in the above description in a loose sense, intended to cover both the enthalpy changes and the changes in the Gibbs free energy. The standard enthalpy, entropy and Gibbs free energy changes involved in these steps[15,16] are summarized in Table 9.7. The standard states involved are the infinite dilutions of all the reactants and products in water and dodecane (the diluent) which are mutually saturated, but because of the low mutual solubilities these solvents may be treated as if they were pure. The results for the entropies and the Gibbs free energies are presented on the mole fraction concentration scale; a change to the molar scale, which should be preferred, would change the figures somewhat, but not the main conclusion, that this extraction process is enthalpy controlled. The overall enthalpy change is several

Table 9.7 The thermodynamic quantities, in kJ mol^{-1}, associated with the hypothetical steps involved in the extraction of uranyl nitrate with TBP at 25 °C (the Gibbs free energy and the entropy on the mole fraction scale)[a]

Step	ΔH^0	$T\Delta S^0$	ΔG^0
$UO_2^{2+}(aq) \rightarrow UO_2^{2+}(g)$	1361	108	1253
$2NO_3^-(aq) \rightarrow 2NO_3^-(g)$	629	88	541
$2TBP(org)^b \rightarrow 2TBP(g)$	123	128	-5^c
$UO_2^{2+}(g) + 2NO_3^-(g) \rightarrow UO_2(NO_3)_2(g)$	-1642	-198 (combined)	-1847 (combined)
$UO_2(NO_3)_2(g) + 2TBP(g) \rightarrow UO_2(NO_3)_2(TBP)_2(g)$	-403		
$UO_2(NO_3)_2(TBP)_2(g) \rightarrow UO_2(NO_3)_2(TBP)_2(org)$	-123	-135	12^c
$UO_2^{2+}(aq) + 2NO_3^-(aq) + 2TBP(org) \rightarrow UO_2(NO_3)_2(TBP)_2(org)$	-55	-9	-46

[a] From Y. Marcus, *J. Inorg. Nucl. Chem.*, **37**, 493 (1975); [b] org = dodecane solvent; [c] the negative (positive) value is due to the correction to the mole fraction scale.

times larger than the overall $T\Delta S^0$ change, and is mainly responsible for the magnitude and sign of the overall ΔG^0 for the extraction of U(VI) by TBP in dodecane from aqueous nitrate solutions.

Ion solvation plays major roles in the energetics of this process: a large amount of enthalpy must be invested in order to free the uranyl cation and the nitrate anions from their hydration shells. If it were altogether desirable (think of the eventual stripping!) and if the technological consideration could allow it, better extraction could have been achieved with less well hydrated anions in the solution, e.g., perchlorate replacing the nitrate. This is indeed the observed fact in laboratory experiments. The enthalpy invested in the dehydration must be returned, but contrary, perhaps, to the expectation, that enthalpy returned by the solvation of the associated ions by the TBP is not large; it constitutes only roughly 20% of the invested amount of enthalpy. Although TBP is a relatively effective electron-pair donor, $DN = 23.7$, it donates its unshared electrons of the phosphoryl group not to the cation UO_2^{2+} but to the already charge-neutralized $UO_2(NO_3)_2$ species, according to the scheme presented here. Still, solvents that are more basic than TBP, i.e., those that have larger donor numbers, such as trialkylphosphine oxides (tri-n-octylphosphine oxide has $DN \approx 32$),[17] should form stronger solvate bonds. Hence they should make the corresponding enthalpy term more negative and lead to better extraction. This, again, is what is observable in the laboratory.[16] It appears that the major cause for the successful extraction in the present commercially applied system is the charge neutralization of the uranyl cation by the two nitrate anions. It is immaterial whether this neutralization takes place in the gas phase, according to the hypothetical scheme outlined above, or in the organic phase, having a low relative permittivity, where it actually seems to occur, since the end effect is the same. This appears to be a general result for the extraction of metal salts from aqueous solutions into organic solvents, or mixtures of extractants with diluents.

The role of a salting agent, such as aluminium nitrate or nitric acid, that may be added to the aqueous phase in order to enhance the extractability of the uranium, may be viewed in the light of what has been said in Chapter 8 regarding the solvation of ion pairs. If the complication introduced by the extractability of the nitric acid by TBP is disregarded (aluminium nitrate is non-extractable, due to the too large investment of hydration enthalpy required), the main roles of these agents are twofold. One is to form $UO_2NO_3^+$ ion pairs and the other is to reduce the activity of the water in the system. The former effect releases some of the water of hydration, since the singly charged ion pair has a smaller hydration sheath than the doubly charged free cation (it also reduces the electrostatic energy released on ion association, but to a lesser extent). The second effect reduces even further the extent of hydration of the uranyl species to be extracted by competition for the available water, thus facilitating its removal from the aqueous phase. A third possible effect could be reduction of the amount of water that hydrates the TBP in the organic phase, hence an increase in the availability of the TBP for the solvation of the extracted uranyl nitrate.

9.4 Applications to separation chemistry

Separations by means of solvent extraction that involves ion solvation are not limited to heavy metal ions as discussed in the previous Section. An interesting, perhaps unique, application of ion solvation is to be the extraction of alkali metal ions by crown ethers. These are cyclic ethers having a ring of $-(C_2H_4O-)_n$ groups, with $4 \leqslant n \leqslant 10$, possibly substituted or replaced by other groupings with the O–C–C–O skeleton that may be incorporated in another ring. This type of solvent extraction is of special interest, since these d^0 monovalent ions do not form extractable chelates or ion pairs with most other (anionic) extractants. An exception to the latter statement is the extractability of the very large caesium ion by certain substituted phenolate anions, on the one hand, and as an ion pair with certain very large anions, such as the triiodide one, I_3^-, into nitrobenzene, on the other. The association of certain crown ethers with alkali metal cations can take place in aqueous solutions, and it is a semantic question whether the reaction is called complex formation or solvation. Indeed, some open-chain analogues of the crown ethers, the multiglymes (multiethyleneglycol dimethyl ethers, e.g., tetraglyme = $CH_3O(C_2H_4O)_4CH_3$), are liquids, and hence can be considered as solvents that solvate the cations dissolved in them by wrapping them in a 'cage' of oxygen atoms that act as donors in the optimal geometry. The closer the size of the alkali metal cation is to the inner diameter of this cage, whether formed by the linear glyme or by the closed ring of the crown ether, the better is its solvation by this reagent.

The extraction of salts of the alkali metals from aqueous solutions by crown ethers (made water-immiscible by proper substitution of benzo-, cyclohexo-, etc. groups for $-C_2H_4-$ groups on the crown ether ring), dissolved in suitable solvents proceeds by the exchange of the oxygen donor atoms of the hydrating water molecules by the same kind of atoms from the crown ether. No great loss or gain of bonding energy is expected in such a process, provided the number of oxygen atoms involved and their mean distances from the cation are similar in the two situations in which the cation finds itself. A further provision is that the anion is solvated by the solvent employed as well as it is in water, and successful extraction depends, therefore, on the readiness of the anion to leave its aqueous environment and enter that of the organic solvent. Common practice is to employ large, poorly hydrated, anions that may interact by polar and dispersion forces with the solvent, picrate being the favourite anion chosen for this purpose, see Table 9.8.

However, the challenge that the extraction of the alkali metal cations by crown ethers poses is the carrying out of this extraction from aqueous solutions containing only relatively small and hydrophilic anions. This has been accomplished by Marcus and Asher[18] by the provision of these anions with a solvating solvent that is able to compete with water as an electron pair acceptor or hydrogen bond donor, that is, a protic solvent, immiscible with water, that has a high E_T value. Most of the better protic acceptor solvents, such as methanol, glycerol, or formamide, are water-soluble, and the choice falls on

Table 9.8 The extraction of alkali metal salts with crown ethers

Alkali metal cations	Anion	Crown ether[a]	Solvent	Ref.
Li^+, Na^+, K^+, Rb^+, Cs^+	Picrate	DC-18-C-6	Hexane, CH_2Cl_2	b
Li^+, Na^+, K^+, Cs^+	Picrate, MnO_4^-	DC-18-C-6	Toluene, cyclohexane, CH_2Cl_2	c
Li^+, Na^+, K^+, Cs^+	Picrate	DB-18-C-6	Benzene, $CHCl_3$, CH_2Cl_2 $C_6H_5NO_2$, C_6H_5Br	c
Na^+, K^+, Rb^+, Cs^+	SCN^-	DC-18-C-6	Toluene + C_4H_9OH	d
Li^+, Na^+, K^+, Cs^+, NH_4^+	Dipicryl-aminate, I^-, ClO_4^-	DB-18-C-6	Isopentyl acetate, PC, CCl_4, CH_2Cl_2, 2,2-di-chloroethylether	e
Na^+, K^+	Picrate	DC-18-C-6	CH_2Cl_2, hexane	f
K^+	Picrate	DB-18-C-6	CH_2Cl_2	f
Li^+, Na^+, K^+, Rb^+, Cs^+, NH_4^+	Picrate	B-18-C-6	CH_2Cl_2	g
Na^+, Cs^+	Dipicryl-aminate	DB-18-C-6	CH_2Cl_2, $CHCl_3$, $C_6H_5NO_2$, CH_3NO_2, PC, C_6H_5Cl	h
K^+	Cl^-	DB-18-C-6	$C_2H_2Cl_4$(membrane)	i
Na^+	$B(C_6H_5)_4^-$	DC-18-C-6	$CHCl_3$	j
Li^+, Na^+, K^+, Rb^+, Cs^+	Picrate	DB-18-C-6	$C_6H_5NO_2$ + toluene	k
Cs^+	Picrate	D(tBu)B-18-C-6	Nitrobenzene	l
Li^+, Na^+, K^+, Rb^+, Cs^+	Picrate	DB-18-C-6	Benzene	m
Na^+, K^+	F^-, Cl^-, Br^-, NO_3^-, SO_4^{2-}, Acetate	DB-18-C-6, DC-18-C-6	m-Cresol	n
Na^+, K^+	Cl^-	DB-18-C-6, others	Substituted phenols	p
Li^+, Na^+, K^+, Rb^+, Cs^+	Picrate	12-C-4	Benzene	q
Rb^+, Cs^+	Picrate	12-C-4, 15-C-6	Tributyl phosphate	r
Li^+, Na^+, K^+	Picrate	B-13-C-4, DB-14-C-4, B-15-C-4	Benzene, CH_2Cl_2 + tricresyl phosphate	s

[a] The symbols for the crown ethers are a sequence of letters and digits, the rightmost digit being the number of oxygen atoms in the ring, the 'C' to its left denotes 'crown', the number to the left of the 'C' is the total number of atoms in the ring, including the oxygen atoms, and the leftmost letters denote substituents: B = benzo, C = cyclohexo, tBu = t-butyl, D = di. Thus DB-18-C-6 is dibenzo-18-crown-6 or octahydrodibenzo-2,5,8,15,18,21-hexaoxacyclooctadecin, in the notation of *Chemical Abstracts*; [b] G. Eisenman, S. M. Ciani, and G. Szabo, *Fed. Proc.*, **27**, 1289 (1968); [c] C. K. Pedersen, *Fed. Proc.*, **27**, 1305 (1968); [d] B. C. Pressman, *Fed. Proc.*, **27**, 1283 (1968); cf. also D. C. Hayness and B. C. Pressman, *J. Membrane Biol.*, **18**, 1 (1974); [e] J. Rais and S. Selucky, *Radiochem. Radioanal. Lett.*, **6**, 257 (1971); [g] S. Kopolow, T. E. Hogen Esch, and J. Smid, *J. Macromol. Sci.*, **6**, 133 (1973); S. Kopolow, Z. Machacek, U. Takaki, and J. Smid, *J. Macromol. Sci.*, **A7**, 1015 (1973); [f] H. K. Frensdorff, *J. Am. Chem. Soc.*, **93**, 4684 (1971); [h] J. Rais, M. Kyrs, and L. Kadlecova, *Proc. Int. Conf. Solvent Extr., ISEC '74, Lyon , 1974*, **2**, 1705 (1974); [i] F. Caracciolo, E. L. Cussler, and J. F. Evans, *AIChE J.*, **21**, 160 (1975); [j] D. W. Mitchell and D. L. Shanks, *Anal. Chem.*, **47**, 642 (1975); [k] P. R. Danesi, H. Meider-Gorican, R. Chiarizia, and G. Scibona, *J. Inorg. Nucl. Chem.*, **37**, 1479 (1975); [l] Lj. Tusek, P. R. Danesi, and R. Chiarizia, *J. Inorg. Nucl. Chem.*, **37**, 1538 (1975); [m] A. Sadakane, T. Iwachido, and K. Toei, *Bull. Chem. Soc. Japan*, **48**, 60 (1975): [n] Y. Marcus and L. E. Asher, *J. Phys. Chem.*, **82**, 1246 (1978); [p] Y. Marcus, L. E. Asher, J. Hormadaly, and E. Pross, *Hydromet.*, **7**, 27 (1980); [q] Y. Takeda, *Bull. Chem. Soc. Japan*, **53**, 2393 (1980); [r] Y. Takeda, *Bull. Chem. Soc. Japan*, **54**, 526 (1981); [s] U. Olsher and J. Jagur-Grodzinski, *J. Chem. Soc., Dalton Trans.*, **1981**, 501.

substituted phenols as the most suitable solvents for the envisaged purpose.[16,18] The requirement that both the cation and the anion of, say, potassium chloride be solvated for this salt to be extractable from aqueous solutions by, say, DB-18-C-6 (for the designation of the crown ethers see Table 9.8) is demonstrated by the non-extractability of this salt when toluene is used as the solvent and its ready extractability when benzyl alcohol is used, although this solvent is far inferior to some substituted phenols.

Once the extraction of the cation is taken care of by the crown ether, the attention can be focused on the extraction of the anion. The order of the extractability of potassium salts with various anions by 0.1 M DB-18-C-6 in 3-methylphenol (m-cresol) is[18]:

$$SO_4^{2+} \ll Cl^- < Br^- < I^- < NO_3^- < CH_3CO_2^- < F^- \qquad (9.18)$$

which is *not* the order of the difficulty of their removal from the aqueous phase (given by ΔG^0_{hydr}, see Table 5.10). The reason for this is that account must be taken also of the ability of the solvent to solvate the anion. The more acidic nature of 3-methylphenol as compared with water favours its interaction with the strongly hydrogen-bond-accepting fluoride anion, although there is space for fewer of the bulkier 3-methylphenol molecules than for water molecules round the anion. For a given anion, in fact, the higher the acidicity or the electron pair acceptance power of the solvent, as measured by its E_T, the better the extraction of the potassium salt with a given crown ether.[19] When, for a given solvent, the anions are compared, then not only their basicities are important but also their sizes. Thus the equilibrium constant K_1 for the reaction of a molecule of phenol in dichloromethane solvent has been measured by infrared spectrophotometry for a series of anions[20]:

$$BF_4^- < ClO_4^- < I^- < HSO_4^- < SCN^- < Br^- < NO_3^- < Cl^- \ll SO_4^{2-} \qquad (9.19)$$

This order follows almost exactly the order of the standard Gibbs free energies of hydration:

$$BF_4^- < ClO_4^- < SCN^- < I^- < NO_3^- < Br^- < Cl^- < F^- \ll SO_4^{2-} \qquad (9.20)$$

No value of log K_1 for the fluoride anion is available for the sequence 9.19, but it is expected to be larger than that for the chloride anion. When the two sequences 9.19 and 9.20 oppose each other in their effects, as they do when solvation in the organic phase by 3-methylphenol competes with hydration when extraction is carried out, the delicate balance of the effects produces the unexpected observed order in sequence 9.18. This simple picture is obscured, however, by an additional effect, that of ion pairing between the potassium ion, solvated in an equatorial plane by the crown ether, with the anion in the axial direction. A water molecule from the potassium ion and a solvent molecule from the anion are probably displaced in this ion pairing, if it leads to a contact pair.[16]

Many compounds of pharmaceutical significance are ionic or at least ionogenic, i.e., ionizable in solvents of sufficiently high relative permeabilities. Their separation from similar compounds for either preparative or analytical purposes is often carried out by liquid–liquid distribution in its HPLC mode, i.e., by high performance liquid chromatography. One of the liquid phases is generally aqueous and the other a practically immiscible organic solvent. If the former is adsorbed on a hydrophilic support, such as silica gel, then normal extraction chromatography will be employed, but if it is the organic solvent that is adsorbed, this time on a hydrophobic support such as polystyrene, then so-called reversed phase extraction chromatography will be used.

The compounds of interest are commonly weak acids and bases, hence an adjustment of the pH of the aqueous phase determines their degree of ionization. It is generally accepted that the ionized forms of the compounds are relatively strongly hydrated, whereas the non-ionized forms are not, though they may be solvated by polar solvents. Hence the affinity of the compound towards the aqueous phase is governed primarily by the extent of its ionization under the prevailing conditions, and secondarily on the relative strengths of the hydration or solvation of the ionized and non-ionized forms by the solvents in the two phases.

One can modify the affinity of the ionized form of the compound of interest for the aqueous phase by the provision of a suitable ion of the opposite charge to form an ion pair that is much more hydrophobic than the original ion. In this case the organic solvent phase will be strongly preferred, even for a solvent of low polarity and negligible solvating ability, such as chloroform. The hydrophobic counter ion may be added directly as its salt, or alternatively it may result from the ionic dissociation of a weak acid or base present in the organic phase, in the presence of an aqueous phase with a suitably adjusted pH. The ionization generally takes place only at the interface, not in the bulk of an organic solvent of low polarity. The net effect is an ion exchange reaction between the ion of interest in the aqueous phase and either a hydrogen ion for a cation, or a chloride ion, say, for an anion, in the organic phase.

These generalities may be illustrated with a large number of examples that have been studied, mainly by Schill and co-workers.[21] With tetra-n-butylammonium as the counter ion and chloroform as the organic solvent, the equilibrium constants for the extraction of some aromatic anions from aqueous solutions increase in the following order: phenolate $<$ benzoate $<$ toluene-4-sulphonate $<$ salicylate $<$ 1-phenylpropyl sulphate. With chloride as the counter ion and chloroform, again, as the organic solvent, the equilibrium constants for the extraction of substituted ammonium ions from aqueous solutions increase in the following order: quaternary $<$ primary $<$ secondary $<$ tertiary, all having one long-chain substituent and from zero to three methyl groups as substituents. This order is changed when a solvent with good donor properties is used instead of the chloroform, e.g., methyl isobutyl ketone. The tertiary ammonium ion is then no longer preferred. Furthermore, if the ion of interest is strongly hydrophilic on account of hydroxylic, carboxylic, or amino groups, solvents that are able to interact with these groups by hydrogen bonding have to be used for obtaining any reasonable extraction into the organic phase, since they can provide solvation to the extracted ion pair.[21]

A specific example for this case is the extraction of the primary octylammonium and the quaternary trimethylnonyl ammonium cations as their ion pairs with the salicylate anion into chloroform, the selectivity of the extraction being rather small. It is increased manyfold by the inclusion in the organic phase of a crown ether, e.g., dibenzo-18-crown-6. The charged group of the primary ammonium cation can be accommodated within the ring of six oxygen donor atoms of the crown ether, but not the bulkier group of the quaternary ammonium cation.[22]

Amino-alcohols and aminophenols of low molecular weight are highly hydrophilic, but their protonated ammonium cations can exchange for the dissociatable acidic hydrogen atom of such a solvent as bis(2-ethylhexyl) phosphoric acid. This solvent solvates the resulting ion pair in the organic phase with further solvent molecules. The extraction of adrenaline (2-hydroxy-2-(3,4-dihydroxyphenyl)ethylmethylamine) is a case[23] in point. Its protonated form is extracted as the bis(2-ethylhexyl) phospate solvated by two further molecules of the undissociated phosphoric acid ester, to produce a highly hydrophobic moiety extractable into chloroform, in spite of the highly hydrophilic nature of the adrenaline. Even enantiomeric α-aminoalcohols can be separated, if optically active esters of, say, tartaric acid, are used.[24]

In summary, it should be noted that although the hydrophilic organic ion of interest may be strongly hydrated in the aqueous phase, the Gibbs free energy of hydration that must be invested in order to remove it from its aqueous environment is compensated by the electrostatic work released when the ion pair is formed and by any extra solvation Gibbs free energy that the solvation of this ion pair in the organic phase may provide. The separation aspect of this compensation manifests itself in the different extents to which it proceeds for different solutes. Even if these differences are small, leading to ratios of the equilibrium constants for extraction that are near unity, the chromatographic process multiplies the extraction and re-extraction (stripping) equilibria manyfold, enhancing the unit effect sufficiently to obtain very effective separations.

The use of ion exchange resins in mixed aqueous–organic solvents for the separation of ions is the last example of separation methods involving ion solvation to be discussed here. The ion exchange resins have a crosslinked polymeric skeleton to which fixed ions are attached by covalent bonds, the necessary counter ions being located in their vicinity in the pores and molecular sized channels of the resin structure. It is possible, thus, to speak of the 'inside' solution phase of the resin, that is in equilibrium with the external solution. Since electroneutrality must prevail in both phases, the ion exchange process is characterized by its stoichiometric nature, on an equivalent basis. The ion exchange resin, in turn, is characterized by its capacity—the number of equivalents of fixed ions (or mobile counter ions) it contains per unit amount of resin—and by its degree of crosslinking, expressed in % of crosslinking agent, that limits the extendability of the polymeric skeleton.

When a dry ion exchange resin is immersed in a solvent it imbibes it and swells. In the case of a mixed aqueous–organic solvent, the composition of the solvent inside the resin differs from that outside it. The resin takes up, and is swollen by, one of the components selectively. If the concentration of the electrolytes added to the system of the ion exchanger + water + organic solvent is low, then the ions distribute between the resin and the outside solution according to their preferential solvation as a function of the different compositions of the solvents in the two phases. If the electrolyte concentration is high, however, it may invade the resin and the solvent composition in the two phases become coupled with the distribution of the electrolyte. Even without this complication, that takes place at high electrolyte concentrations, the solvent composition inside the resin depends on the natures of both the fixed ion and the counter ion.

The swelling of ion exchange resins in the pure nonaqueous solvents[25] is presented in Table 9.9. Swelling is seen to be highest in water and in highly polar hydrogen-bonding solvents, such as methanol and formamide. It is lower in polar liquids that have poorer ion-solvating abilities and negligibly low in sol-

Table 9.9 The swelling of typical ion exchange resins by solvents, $\bar{n}_S$, in moles per equivalent of resin[a]

Solvent	Cation exchanger[b]	Anion exchanger[c]
Water	12.9	14.1
Methanol	3.5	6.8
Ethanol	2.3	2.9
1-Propanol		2.4
1-Butanol	1.9	
Ethylene glycol	3.0[d]	3.3[i]
Glycerol	0.8[e]	
Tetrahydrofuran	1.6	
Dioxane	0.1	0.2
Acetone	0.6[e]	0.5
Acetic acid	1.9	3.1[g]
Dimethylsulphoxide		2.0[h]
Pyridine	0.2[f]	1.3[j]
Acetonitrile	1.4	2.5
Formamide		8.2
N,N-Dimethylformamide	1.6	0.3[g]
Heptane	0.0	0.1[j]
Benzene	0.0	0.5[j]

[a] From Y. Marcus, ref. 25; [b] Polystyrene sulphonate, 8% crosslinked (Amberlite IR 120), unless otherwise noted, in H^+-form, from D. J. Pietrzyk, *Talanta*, **16**, 169 (1969); [c] Polystyrene methylenetrimethylammonium, 8% crosslinked (Dowex-1), unless otherwise noted, in Cl^--form, from Y. Marcus and J. Naveh, ref. 27; [d] Na^+-form of resin; [e] G. W. Bodamer and R. Kunin, *Ind. Eng. Chem.*, **45**, 2577 (1953); [f] Dowex-50 resin, J. Inczedy and E. Pasztler, *Acta Chim. Acad. Sci. Hung.*, **56**, 9 (1968); [g] Lewatit 500 resin, J. Inczedy and I. Hogye, *Acta Chim. Acad. Sci. Hung.*, **56**, 109 (1968); [h] A. M. Phipps, *Anal. Chem.*, **40**, 1769 (1968); [i] 16% crosslinked resin; [j] Amberlite IRA-400 resin, Bodamer and Kunin (*loc. cit.*).

vents of low relative permittivity and polarity: heptane, benzene, and dioxane. The hydrogen-form of a cation exchange resin of the polystyrene sulphonate type is generally somewhat more highly swollen than the sodium-form, and for some solvents (e.g., ethanol and N,N-dimethylformamide) considerably more. In general, the swelling of these cation exchangers decreases with increasing size of the cation. Anion exchange resins of the polystyrene methylenetrimethylammonium type swell as a rule somewhat more in a given solvent than cation exchange resins of similar crosslinking. Again, swelling in protic solvents decreases with increasing size of the anion (e.g., perchlorate versus chloride), but the reverse is true for aprotic solvents, where the perchlorate-form is the more highly swollen form.

These phenomena result from the combined action of several effects, including ion solvation. The electrostatic repulsion between the like-charged fixed ions (and between the like-charged counter ions), working against the tension of the crosslinked polymer network, decreases as the relative permittivity of a series of solvents increases. This effect is compensated by the decreased ion-pairing of the fixed with the counter ions, that permits better solvation of the individual ions. The fixed sulphonate group of the cation exchange resins is strongly hydrophilic, the fixed methylenetrimethylammonium group of the anion exchange resins considerably less so, perhaps somewhat hydrophobic. The mobile cations are solvated best by solvents with good donor properties, the mobile anions with ones with good acceptor properties, see Sections 6.1 and 6.5. However, bulky solvents must expend an appreciably larger amount of work against the tension of the polymer network than solvents of smaller molecular size, and are at a relative disadvantage.

The swelling behaviour in mixed solvents of cation exchange resins[26] and anion exchange resins[27,28] is most readily represented as diagrams of the total amount, $\bar{n}_W + \bar{n}_S$, of solvent imbibed by a unit amount of dry resin on the one hand, and as composition isotherms, $\bar{x}_S$, on the other, in both cases plotted versus the composition of the external mixture of solvents, x_S, see Figures (9.2) and (9.3), respectively. It is interesting to note that for an anion exchange resin in the chloride form the quantity $\bar{n}_W + \bar{n}_S$ decreases monotonously (nearly linearly, but see below) with the mole fraction of solvent x_S in the external mixture, from the value $\bar{n}_W$ it has in water to the value $\bar{n}_S$ it has in the nonaqueous solvent for methanol, ethanol, 1-propanol, and formamide. However, the perchlorate-form of the resin exhibits a definitely non-monotonous course of the curves, most pronouncedly for formamide and N,N-dimethylformamide, but also for the alcohols. This effect increases with decreasing cross-linking of the resin.[27]

The composition isotherms (Figure 9.3) show that water is preferred inside the resin. Moreover, this is true not only for the few resin forms and solvents shown but generally. At very low concentrations of some solvents, however, it is not the water that is preferred but the organic solvent. These cases are discussed further below, but the bulk of the data can be described in terms of the equation[25,27]

$$\log[(1 - \bar{x}_S)/\bar{x}_S] = \log k + p \log[(1 - x_S)/x_S] \qquad (9.21)$$

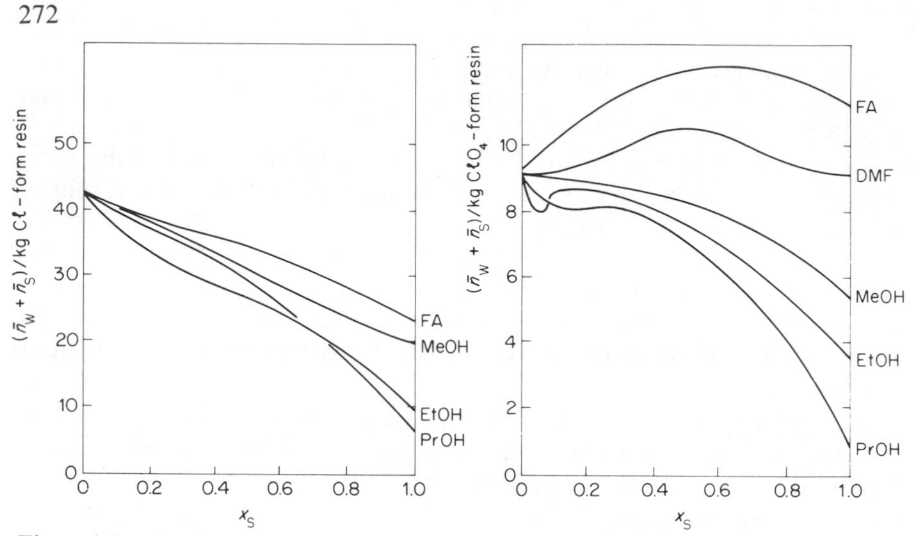

Figure 9.2 The total swelling of an 8%-cross-linked anion exchange resin, in moles total solvents per kg dry resin, in various aqueous solvents as a function of the composition of the latter[27]. (Reprinted with permission from Y. Marcus and J. Naveh. *J. Phys. Chem.*, **73**, 591 (1969). Copyright 1969 American Chemical Society)

with two characteristic parameters, p and k, the values of which are shown in Table 9.10 The bar above a symbol in equation 9.21, as elsewhere in the present discussion, denotes the inside of the resin phase. Positive values of log k mean that water is preferred inside the resin, hence the negative values noted for the perchlorate-form of the anion exchange resin with formamide and N,N-dimethylformamide describe the unusual affinity in these cases for the organic solvents. The solvents that have a low value of p have in their isotherms a region of considerable extent where $\bar{x}_S$ is not only lower than x_S but also practically independent of it, see Figure 9.3 for examples.

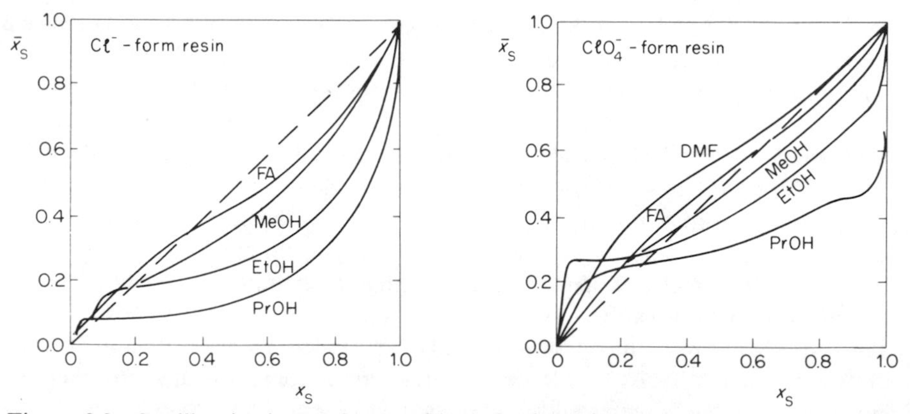

Figure 9.3 Swelling isotherms (composition of solvent in the resin versus that in the external equilibrium solvent) of 8%-cross-linked anion exchange resins in various aqueous solvent mixtures[27]. (Reprinted with permission from Y. Marcus and J. Naveh, *J. Phys. Chem.*, **73**, 591 (1969). Copyright 1969 American Chemical Society)

Table 9.10 Parameters describing the selective swelling of typical ion exchange resins in mixed aqueous-organic solvents according to equation 9.21[25]

Solvent	Cation exchanger[a]		Anion exchanger[b]			
	Na^+-form		Cl^--form		ClO_4^--form	
	p	log k	p	log k	p	log k
Methanol			0.75	0.21	0.75	0.06
Ethanol	0.60[c]	0.71[c]	0.55	0.46	0.55	0.14
1-Propanol			0.25	0.74	0.25[f]	0.27[f]
2-Propanol			0.34[g]	0.28[g]		
Allyl alcohol			0.42[g]	0.07[g]		
Acetone			0.29[g]	0.66[g]		
Ethanolamine	1.18[d]	0.28[d]				
Dimethylsulphoxide			1.00[e]	0.22[e]		
Formamide			0.80	0.09	0.80	−0.05
N,N-Dimethylformamide					0.75	−0.16

[a] Polystyrene sulphonate type, 8% crosslinked; [b] Polystyrene methylenetrimethylammonium type, 8% crosslinked; [c] H. Rückert and O. Samuelson, in ref. 28; [d] R. Arnold and S. C. Churns, *J. Chem. Soc.*, **1965**, 325; [e] L. S. Frankel, in ref. 28; [f] 7.5% crosslinked resin; [g] from Y. Marcus and J. Naveh, *J. Phys. Chem.*, **82**, 858 (1978).

As already mentioned above, another phenomenon may occur in the water-rich part of the isotherm, $x_S < 0.1$, in particular for the perchlorate-form of the anion exchange resin and for solvents with a bulky nonpolar group attched to a polar head, e.g., 1-propanol. In such cases the resin imbibes more of the organic solvent than its share in the external mixture of solvents. This is ascribed to the presence of vestiges of the water-structure in the water-rich external solution, that the nonpolar groups tend to break. Since inside the resin, in particular the poorly swollen perchlorate-form, no such structure can prevail, due to the tightness of this kind of resin the result is that the organic solvent passes into the resin. It is pushed from a region of much water-structure to one of little water-structure. At higher mole fractions of the solvent, however, all the water-structure in the external solution is destroyed completely, so that this effect no longer operates. The relative strength of the hydration and the solvation of the ions inside the resin is then the main driving force, and selective hydration of the ions is generally the predominant phenomenon. This latter effect is manifested not only in the distribution isotherms of the two solvents, but also in the total solvation, which decreases more slowly at the intermediate compositions than at high organic solvent contents, the curves of $\bar{n}_W + \bar{n}_S$ versus x_S being concave downwards.[27]

The volumetric and calorimetric behaviour of these systems show similar effects. The former can be described in terms of the contraction of the solvent inside the resin due to electrostriction by the fixed ions and the counter ions. The partial molar electrostriction of the water component in the mixture is much larger than that of the organic solvent, and this is an indication of strong preferential hydration. Similarly, the calorimetric data show that for the swelling of an anion exchange resin in aqueous acetone, for instance, the total enthalpy of swelling practically equals the partial molar heat of swelling by the water, up to $x_S = 0.85$. This, again, is an indication of strong preferential hydration. Only at still higher acetone concentrations, where the total swelling is rather less than one mole each of water and acetone per equivalent of resin, is an enthalpy effect due to solvation of the ions inside the resin by acetone at all perceptible.[29]

Once the composition of the solvent inside the ion exchange resin is known as a function of the composition of the external mixed aqueous–organic solvent and well understood, as is also the total swelling of the resin in the mixture, then the ion exchange selectivities for particular cations or anions can be understood too. The exchange of two ions (the following discussion is restricted to univalent ones) A and B can be described by the equilibrium

$$A + \overline{B}\overline{R} \rightleftharpoons \overline{A}\overline{R} + B \quad \text{(charges are omitted)} \tag{9.22}$$

where R is an equivalent of resin, a bar above a symbol signifies, as before, the resin phase, and its absence the external solution. The selectivity quotient for A over B is

$$Q_B^A = \bar{x}_A x_B / x_A \bar{x}_B = (\bar{x}_A/\bar{x}_B)/(c_A/c_B) \tag{9.23}$$

where the mole fractions can be changed to other convenient concentration scales, as is done in the second equality on the right-hand side for the external solution phase. The selectivity quotient Q_B^A generally depends on the ionic composition of the resin, i.e., on $\bar{x}_A$, but at the point $\bar{x}_A = \bar{x}_B = 0.5$ (for univalent ions) it approximates well the thermodynamic selectivity constant $K_B^A = \int_0^1 Q_B^A \, d\bar{x}_A$. In the following Q_B^A stands for this respresentative $Q_B^A(0.5)$.

The following generalizations can be made concerning the cation exchange selectivities.[25] In water-rich mixtures, down to a relative permittivity of 50, log Q_B^A is linear with $1/\varepsilon$, and if the ion preferred is designated by A, then the slope of this plot is positive and the preference is enhanced as the content of the organic solvent in the mixture is increased. In exchanges involving the hydrogen ion, however, the curves may show an extremum, and in exchanges where $Q_B^A \sim 1$, an inversion of the selectivity may occur when the organic solvent is added to water. The linear dependence of log Q_B^A on $1/\varepsilon$ can be ascribed to the Born term in the solvation Gibbs free energies of the ions A and B, see Section 3.4, although the slopes do not agree with the predicted values. Site-binding, i.e., ion pairing of the mobile counter ion with the fixed ion, is a factor that must be reckoned with too, in addition to ion solvation, for an understanding of the observed selectivities. The extremum observed in the exchanges involving hydrogen ions is ascribed to a change in the mean basicity of the solvent that does not follow the change in the relative permittivity. Some data for the selectivity quotients in different solvents are presented for the sake of illustration in Table 9.11. Since the data do not pertain to the same resins and are not from the same sources for

Table 9.11 Illustrative data on the selectivity quotients of ion exchange in mixed aqueous organic solvents[25]

Solvent	Ion A	Ion B	$\log Q_B^A(W)$	$\log[Q_B^A(W+S)/Q_B^A(W)]$		
				$x_S = 0.25$	$x_S = 0.50$	$x_S = 0.75$
8–10% Crosslinked polystyrene-sulphonate-type cation exchangers						
Methanol	Li^+	H^+	−0.19	0.19	0.39	0.58
	Na^+	H^+	0.11	0.23	0.46	0.69
	K^+	H^+	0.35	0.71	1.22	1.59
	Cs^+	H^+	0.36	0.61	1.04	1.38
	Na^+	Li^+	0.25	0.20	0.40	0.68
	NH_4^+	Li^+	0.31	0.09	0.20	0.34
	NH_4^+	Na^+	0.10	−0.17	−0.38	
Ethanol	Li^+	H^+	−0.20	0.06	0.22	0.54
	Na^+	H^+	−0.04	0.04	0.28	0.79
	K^+	H^+	0.12	0.10	0.37	0.90
	NH_4^+	Li^+	0.28	0.03	0.06	0.08
	NH_4^+	Na^+	0.11	−0.27	−0.52	−0.77
1-Propanol	Li^+	H^+	−0.01	0.50		
	Na^+	H^+	0.23	0.72	0.93	
	NH_4^+	Li^+	0.28	0.03	0.05	
	NH_4^+	Na^+	0.11	0.27	0.53	
2-Propanol	Li^+	H^+	−0.01	0.55	0.74	
	Na^+	H^+	0.23	0.82	1.08	
	NH_4^+	Li^+	0.28	0.04	0.09	
Acetone	Li^+	H^+	−0.19	0.39	0.59	
	Na^+	H^+	0.19	0.35	0.71	
	K^+	H^+	0.52	0.72	0.92	
	NH_4^+	Li^+	0.28	−0.05	−0.08	0.08
	NH_4^+	Na^+	0.11	−0.18	−0.35	−0.53

Table 9.11—*continued*

Solvent	Ion A	Ion B	$\log Q_B^A(W)$	$\log[Q_B^A(W+S)/Q_B^A(W)]$ $x_S = 0.50$
8–10% Crosslinked polystyrene-methylenetrimethylammonium-type anion exchangers				
Methanol	Cl^-	OH^-	1.02	0.38
	NO_3^-	Cl^-	0.57	−0.10
	SCN^-	Cl^-	1.40	−0.74
Ethanol	F^-	OH^-	−1.30	−1.47
	Cl^-	OH^-	−0.40	−1.50
	Br^-	OH^-	−0.25	−1.55
	I^-	OH^-	−0.15	−1.60
	NO_3^-	Cl^-	0.57	−0.28
Acetone	NO_3^-	Cl^-	0.57	−0.92
Dimethyl-sulphoxide	NO_3^-	Cl^-	0.51	−0.71
	Br^-	NO_3^-	−0.05	0.33
	SCN^-	NO_3^-	0.89	−0.89
	ClO_4^-	NO_3^-	1.28	−1.06

the various ions and solvents, they are not strictly comparable and yield only general trends.

Fewer data are available on the exchange of anions on anion exchange resins in mixed aqueous–organic solvents. Some illustrative data are shown in Table 9.11. Exchanges involving the hydroxide anion differ from those that involve only other anions in that $\log[Q_B^A(W + S)/Q_B^A(W)]$ for the former exchanges has the same sign as $\log Q_B^A(W)$ but has the opposite sign for the latter ones. In the exchanges involving the hydroxide anion the anion preferred in purely aqueous solutions is even more preferred in the mixed solvents. In the other kind of exchanges the preference for a given ion in water decreases as the organic component is added. This may be explained[30] in terms of the water-structure-breaking effects of the anions (see Section 5.4). If the swelling of the resin is appreciable and a considerable amount of solvent is present as a 'free' solvent, i.e., not bound in the solvation shells of the ions in the resin, then this solvent is generally richer in water than the external solution, as discussed above. Hence the water in this interior solvent will be more structured than the external solvent. This is more noticeable in an anion exchange resin than in a cation exchange resin, since in the former the fixed ion is the poorly hydrophilic methylenetrimethylammonium ion that promotes water structure, whereas in the latter it is the strongly hydrated sulphonate ion that orients the water around it but not in the tetrahedral structure of bulk water. The more structure-breaking anion, that is preferred inside the resin in pure water, will be less preferred when the external solution becomes more rapidly 'destructured' than the internal one, on the addition of the organic component, the latter keeping a more or less constant composition, see above.

9.5 Applications to organic synthesis

Since organic synthetic reactions are generally carried out in solution, it is natural to assume that the solvent plays some role, with respect both to the solubilities of the reactants and the products and to the rates of the reactions. In the present context only some examples of reactions that involve *ions* as reactants, products, or activated states are discussed. Typical reactions that involve ions are the S_N1 solvolysis reaction

$$RX \rightleftharpoons [R^{\delta+} \cdots X^{\delta-}]^{\neq} \rightarrow R^+ + X^- \rightarrow \text{solvolysis products} \tag{9.24}$$

and the S_N2 nucleophilic substitution reaction

$$Y^- + RX \rightarrow [YRX^-]^{\neq} \rightarrow YR + X^- \tag{9.25}$$

The transition state of the S_N1 reaction 9.24 is not an ion nor an ion pair, but a highly polarized dipole, where the exposed X atom that carries a partial negative charge can be solvated by protic solvents. Since the primary products of the solvolytic dissociation of this transition state are ions, the role of the solvent in promoting this dissociation is readily understood in terms of the solvation of the ions produced. The carbonium ion then undergoes further reactions to give the final solvolysis product. In the S_N2 reaction 9.25 one of the reactants, Y^-, and one of the products, X^-, are ions, specifically anions, as is the transition state. Their solvation is thus of direct consequence on the rate of the reaction and its direction.

Because of solubility limitations most organic reactions are carried out in nonaqueous solvents. It is common practice to use methanol, symbolized by M in the following, rather than water, as a reference solvent for the expression of the solvent effect on the rate of the reaction. The expression

$$k_S = k_M \exp(-\Delta G_t^{\neq}/\mathbf{R}T) \tag{9.26}$$

describes the rate constant k_S in any solvent S (unimolecular for reaction 9.24, involving given R and X, and bimolecular for reaction 9.25, involving given R, X, and Y) in terms of the rate constant in methanol, k_M, for the same reaction, and a transfer Gibbs free energy of activation, $\Delta G_t^{\neq}$. It has been shown by Parker and co-workers[31,32] that the latter quantity is given by the linear solvation energy relationships

$$\Delta G_t^{\neq}(R\cdots X) = p\Delta G_t^{\infty}(K^+) + n\Delta G_t^{\infty}(Cl^-) - \Delta G_t^{\infty}(RX) \tag{9.27}$$

$$\Delta G_t^{\neq}(YRX) = -n\Delta G_t^{\infty}(Cl^-) - \Delta G_t^{\infty}(RX) \tag{9.28}$$

for reactions 9.24 and 9.25, respectively, where the transfers of the ions K^+ and Cl^- and of the nonelectrolyte RX are from methanol to the solvent S. The so-called 'sensitivity parameters' p and n pertain to a given reaction characterized by RX or by RX and Y, but are solvent-independent.

The standard molar Gibbs free energies of transfer of the ions K^+ and Cl^- from methanol to the solvent S are obtained from the entries for the transfer from water presented in Table 6.8:

$$\Delta G_t^{\infty}(K^+ \text{ or } Cl^-, M \to S) = \Delta G_t^{\infty}(K^+ \text{ or } Cl^-, W \to S) - \Delta G_t^{\infty}(K^+ \text{ or } Cl^-, W \to M) \tag{9.29}$$

Since the standard molar Gibbs free energies of transfer are directly related to the donor and acceptor properties of the solvents (and to a lesser degree to some other properties, as discussed in Section 6.5), equations 9.27 to 9.29 can also be formulated in terms of these properties, say DN and AN, with appropriately modified sensitivity parameters p' and n'.[32] The Gibbs free energies of transfer of the ions depend, of course, on an extrathermodynamic assumption, the tetraphenylarsonium tetraphenylborate one in the case of the entries in Table 6.8. On the other hand, the standard molar Gibbs free energies of the nonelectrolyte RX, that are also required in equation 9.27 and 9.28 and also pertain to transfer from methanol to the solvent S, are directly observable quantities.

Table 9.12 presents data for the transfer of some model compounds RX from methanol to other polar solvents and Table 9.13 presents some typical values of the sensitivity parameters p and n. The former table shows the expected trends: the bulkier 4-nitro-iodobenzene has a larger standard molar Gibbs free energy

Table 9.12 Standard molar Gibbs free energies of transfer of some model compounds from methanol to other polar solvents[32]

Solvent, S	ΔG_t^{∞}(RX, M → S)/kJ mol^{-1}		
	CH_3I	1-C_4H_9Br	4-$NO_2C_6H_4I$[a]
Water	8	14	
Ethanol	−1		
Acetone	−5	−1	
Dimethylsulphoxide	−3	0	−6
Nitromethane	−1		−3
Acetonitrile	−2	−1.5	−2
Formamide	3	3	2
N,N-Dimethylformamide	−3	−1	−7
N,N-Dimethylacetamide	−3	−1	
N-Methylpyrrolidinone	−4		−8
Hexamethyl phosphoric triamide	−4[a]	−2.5	−9

[a] R. Alexander, E. F. C. Ko, A. J. Parker, and T. J. Broxton, *J. Am. Chem. Soc.*, **90**, 5049 (1968).

of transfer than the less bulky iodomethane or 1-bromobutane. Also, the transfer from methanol to non-hydrogen-bonded solvents is favourable, but that to the solvents that are more hydrogen-bonded and structured than methanol—water and formamide—is unfavourable. The sensitivity parameters presented in Table 9.13 for the ions can be understood in terms of the analysis presented in equation 6.28, concerning the relation of the standard molar Gibbs free energies of transfer of ions on their properties. The sensitivity parameters for the reactants of the S_N2 nucleophilic substitution reaction 9.25 also presented in Table 9.12 are seen to depend mainly on the anions involved and to be nearly independent of the uncharged molecular reactant.

Table 9.13 Some sensitivity parameters for linear solvation energy relationships pertinent to S_N1 solvolysis and S_N2 nucleophilic substitution reactions[32]

Ion	p	n	Reactants (equation 9.25)	n
Li^+	2.1		$CH_3Br + Cl^-$	0.80
K^+	(1.0)[a]		$CH_3Br + SCN^-$	0.35
$(CH_3)_4N^+$	0.5		$CH_3I + Cl^-$	1.0
$(C_6H_5)_4As^+$	0.25		$CH_3I + Br^-$	0.7
$CH_3CO_2^-$		1.33	$CH_3I + SCN^-$	0.35
F^-		1.25	1-$C_4H_9Br + N_3^-$	0.55
Cl^-		(1.0)[a]	4-$NO_2C_6H_4F + N_3^-$	0.75
N_3^-		0.75	4-$NO_2C_6H_4I + N_3^-$	0.75
Br^-		0.74	2,4-$(NO_2)_2C_6H_3I + Cl^-$	1.0
SCN^-		0.35	2,4-$(NO_2)_2C_6H_3I + SCN^-$	0.35
I^-		0.35	$(CH_3)_3CCl$ ($p = 0.20$)[b]	0.50
$(C_6H_5)_4B^-$		0.05	$(CH_3)_3CBr$ ($p = 0.20$)[b]	0.35
ClO_4^-		0		

[a] Reference values; [b] Reaction 9.24.

The success of the linear solvation energy relationships 9.27 and 9.28 in describing the solvent effects on the rates of the S_N1 solvolysis and the S_N2 nucleophilic substitution reaction, respectively, depends on the account they take of the (unmeasurable) standard molar Gibbs free energy of transfer of the activated transition state.

In equation 9.27 this quantity is estimated by simulating the dipolar activated state by the pair of ions K^+ and Cl^- (but not with the ion pair K^+Cl^-). The sensitivity of the dipolar transition state to the solvents is, however, less than that of the free ions K^+ and Cl^-. The role of the solvent lies mainly in the solvation of the leaving anion X^-, if the resulting carbonium ion is not particularly accessible to solvation, as is, for instance the *tert*-butyl carbonium ion. Thus in some carbonium ions the positive charge is dispersed over a large surface and the ionic analogue may be $(C_6H_5)_4As^+$ rather than K^+.

In equation 9.28 the standard molar Gibbs free energy of the transition state, $[YRX^-]^‡$, is neglected altogether. This tight transition state anion is large and has a low surface density of negative charge. Its ionic analogue is $B(C_6H_5)_4{}^-$ or $ClO_4{}^-$, which have a negligibly small value of the sensitivity parameter n. Hence the rate of the reaction is insensitive to the Gibbs free energy of transfer of this transition state anion from methanol to the other solvent. However, when the activated transition state anion is 'loose', the individual natures of X^- and Y^- are manifested more distinctively, and the solvation of this state by good anion solvating solvents, i.e., protic solvents, does play some role in determining the rate.[33] 'Loosening' of the transition state may be due to electronic or steric factors or both. For instance, when two methyl groups are situated on the carbon atom at which the substitution takes place, more loosening is caused than when one propyl group is located there, in which case, in turn, more loosening is caused relative to the unsubstituted carbon atom in the reaction of bromomethane with azide anions. Similarly, a para-substituted methoxy group causes loosening, a para-substituted nitro group causes a tightening of the transition state of α-bromotoluene reacting with azide anions, through an electronic induction effect.

The two types of reactions discussed here as illustrations for ion solvation as the major cause of the solvent effects on the rates of some organic reactions are, of course, only examples of the many more that have been studied. These may involve not only anions but also cations as reactants or transition states, and are described in detail in standard works on solvent effects, e.g. ref. 34.

Another area of organic reaction chemistry where ion solvation plays an important role is phase transfer catalysis. Within the scope of this book, again, only one example illustrating the principles involved can be discussed; further information is available elsewhere, e.g. ref. 35. A typical organic substitution reaction, where aqueous sodium cyanide reacts with 1-bromo-octane to yield nonanitrile (and aqueous sodium bromide), formulated as

$$NaCN(aq) + C_8H_{17}Br(org) \rightarrow C_8H_{17}CN(org) + NaBr(aq) \qquad (9.30)$$

proceeds at a negligible rate, even at elevated temperatures. This is because the reactant cyanide anion is not present at the reaction site of the bromooctane. If a quaternary ammonium salt, such as tetrabutylammonium bromide, is added to the heterogeneous reaction mixture, fast reaction with a high yield of nonanitrile is observed. This and similar observations on related reactions led Starks[36] to coin the name 'phase transfer catalysis' for these phenomena, in view of the obvious role of the quaternary salt, added at catalytic concentrations, to transfer the reactant anion (cyanide) from the aqueous phase, where it normally finds its preferred environment, into the organic phase, where it can react.

The rate determining step is generally the substitution reaction that takes place in the organic phase. This phase consists of either the neat reactant, e.g., 1-bromooctane in the example given, or its solution in an inert and not very polar solvent, such as dichloromethane. On this relatively slow substitution reaction is superimposed a generally fast liquid ion exchange reaction involving the quaternary ammonium salt, the cation of which may be symbolized by Q^+:

$$Q^+Br^-(org) + CN^-(aq) \rightleftharpoons Q^+CN^-(org) + Br^-(aq) \tag{9.31}$$

This equilibrium provides a sufficiently high concentration of cyanide in the organic phase for the homogeneous reaction

$$Q^+CN^-(org) + C_8H_{17}Br(org) \rightarrow C_8H_{17}CN(org) + Q^+Br^-(org) \tag{9.32}$$

to proceed at a reasonable rate. The formulation 9.31 presumes the quaternary salt to reside practically exclusively in the organic phase. This is the case when its alkyl chains consist altogether of some 24 carbon atoms or more, all of the chains being of similar length. If a quaternary ammonium salt is employed that has fewer carbon atoms in its alkyl chains and if these are of grossly unequal length, then the phase transfer catalyst itself partitions between the two phases. The liquid anion exchange equilibrium then proceeds as the coupled reactions:

$$Q^+(aq) + CN^-(aq) \rightleftharpoons Q^+CN^-(org) \tag{9.33a}$$

$$Q^+(aq) + Br^-(aq) \rightleftharpoons Q^+Br^-(org) \tag{9.33b}$$

The net result of the two reactions 9.33 is reaction 9.31, but the difference lies in the presence of Q^+ at appreciable concentrations in the aqueous phase. Both mechanisms, in fact, yield the same rate equation.

Ion solvation comes into play in phase transfer catalysis by determining the relative extents of reactions 9.33a and 9.33b or the position of the equilibrium 9.31. Since the anions, in order to be reactive, may not be appreciably solvated in the organic phase, the main solvation effect is the requirement for the provision of the Gibbs free energy of dehydration for the equilibria 9.33 to proceed from left to right. For values of this quantity see Table 5.10. Two complications must, however, be considered: one is incomplete dehydration, the other is high aqueous salt concentrations.

Under certain circumstances the quaternary salts carry some water with them into the organic phase. Tetraoctylammonium salts carry, per mole of salt, into toluene 1.5 moles of water with the nitrate, 3.2 with the chloride, 2.4 with the bromide, and 18 with the sulphate.[37] The actually observed equilibrium quotients for the liquid anion exchange reactions of the type 9.31 for several anions, besides the cyanide, exchanging with bromide are presented in Table 9.14. These equilibria have been measured at low aqueous concentrations of the salts.

Reactions carried out with phase transfer catalysis generally employ very concentrated aqueous solutions of the inorganic salt, even saturated ones (up to the point of using solid salts just wetted by water). This introduces the complication of the salting-out of the quaternary ammonium salt induced by the paucity of free water in the aqueous phase, a phenomenon that itself acts in an anion-selective manner. Thus the position of the equilibrium 9.31 or the pair of equilibria 9.33 cannot be predicted accurately from the Gibbs free energies of hydration given in Table 5.10 under actual reaction conditions, and even Table 9.14 gives only a rough guide.

If the aqueous phase is saturated with respect to both the reactant salt (NaCN in the example treated here) and the product salt (NaBr), then the ratio of the activities of these salts is a constant, irrespective of how far the substitution reaction has proceeded. Then also the fraction of the quaternary ammonium salt in the organic phase that is in the form of the reactant (Q^+CN^-) is a constant. For a given amount of phase transfer catalyst added to the system the concentration of the anionic reagent in the organic phase is then also constant. Rates of phase transfer catalysed reactions have often been studied under precisely these conditions, in order to simplify the kinetic equations.

A further, though minor, complication that occurs when the neat organic reactant is used as the organic phase is due to the gradual change of the nature of this phase as the reaction proceeds. In the example given here, bromooctane has a very low polarity but nonanitrile has a considerably higher one. These changes in the polarity of the organic phase cause a gradual decrease of the

Table 9.14 Equilibrium quotients for liquid anion exchange reactions of quaternary salts of various anions with bromide

Anion, X^-	Q^X_{Br} [a]	Q^X_{Br} [b]	Anion, X^-	Q^X_{Br} [a]	Q^X_{Br} [a]
F^-	0.0012[c]	<0.0021	IO_3^-		0.0008
Cl^-	0.061	0.040	ClO_4^-	30	>40
Br^-	(1.00)	(1.00)	MnO_4^-	>6	>60
I^-	300	>60	IO_4^-		0.004
CN^-	0.06		SO_4^{2-}	0.0008[d]	
SCN^-		36[e]	HSO_4^-	0.013[d]	
OH^-	0.0006		CO_3^{2-}	0.0003[d]	
NO_3^-		4.2	HCO_3^-	0.003[d]	
NO_2^-		0.04	HCO_2^-	0.005[d]	
ClO_3^-		>30	$CH_3CO_2^-$	0.007	
BrO_3^-	0.013[d]		$CH_3SO_3^-$	0.015	

[a] From ref. 35, p. 24, $Q^+ = (C_{10}H_{21})_3CH_3N^+$, unless otherwise noted, solvent is toluene; [b] R. Bock and J. Jainz, *Z. Anal. Chem.*, **198**, 316 (1963), $Q^+ = (C_6H_5)_4As^+$, solvent is chloroform; [c] $Q^+ = (C_{10}H_{21})_3(C_3H_7)N^+$; [d] $Q^+ = (C_{18}H_{37})_2(CH_3)_2N^+$; [e] Q^+ = N-hexadecylpyridinium.

selectivity for bromide versus cyanide that the organic phase generally has (see Table 9.14), and an increase in the ionic dissociation of the ion pairs formed between the quaternary cation with the anions. The result is an autocatalytic reaction, becoming faster the further it proceeds. This behaviour is manifested by a phase transfer catalyst that is appreciably soluble in the aqueous phase, such as tetrabutylphosphonium bromide, but not by one that resides totally in the organic phases, such as hexadecyltributylphosphonium bromide. Only for the former are there equilibria of the type 9.33 to be affected by the changing nature of the organic phase.[38]

Ion solvation in the organic phase is of little importance in organic reactions carried out by phase transfer catalysis as ordinarily practiced, since conditions are generally selected that avoid extensive solvation of the reactant anions. Protic solvents are thus normally excluded, but dipolar aprotic solvents have also relatively little use, nor very important effects, if used, since they solvate neither the anions nor the higher symmetrical tetraalkylammonium cations effectively. They do solvate small cations, if used, as the relative rates of the reaction of 1-chlorobutane with potassium or tetrabutylammonium phenoxide in homogeneous solutions, presented in Table 9.15, show.[39] The rates are seen to increase with both the relative permittivity and the donor number of the solvent, both effects, promoting the release of the anion from association with the cation, and hence greater reactivity of the anion. The effect is seen to be manyfold larger for the small, well solvated, potassium cation than for the poorly solvated tetrabutylammonium cation.

The effectivity of the phase transfer catalysts thus depends not only on their ability to carry reactant anions from the aqueous phase into the organic one and to remove the product anion from the latter back into the former, but also on their ability to provide the anion in the organic phase in a reactive form. This requirement precludes the use of, e.g., amines substituted by fewer than four alkyl groups, since the hydrogen atoms attached to the nitrogen atom in tertiary or lower ammonium cations form hydrogen bonds with the anion and confer on it a low reactivity. Even quaternary ammonium cations with one or more methyl substituents on the nitrogen atom are prone to ion pairing with the anion and are to be avoided. It is therefore preferable to employ symmetrical tetra-alkyl ammonium salts as phase transfer catalysts, since for them the anion is at the greatest distance away from the centre of positive charge, ion pairing is at a

Table 9.15 Rate constants (in $dm^3 mol^{-1} s^{-1} \cdot 10^5$) for the reaction of 1-chlorobutane at 25 °C with potassium or tetrabutylammonium phenoxide in homogeneous solutions[39]

Solvent (ε, DN)	$K^+C_6H_5O^-$	$(C_4H_9)_4N^+C_6H_5O^-$
Tetrahydrofuran (7.6, 20.0)	0.0023	4.9
50% Dioxane + 50% acetonitrile (20, ?)	0.084	4.0
Acetonitrile (37.5, 14.1)	0.33	2.2
N,N-Dimethylformamide (36.7, 31.0)	12	17

minimum, and anion reactivity is maximal. For this reason it may be advisable to dilute an organic reactant that has a very low polarity (e.g., bromooctane) with an inert diluent of sufficiently high relative permittivity but low solvating power (such as the volatile, hence readily removable, dichloromethane). Such a mixture permits sufficient ionic dissociation of the quaternary ammonium salt in the organic phase while it does not bind the anion by solvation.

The nature of the cation in the aqueous phase used in the substitution reaction, i.e., the sodium ion in reaction 9.30, is generally immaterial with the phase transfer catalysts discussed above. A variant phase transfer catalytic method, however, makes use of crown ethers to transfer both cation and anion from the aqueous to the organic phase. Such transfers have been discussed in Section 9.4. The crown ethers, of course, require a cation that fits into the right opening, but otherwise the combination of crown ether (e.g., dibenzo-18-crown-6) and cation (e.g., potassium) acts very much like the quaternary ammonium cations discussed above.[40]

References

1. R. G. Bates, *Pure Appl. Chem.*, **54**, 229 (1982); A. K. Covington, *ibid.*, **55**, 1467 (1983).
2. T. Mussini, A. K. Covington, P. Longhi, and S. Rondinini, *Report to IUPAC Comm. V5*, 1984, to be published in *Pure Appl. Chem.*
3. C. L. DeLigny and M. Rehbach, *Rec. Chim.*, **79**, 727 (1960); W. J. Gelsema, C. L. DeLigny, A. G. Remijnse, and H. A. Blijleven, *ibid.*, **85**, 647 (1966).
4. R. G. Bates, M. Paabo, and R. A. Robinson, *J. Phys. Chem.*, **67**, 1833 (1963.
5. I. M. Kohltoff and F. G. Thomas, *J. Phys. Chem.*, **69**, 3049 (1965).
6. Y. Marcus, *Pure Appl. Chem.*, **57**, July (1985).
7. J. Barthel, H. J. Gores, G. Schmeer, and R. Wachter, *Topics Curr. Chem.*, **111**, 33 (1983); H. J. Gores and J. Barthel, *J. Soln. Chem.*, **9**, 939 (1980).
8. B. Scrosati, *Electrochim. Acta*, **26**, 1559 (1981).
9. A. J. Parker, *Pure Appl. Chem.*, **53**, 1437 (1981).
10. I. D. McLeod, D. M. Muir, A. J. Parker, and P. Singh, *Austr. J. Chem.*, **30**, 1423 (1977); A. J. Parker and D. M. Muir, *Hydromet.*, **6**, 239 (1980); A. J. Parker, D. M. Muir, Y. C. Smart, and J. Avraamides, *ibid.*, **7**, 213 (1981).
11. A. J. Parker, B. W. Clare, and R. P. Smith, *Hydromet.*, **4**, 233 (1979).
12. D. E. Couch and A. Brenner, *J. Electrochem. Soc.*, **99**, 234 (1952); J. H. Connor and A. Brenner, *ibid.*, **103**, 657 (1956).
13. M. Yoshio and N. Ishibashi, *J. Appl. Electrochem.*, **3**, 321 (1973); N. Ishibashi and M. Yoshio, *Electrochim. Acta*, **17**, 1343 (1972); M. Yoshio, N. Ishibashi, W. Waki, and T. Seiyama, *J. Inorg. Nucl. Chem.*, **34**, 2439 (1972).
14. E. Peled and E. Gileadi, *J. Electrochem. Soc.*, **123**, 15 (1976); A. Reger, E. Peled, and E. Gileadi, *ibid.*, **123**, 638 (1976); S. Ziegel, E. Peled, and E. Gileadi, *Electrochim. Acta*, **23**, 363 (1978).
15. Y. Marcus and Z. Kolarik, *J. Chem. Eng. Data*, **18**, 155 (1973); Y. Marcus, *J. Inorg. Nucl. Chem.*, **37**, 493 (1975); Y. Marcus, *CIM Spec. Publ.* No. **21** (Proc. ISEC '77), 154 (1979).
16. Y. Marcus, *Pure Appl. Chem.*, **54**, 2327 (1982).
17. Y. Marcus, *J. Soln. Chem.*, **13**, 599 (1984).
18. Y. Marcus and L. E. Asher, *J. Phys. Chem.*, **82**, 1246 (1978); Y. Marcus, L. E. Asher, J. Hormadaly, and E. Pross, *Hydromet.*, **7**, 27 (1981).
19. J. Hormadaly and Y. Marcus, *J. Phys. Chem.*, **83**, 2843 (1979).

20. R. P. Taylor and I. D. Kuntz, Jr., *J. Phys. Chem.*, **74**, 4573 (1970).
21. G. Schill, in *Ion Exchange and Solvent Extraction*, J. A. Marinsky and Y. Marcus, eds., Dekker, New York, Vol. 6, p. 1 (1974); R. Modin and G. Schill, *Talanta*, **22**, 1017 (1975).
22. M. Schröder-Nielsen, *Acta Pharm. Suecica*, **11**, 541 (1974).
23. R. Modin and M. Johansson, *Acta Pharm. Suecica*, **8**, 561 (1971); G. Hoogewijs and D. L. Massart, *Anal. Chim. Acta*, **106**, 271 (1979).
24. V. Prelog, Z. Stojanac, and M. Kovacevic, *Helv. Chim. Acta*, **65**, 377 (1982).
25. Y. Marcus, in *Ion Exchange and Solvent Extraction*, J. A. Marinsky and Y. Marcus, eds., Dekker, New York, Vol., 3, p. 1 (1973).
26. O. D. Bonner and J. C. Moorefield, *J. Phys. Chem.*, **58**, 555 (1954); C. W. Davies and B. D. R. Owen, *J. Chem. Soc.*, **1956**, 1676; C. W. Davies and V. C. Patel, *J. Chem. Soc.*, **1962**, 880; H. Ohtaki, K. Gonda, and H. Kakihana, *Ber. Bunsenges. Phys. Chem.*, **67**, 87 (1963); M. Mattison and O. Samuelson, *Acta Chem. Scand.*, **12**, 1386 (1958).
27. Y. Marcus and J. Naveh, *J. Phys. Chem.*, **73**, 591 (1969).
28. H. Rückert and O. Samuelson, *Acta Chem. Scand.*, **11**, 315 (1957); L. S. Frankel, *Canad. J. Chem.*, **48**, 2432 (1970); J. I. Kim, H. J. Born, and H. Lagally, *J. Inorg. Nucl. Chem.*, **37**, 1259 (1975).
29. Y. Marcus, J. Naveh, and M. Nissim, *J. Phys. Chem.*, **73**, 4415 (1969); Y. Marcus and J. Naveh, *Israel J. Chem.*, **10**, 899 (1972).
30. G. E. Janauer and I. M. Turner, *J. Phys. Chem.*, **73**, 2194 (1969).
31. A. J. Parker, *Chem. Rev.*, **69**, 1 (1969); M. H. Abraham, *Progr. Phys. Org. Chem.*, **11**, 1 (1974).
32. A. J. Parker, U. Mayer, and V. Gutmann, *J. Org. Chem.*, **43**, 1843 (1978).
33. A. J. Parker and E. C. F. Ko, *J. Am. Chem. Soc.*, **90**, 6447 (1968).
34. C. Reichardt, *Solvent Effects in Organic Chemistry*, Springer, Berlin (1979).
35. C. M. Starks and C. Liotta, *Phase Transfer Catalysis*, Academic Press, New York (1978).
36. C. M. Starks, *J. Am. Chem. Soc.*, **93**, 195 (1971).
37. V. L. Heifets, N. A. Yakovleva, and B. Ya. Krasil'shchik, *Zh. Prikl. Khim.*, **46**, 549 (1973).
38. C. M. Starks and R. M. Owens, *J. Am. Chem. Soc.*, **95**, 3613 (1973).
39. J. Uglestad, T. Ellingsen, and A. Beige, *Acta Chem. Scand.*, **20**, 1593 (1966).
40. C. L. Liotta and H. P. Harris, *J. Am. Chem. Soc.*, **96**, 2250 (1974).

Author Index

Subject Index

The names of ions and of solvents are entered in this index only if they are specifically mentioned in the text or in the heading of a Table or Figure. They are not included if they are entries in extensive Tables dealing with many ions or solvents, and such Tables should also be consulted.